# *Human* PHYSIOLOGY

## LAB MANUAL AND STUDY GUIDE

Fourth Edition

PAT CLARK

Indiana University – Purdue University Indianapolis

Kendall Hunt
publishing company

**Kendall Hunt**
publishing company

www.kendallhunt.com
*Send all inquiries to*:
4050 Westmark Drive
Dubuque, IA 52004-1840



ISBN 978-1-5249-8511-0



Printed in the United States of America

# CONTENTS

# Exercise 1

# Introduction to N217: Human Physiology Lab

## Information to Know

Lab Section Number 23642
Lab Instructor Name Brandon Wysong
Lab Instructor Contact Information
email bwysong@iuk.edu
Phone ____________

## Student Responsibilities

- Bring the necessary materials to lab including the appropriate lab manual(s) and other items that may be of use for note-taking and calculations.
- Be prepared for lab; **read the lab exercise before coming to lab**.
- Know and understand the procedures for each exercise or experiment.
- Understand what happened in each exercise or experiment.
- Understand the results of each exercise or experiment.
- Ask questions when you do not understand concepts.
- Relate lab activities with lecture topics.
- Return lab materials to their proper locations and in the proper condition.
- Follow all lab safety procedures, especially the proper disposal of waste material.
- Leave the lab cleaner than when you entered it.
- Turn in your lab reports on time.

## Emergency Procedures

- In case of a fire, immediately exit and move away from the building.
- In case of a tornado, move to the lower level and seek safety away from windows and glass.
- For other emergencies, call the campus police, 9-911 or 4-7911.

## Lab Safety

- biohazard

You **must** follow the lab safety instructions and guidelines printed (following) as well as those specified by your lab instructor. Failure to follow the printed guidelines or instructions given in the lab may result in the loss of lab points. Remember that you are responsible for your own safety.

- Familiarize yourself with the location and operation of the safety equipment, including the Material Safety Data Sheets (MSDSs).
- Do not eat, drink, smoke, or chew in the lab. This includes application of lip balm and cosmetics. Food items must remain outside the lab.
- Lab attire: no sandals or open toed shoes are allowed in lab and no shorts are allowed in lab. These are University safety guidelines that we must adhere to. You may be required to leave lab if you are not dressed appropriately.
- Wear safety glasses and lab coats when handling dangerous chemicals or when heating materials. Always wear gloves when working with body fluids, chemicals, or when dissecting specimens.
- If body fluids are being used in lab, use only your own and follow all special procedures as directed below and/or by your lab instructor.
- Keep your work area clear of unnecessary equipment. Keep long hair and loose-fitting clothing restrained and away from the work area.
- Spills, accidents, and damaged equipment should be reported to your lab instructor immediately.
- Use only the appropriate material and only as directed in the lab instructions and by your lab instructor.
- Dispose of all used disposable materials properly and immediately. Disposable contaminated materials should be placed in the biohazard boxes or bags provided in the lab. Reusable items contaminated with body fluids should be placed in a disinfectant solution and autoclaved. Uncontaminated sharp objects should be placed in the sharps boxes or if glass, in the glass box provided. Never force an object into a full box. One of the most common causes of an accidental needle puncture or cut is due to the overfilling of a disposal container. If disposal boxes appear full, inform the instructor and new boxes will be provided.
- All disposable lab equipment is to be used only once and then properly discarded.
- All equipment and work surfaces must be cleaned and decontaminated immediately after your procedures are completed.
- Wash your hands before leaving the lab. Hand washing is very important in reducing the possibility of spreading infectious agents.
- Failure to follow the safety rules and regulations of the lab may result in loss of lab points and your dismissal from lab.

## Blood-Borne Pathogens Facts

- Gloves reduce exposure to many pathogens, but they cannot prevent penetration by infected sharps or needles.
- HIV is susceptible to most common disinfectants, including bleach and ethanol.
- Dried-out HIV suspensions can survive for several days and remain potentially infectious.
- A single milliliter of hepatitis B or C virus (HBV, HCV) infected blood may contain more than 1 million infectious viral particles.
- The risk of infection from a single needle stick of HBV- or HBC-infected blood can be as high as 1 in 3.
- HBV and HCV can remain in dried blood on surfaces and likely remain infectious for several weeks.
- HBV and HCV infection can occur through direct or indirect exposure of dried blood into abrasions, lesions, or any mucosal surface.
- HBV and HCV are the leading causes of chronic liver infections and diseases including cancer.

## In the Event of a Needle Stick

- Report the incident to your lab instructor or a lab supervisor and call the campus police, 9-911 or 4-7911.
- Make the site bleed.
- Wash or flush the affected area with running water for at least 15 minutes.
- Seek treatment immediately.

## Precautions to be Taken with Human Subjects

- Always use care when utilizing a human subject in a lab. Treat the subject in the same manner that you would like to be treated. Stop any procedure immediately if the subject is demonstrating any distress.
- Do not act as the subject in any lab that requires strenuous exercise if you have an abnormal heart beat, have hypertension, or have any other cardiovascular disease.
- Do not act as a blood donor if you have any known blood abnormalities such as hemophilia.
- Do not act as the subject in respiratory experiments if you have a respiratory infection or any respiratory system associated disease.
- Do not act as the subject in any lab requiring hyperventilation or the breathing of carbon dioxide if you have epilepsy.
- Do not act as the subject in the oral glucose tolerance lab if you know you are prediabetic, have diabetes type I or type II, or if you have hypoglycemia. Do not act as the subject in the oral glucose tolerance test if your fasting blood glucose concentration is above 120 mg/dL of whole blood.

## Lab Format

The exact format of your lab will depend in part on your lab instructor. However, the following are the general components of the lab you can expect to encounter in each lab.

- Review of the previous lab
  Time will be available to answer questions regarding the previous lab, including lab report questions due at the current lab. This is also the time lab quizzes will be given.
- Introduction of the current lab
  The material and concepts covered in the current lab will be explained. Special procedures used in the lab will also be explained. However, the student is expected to have read and to be prepared for the current lab.
- Setting up the equipment and experimentation
  Be sure that you have all the necessary equipment available before you begin the experimental procedure. Be clear regarding the experimental procedure before you begin the experiment to avoid having to repeat the experiment because of procedural error. Take the time to record your observations during the experimentation. These observations may help you with the interpretation of the data collected. Follow all instructions, whether written in the lab manual or given by your lab instructor. You should immediately report any accidents, spills, or unexpected outcomes to your lab instructor.

- Lab reports

  The lab reports are designed to stimulate critical thought. Reports may include questions concerning the analysis of the data you collected, the application of concepts from lecture, and applications of concepts and results to the health field. Lab reports are handed in for points and are expected to be neat and answered in your own words. Although discussion among students is encouraged, you are responsible for understanding the information and lab exercises and writing your own answers.
- Lab cleanup

  When you enter the lab, you may expect the lab to be neat and orderly. You are expected to leave the lab neater than when you entered it. You are expected to dispose of all waste materials appropriately (see the lab safety information listed earlier). Equipment should be cleaned as instructed and returned to its proper location. Your lab space should be properly cleaned (with the use of disinfectant wipes when instructed. Failure to clean and return equipment and supplies to their original condition and location may result in the loss of lab points and dismissal from the lab).
- Use of lab computers

  Computers will be used for several of the labs. These computers are to be considered lab equipment and are not to be used for any other activities nor are they to be altered in any way. Violation may result in loss of lab points and dismissal from the lab.

# Exercise 2

# Metrics and Common Computations

## Objectives

- Know the units of length, weight, volume, and temperature in the metric system.
- Be able to make conversions within the metric system.
- Be able to perform proportional calculations.
- Be able to write numbers in *scientific notation*.
- Be able to determine the *molecular weight* of a given molecule.
- Be able to determine the components of *percent* and *molar solutions* and determine the *molarity* and *osmolarity* of a solution.
- Be able to calculate *means*.
- Be able to identify *dependent* and *independent experimental variables*.
- Be able to graph data.
- Be able to interpret graphs.

## Introduction

In this lab you will practice using the metric system of measure and learn how to convert between units within the metric system. You will learn how to write and understand numbers using scientific notation. You will also practice solving for unknowns in proportional relationships. This type of calculation would be used in determining correct drug dosage for various body weights, determining a patient's rate of function compared to a standard value, and many other situations. You will also be looking at the composition of mixtures, specifically solutions. It is important to understand the normal concentrations of solutions in the body to understand the effects and impacts alteration of the makeup of the body's fluid may have (i.e., intravenous solutions and transfusions).

## Metrics - being able to convert

Although we still utilize the English or British-American system of measurements in our everyday lives, the professional world is converting to the metric system of measurement.

| | English System | Metric System |
|---|---|---|
| Length | Inch, foot, yard, mile | Millimeter, centimeter, **meter**, kilometer |
| Volume | Pint, quart, gallon | Milliliter, deciliter, **liter**, kiloliter |
| Mass | Ounce, pound, ton | Milligram, **gram**, kilogram, metric ton |
| Temperature | Degree Fahrenheit | Degree Celsius or centigrade |

This is especially prevalent in the health field and in international industry. The benefit of the metric system is that it is based on the decimal system and units of 10. The difference between metric units will always be a difference in powers of 10. For example, a decimeter is 1/10 of a meter (or $10^{-1}$) and a centimeter is 1/100 of a meter (or $10^{-2}$). Because the difference between metric units is in the power of 10, all you need to do to convert from one unit to another is to move the decimal point in the correct direction and the correct number of places. The decimal point will move to the right when you convert from a larger unit to a smaller unit, and the number of units will get larger. The decimal point will move to the left when you convert from a smaller unit to a larger unit, and the number of units will get smaller. This should make intuitive sense when you stop to think about it. Consider these examples. If you take a pie and consider it to be one unit and then cut it into 10 pieces, you now have 10 smaller pieces or subunits. In counting the pieces, you just moved the decimal point to the right one place or by a power of 10, from 1.0 unit to 10.0 units.

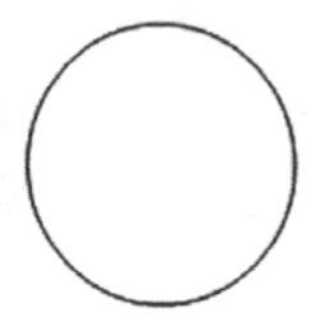

1 large unit

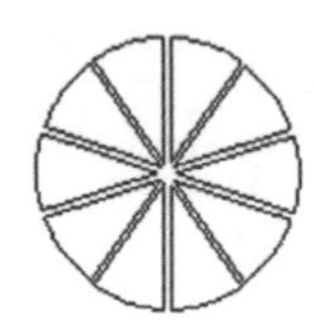

1 large unit divided into 10 equal pieces
1 large unit • $10^1$ = 10 small units

If you take 10 pieces of a metal rod and solder them together to make a one longer rod, you now have one large unit made up of the 10 original pieces or subunits. In counting the individual pieces, you just moved the decimal point to the left one place or by a power of 10, from 10.0 units to 1.0 unit.

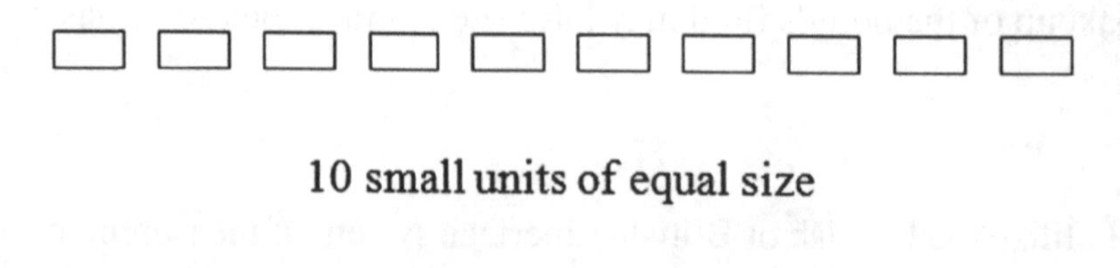

10 small units of equal size

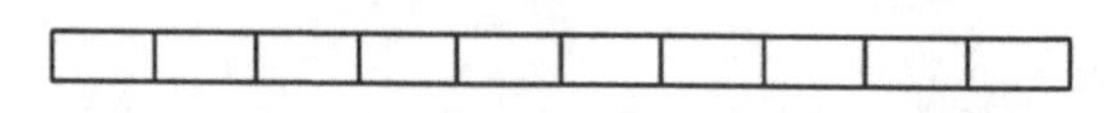

10 small units combined to make 1 large unit
10 small units • $10^{-1}$ = 1 large unit

If each of the small units was further divided into 10 very small units for a total of 100 very small units, 10 small units, and 1 large unit, the following conversions would apply:

conversion from very small units to small units
10 very small units combined to make 1 small unit
10 very small units • $10^{-1}$ = 1 small unit

conversion from very small units to large units
100 very small units combined to make 1 large unit
100 very small units • $10^{-2}$ = 1 large unit

Also, the prefixes used in the metric system are very descriptive. For example, centi- means 1/100. Thus one centimeter is 1/100 of a meter.

- **Metric Units**
  - Measure of Length
    In the metric system, the standard unit of length is the meter. To give you a comparison with the more familiar English system, 1 meter is equal to 39.37 inches or 1.094 yards.
  - Measure of Mass
    In the metric system, the base unit of mass is the gram. To give you a comparison with the more familiar English system, 1 g is equal to 0.035 ounces. The International System of Units (SI) uses the kilogram as the base unit of mass.
    The mass of an object is technically not the same as the weight of an object. The mass of an object is a measure of the quantity of matter in the object. The weight of an object is the force exerted by an object as a result of the gravitational force or gravitational pull on the matter of the object. When we determine the weight of an object, we are technically measuring the object's mass and its associated gravitational pull, and we record that weight in kilograms, grams, etc.
  - Measure of Volume
    In the metric system, the standard unit of volume is the liter. To give you a comparison with the more familiar English system, 1 L is equal to 1.057 quarts. Another measure of volume is based on linear measurement. When any linear measurement is cubed, it becomes a volume. For example, 1 meter length • 1 meter width • 1 meter height = 1 meter cubed (or 1 cubic meter).
  - Temperature
    In the metric system, the standard unit of temperature is the degree on the Celsius or centigrade (°C) scale. To give you a comparison with the more familiar English (Fahrenheit, °F) system, the freezing point of water at 32°F is equal to 0°C, and the boiling point of water at 212°F is equal to 100°C.
- **Metric Prefixes**
  It is important that you memorize the more commonly used prefixes in the metric system. Again, these prefixes are descriptive and identify the power of 10 used for that particular unit. Also, these prefixes are the same for all measures: length, weight, and volume. The following

table lists the more commonly used metric prefixes along with the abbreviation, definition, power of 10, and decimal notation.

| **Prefix** | **Abbrev.** | **Definition** | **Power** | **Decimal** |
|---|---|---|---|---|
| Giga | G | 1 billion | $10^9$ | 1,000,000,000.0 |
| | | | $10^8$ | |
| | | | $10^7$ | |
| Mega | M | 1 million | $10^6$ | 1,000,000.0 |
| | | | $10^5$ | |
| Myria | My | 10 thousand | $10^4$ | 10,000.0 |
| **Kilo** | k or K | 1 thousand | $10^3$ | 1000.0 |
| Hecto | h | 1 hundred | $10^2$ | 100.0 |
| Deka | da | 10 | $10^1$ | 10.0 |
| **Uni** | — | 1 | $10^0$ | 1.0 |
| Deci | d | 1/10 | $10^{-1}$ | 0.1 |
| **Centi** | c | 1/100 | $10^{-2}$ | 0.01 |
| **Milli** | m | 1/1000 | $10^{-3}$ | 0.001 |
| | | | $10^{-4}$ | |
| | | | $10^{-5}$ | |
| **Micro** | μ or mc | 1/1,000,000 | $10^{-6}$ | 0.000001 |
| | | | $10^{-7}$ | |
| | | | $10^{-8}$ | |
| **Nano** | n | 1/1,000,000,000 | $10^{-9}$ | 0.000000001 |
| | | | $10^{-10}$ | |
| | | | $10^{-11}$ | |
| **Pico** | p | 1/1,000,000,000,000 | $10^{-12}$ | 0.000000000001 |

## Metric Conversions

Suppose you want to convert one meter (m) to centimeters (cm). First think about whether the number of centimeters should be a number that is larger or a number that is smaller than the number of meters you are starting with. A question to ask is, "Am I cutting the length into smaller pieces, or am I trying to determine how much of a larger unit my small units make up?" Because a centimeter is smaller than a meter, for this particular question you are asking how many smaller pieces (the cm) you will have if you cut the larger length (the m) into centimeter-sized pieces. At this point, you know your answer should be greater than the one meter you are starting with. There are a couple of ways you can now do your calculation.

Method 1: Moving Decimal Points

*Example:* You know that 1 m = $10^0$ and 1 cm = $10^{-2}$ m. With this information, you can compare powers of 10 between the two units.

| km | hm | dam | m | dm | cm | mm | | | um | | | nm |
|---|---|---|---|---|---|---|---|---|---|---|---|---|
| $10^3$ | $10^2$ | $10^1$ | $10^0$ | $10^{-1}$ | $10^{-2}$ | $10^{-3}$ | * | * | $10^{-6}$ | * | * | $10^{-9}$ |

You are moving from large unit to a small unit, so you are moving to the right on the scale, and your decimal point must move to the right.

With this information you also know that there is a difference of two powers of 10 between 1 meter and 1 centimeter, so you move the decimal point in your 1 meter two places to the right: 1.0 m = 100.0 cm

How would this change if you were converting from 1 centimeter to meters?
You are now moving from a small unit to a large unit, so you are moving to the left on the scale, and your decimal point must move two places to the left: 1.0 cm = 0.01 m.

How would you use this method to make even larger conversions, such as converting from centimeters to kilometers? You must now count all the decimal places between centimeters and kilometers. You are moving from a small unit to a large unit, so you are moving to the left on the scale, and your decimal point must move to the left. With this information, you know that there is a difference of two powers of 10 between 1 centimeter and 1 meter and a difference of three powers of 10 between 1 meter and 1 kilometer for a total difference of five powers of 10.
You must move the decimal point five places to the left: 1.0 cm = 0.00001 km.

Method 2: Dimensional Analysis

*Example:* You know that (1 m/100 cm) = 1, and you know that (100 cm/1 m) = 1.
With this information, you can set up the following equivalent ratios:

$$1 \text{ m} \cdot (100 \text{ cm}/1 \text{ m}) = 100 \text{ cm}$$

Notice that the unit you are converting to is the numerator in the ratio, and the unit you are converting from is the denominator in the ratio. When you work through the equation, the units of the denominator will cancel out, and you will be left with the units of the numerator. How would this equation change if you were converting from 1 centimeter to meters?

$$1 \text{ cm} \cdot (1 \text{ m}/100 \text{ cm}) = 0.01 \text{ m}$$

How would you use this method to make even larger conversions? For example how would you convert 1 centimeter to kilometers?

$$1 \text{ cm} \cdot (1 \text{ m}/100 \text{ cm}) \cdot (1 \text{ km}/1000 \text{ m}) = 0.00001 \text{ km}$$

- Using either method, try the following conversions:

| | |
|---|---|
| 1 m = 1000 mm | 1 g = 1000000 µg |
| 1 m = .001 km | 1 g = .01 dag |
| 1 dm = 0.01 dam | 1 kg = 10000 dg |

## Scientific Notation

You have already been looking at some numbers that have been written using scientific notation. Many of the numbers used to describe or define the various metric units are written as a form of scientific notation. The point of scientific notation is to simplify the expression of the number. This is done by writing the number as a power of 10.

- **Numbers greater than 10**

  If the number is greater than 10, move the decimal point to the left so the number is between 1 and 10 and then multiply the number by the appropriate power of 10. For each place the decimal point is moved to the left, multiply by a positive power of 10. So if the decimal point is moved two places to the left, multiply by $10^2$.

  23.465 written in scientific notation would be $2.3465 \cdot 10^1$
  234.65 written in scientific notation would be $2.3465 \cdot 10^2$
  2346.5 written in scientific notation would be $2.3465 \cdot 10^3$

- **Numbers less than 1**

  If the number is less than 10, move the decimal point to the right so the number is between 1 and 10 and then multiply the number by the appropriate power of 10. For each place the decimal point is moved to the right, multiply by a negative power of 10. So if the decimal point is moved two places to the right, multiply by $10^{-2}$.

  0.2346 written in scientific notation would be $2.346 \cdot 10^{-1}$
  0.02346 written in scientific notation would be $2.346 \cdot 10^{-2}$
  0.002346 written in scientific notation would be $2.346 \cdot 10^{-3}$

  - Write the following numbers in scientific notation.

| | |
|---|---|
| 31.456 = $3.1456 \cdot 10^1$ | 65378.2 = $6.53782 \cdot 10^4$ |
| 0.0356 = $3.56 \cdot 10^{-2}$ | 0.00046 = $4.6 \cdot 10^{-4}$ |

## Proportions

A proportion shows the equivalency between two ratios. For example, 2/5 = 4/10. In the case of proportions, if you know three of the four numbers, you can calculate the fourth number.

For example, 2/5 = X/20. You can solve for X by cross multiplying.
It is easier to see if we replace the numbers with letters and solve for X.

If A/B = X/Y, then we can rearrange the equation to solve for X:

$$Y \cdot A/B = X.$$

If we put our numbers (2/5 = X/20) into the equation and rearrange the equation to solve for X:

$$20 \bullet 2/5 = X$$
$$X = 8.$$

- Solve the following proportion problems.

3/6 = X/8  X= 4____
Hint: 8 • 3/6 = X

1/3 = X/96 X= 32____
Hint: 96 • 1/3 = X

12/20 = X/8 X= 24/5____

5/7 = X/28 X= 20____

## Mixtures

A mixture is a general term used to describe any two substances mixed together. The substances could be liquid, solid, or gas. We will often use a more specific type of mixture referred to as a **solution**. A solution is specifically a **solute** dissolved in a **solvent**. If you think of salt dissolved in water, salt is the solute, water is the solvent, and saltwater is the solution. When water is the solvent, the solution is specifically referred to as an **aqueous solution**. Because the body is close to 70% water, water serves as the solvent for our bodies. Some mixtures are not solutions but rather suspensions. In a **suspension**, the particles are insoluble, meaning that they will not dissolve but may stay suspended in the mixture. However, these suspended particles will eventually settle to the bottom. In your blood, the cellular components are particles that will settle out of the plasma if the blood is allowed to sit in a stationary environment without disturbance. In a **colloid suspension** the colloids are very small particles that do not dissolve but are small enough to stay in suspension without periodic mixing. The plasma proteins in your blood form a colloid suspension.

There are several descriptive terms that are used in conjunction with the term solution. If a solution is **homogeneous,** the solutes are evenly dispersed throughout the solvent, and the solution is uniformly mixed. If a solution is **heterogeneous**, the solutes are not evenly dispersed throughout the solvent. A **saturated** solution contains as many solute particles as it can possibly hold in solution under normal conditions. If more solute is added to a saturated solution, the additional solute will not be able to be dissolved and will **precipitate** to the bottom.

- **Types of solutions**
  - Percent solution

    A percent solution is determined by the number of parts of solute per 100 parts of solution. Physiologic saline is approximately 0.87% NaCl or 0.87 g of NaCl mixed in enough water to make 100 mL of physiologic saline solution.
    - Given the following % solutions, determine the amount of solute and the approximate amount of solvent in each.

1 L of physiologic saline solution:

8.7 g NaCl and enough water to make 1000 mL of the solution

100 mL of 10% glucose solution:

__________ g glucose and enough water to make __________ mL of the solution

- **Molarity**

  Molarity is the number of moles of solute dissolved in 1 L of solution (or moles per liter). To determine the weight of a specific number of moles of a solute, you must know the molecular weight of the solute. The molecular weight of a solute is calculated by adding the atomic weight of all the atoms in the solute. If you consider a 1-molar aqueous solution of NaCl, the molecular weight of 1 mole (M) of NaCl is the sum of the atomic weight of 1 mole of $Na^+$ (23 g) and 1 mole of $Cl^-$ (35.5 g). A 1-molar solution of NaCl is made up of 23 g ($Na^+$) + 35.5 g ($Cl^-$) = 58.5 g of NaCl dissolved in enough water to make 1 L of solution. A 2-molar solution of NaCl will consist of 117 g (2 M • 58.5 g/M) of NaCl dissolved in enough water to make a 1-L solution.

  - Given the following atomic weights, determine the amount of solute in 1 L of the following solutions.

    H = 1 g, C = 12 g, O = 16 g, Na = 23 g, Mg = 24.3 g, and Cl = 35.5 g

    4-molar solution of NaCl 234 g

    2-molar solution of $MgCl_2$ 190.6 g

    1-molar solution of $C_6H_{12}O_6$ 180 g

    72 12 96

  - How many grams of the solute would be required to make each of the following solutions?

    2 L of a 4-molar solution of NaCl __________ g

    3 L of a 2-molar solution of $MgCl_2$ __________ g

    5 L of a 1-molar solution of $C_6H_{12}O_6$ __________ g

- **Osmolarity**

  Osmolarity is similar to molarity, except that it is a measure of the number of particles in solution. Although adding 1 mole of NaCl to enough water to make 1 L of solution makes a 1-molar solution, it makes a solution with an osmolarity of 2. The reason the osmolarity and the molarity of this solution is not the same is because NaCl dissociates when dissolved in water. This means that although when dry, one molecule of NaCl is one particle, when dissolved in water, NaCl splits apart (dissociates) into two particles, one $Na^+$ ion and one $Cl^-$ ion. Because we are measuring osmolarity we must count each ion as a separate particle (see the following figure).

**Molarity: NaCl molecules - 7 particles**

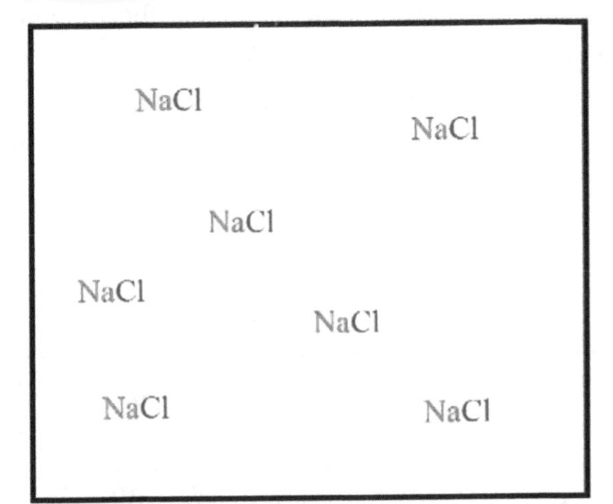


**Osmolarity: Na+ and Cl- ions - 14 particles**

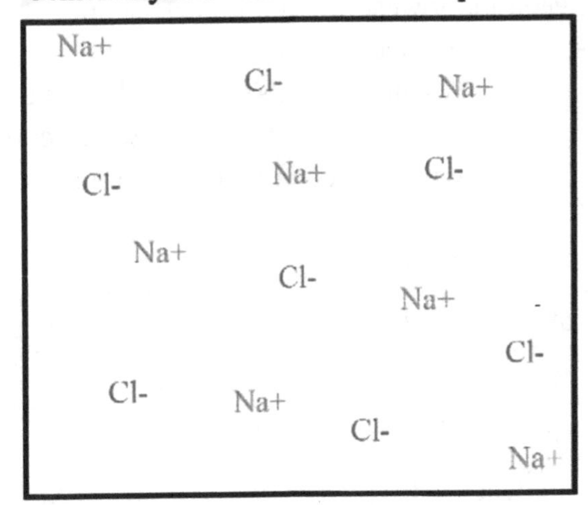


- Determine the osmolarity of the following solutions.

  2 moles of NaCl in a 1-L solution __4__ osmoles
  (NaCl dissociates completely to $Na^+ + Cl^-$)

  2 moles of $C_6H_{12}O_6$ in a 1-L solution __2__ osmoles
  (glucose does not dissociate) does not break apart

  1 mole of $MgCl_2$ in a 1-L solution __3__ osmoles
  ($MgCl_2$ dissociates completely to $Mg^{++} + Cl^- + Cl^-$)

## Means

The arithmetic mean or average of a set of numbers is calculated by summing all the numbers in the set and then dividing that sum by the number of individual values added together. This is represented by the equation:

$\overline{X} = \Sigma X/N$
$\overline{X}$ is the mean.
$\Sigma X$ is the sum of all the values of X.
N is the number of individual values of X.

The following numbers are the scores on a physiology lab quiz.

8, 9, 7, 6, 8, 9, 7, 7, 7, 8, 9, 10, 10, 7

The mean score for the quiz is calculated as:

$\Sigma X = 8+9+7+6+8+9+7+7+7+8+9+10+10+7 = 112$ with $N = 14$
$\overline{X} = 112/14 = 8$

When calculating a mean, the mean value should always fall within the range of the scores (between the lowest and the highest score). If you calculate a mean that is outside the range of the scores, you have done something wrong, and you should recalculate the mean.

- Count your pulse for 1 full minute. Write your pulse rate (bpm) and age on the chart provided by your lab instructor. Record the class data in the following table. These numbers will be used in your lab report.

Chart for pulse rate (bpm) and student age

| Student | 1 | 2 | 3 | 4 | 5 | 6 | 7 | 8 | 9 | 10 | 11 | 12 | 13 | 14 | 15 | 16 |
|---|---|---|---|---|---|---|---|---|---|---|---|---|---|---|---|---|
| bpm | 88 | 68 | 68 | 80 | 91 | 78 | 67 | 84 | 86 | 68 | 80 | 74 | 86 | 76 | 70 | 68 |
| Age | 20 | 19 | 19 | 18 | 17 | 18 | 18 | 18 | 18 | 18 | 20 | 17 | 20 | 18 | 19 | 18 |
| Student | 17 | 18 | 19 | 20 | 21 | 22 | 23 | 24 | 25 | 26 | 27 | 28 | 29 | 30 | 31 | 32 |
| bpm | 72 | 70 | 72 | 72 | 86 | 94 | 60 | 60 | 88 | 74 | 62 | 68 | 72 | | | |
| Age | 18 | 20 | 19 | 20 | 20 | 23 | 20 | 19 | 19 | 24 | 19 | 18 | 18 | | | |

## Graphs

A graph gives a visual representation of data. A graph shows the relationship between two sets of experimental values or **variables**, often comparing a dependent variable against an independent variable. An **independent variable** is the variable that you manipulate in an experiment. An example could be temperature or surface area. Think about the independent variable as the experimental variable that causes something to change or to occur. A **dependent variable** is the variable that you are monitoring in an experiment. An example could be rate of diffusion. Think about the dependent variable as the experimental variable that changes due to a change in the independent variable. When graphing data, the independent variable (the $X$ value) is plotted along the $X$-axis, and the dependent variable (the $Y$ value) is plotted along the $Y$-axis. On the following graph, for one point (2,1) the independent value ($X$) is 2 and the dependent value ($Y$) is 1, and for the second point (4,3) the independent value is 4 and the dependent value is 3.

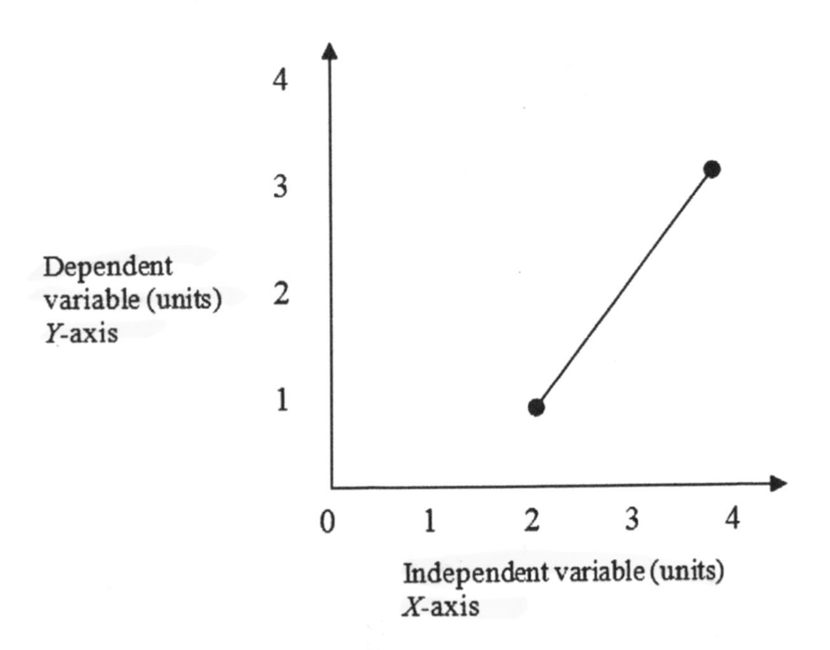


In the relationship illustrated in the graph, the dependent variable increases with the independent variable; this relationship is a **direct relationship**. If the dependent variable decreased as the independent variable increased, or the dependent variable increased as the independent variable decreased, the relationship would be an **indirect relationship**. The relationship illustrated in the graph is also a **linear** relationship. When the points graphed for a relationship do not form a straight line but rather a curved line, that relationship is **curvilinear**.

Name: Heidi Scoggins

Date: 2/8/21

## Exercise 2 Review Questions: Metrics and Common Computations

1. Using a metric ruler, measure the diameter of the circle. Give the diameter in the appropriate units.

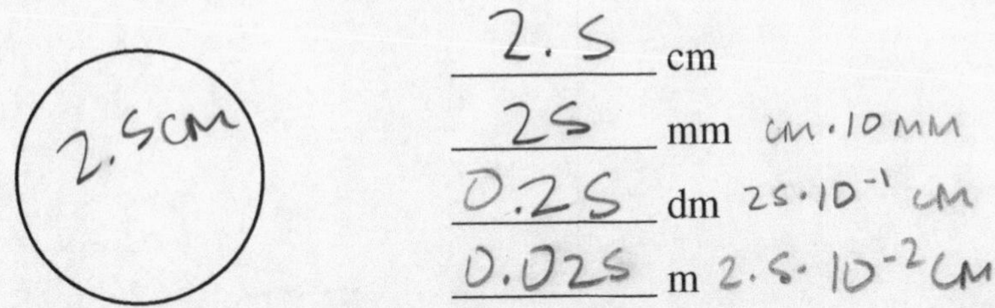


2. Make the following conversions and then write your answers in scientific notation.

| | Conversion | Scientific notation | | Conversion | Scientific notation |
|---|---|---|---|---|---|
| 5 cm = | 50 mm | $5 \times 10^1$ mm | 6892 dm = | 0.6892 km | $6.8 \times 10^{-1}$ km |
| 3.69 g = | 0.00369 kg | $3.69 \times 10^{-3}$ kg | 67 dag = | 670000 mg | $6.7 \times 10^{-5}$ mg |
| 0.0573 L = | 0.0000000573 µL | $5.73 \times 10^{-8}$ µL | 327 µL = | 0.327 mL | $3.27 \times 10^{-1}$ mL |

3. Solve for X in the following proportions.

3/8 = X/24 X = 9 15/45 = X/12 X = 4

X/30 = 1/6 X = 6 X/19 = 5/95 X = 1

4. A man weighs 85.5 kg. The standard dosage for a particular drug prescribed for this man is 2 mg per kg, twice a day. How much of this drug, in grams, should the man be given in each dose? How much of this drug, in grams, should the man be given per day?

Amount per dose = 0.171 g Daily amount = 0.342 g

5. Determine the molecular weight of NaCl and calculate the number of grams needed to make a 2-L solution of 2-molar NaCl (see the lab for atomic weights).

234 g NaCl in 2-liters of a 2-molar solution

6. A total of 360 g of glucose ($C_6H_{12}O_6$) is dissolved in enough water to make 1 L of the glucose solution. What is the molarity of the solution? What is the osmolarity of the solution? (see the lab for atomic weights and dissociation information)

Molarity = 2 mol/L Osmolarity = 2 osmol/L

7. A total of 381.2 g of magnesium chloride ($MgCl_2$) is dissolved in enough water to make 1 L of $MgCl_2$ solution. What is the molarity of the solution? What is the osmolarity of the solution? (see the lab for atomic weights and dissociation information)

Molarity = 4 mol/L Osmolarity = 12 osmol/L

381.2 / 95.39/mol = 4 mol 3·4 = 12

8. Calculate the average pulse rate for your class.

N = 32

ΣX = 2444

$\bar{X}$ = 76

$\bar{x} = \Sigma X / N$

9. Construct a graph of the student pulse rates versus the age of the student. Which variable is the independent variable? Which variable is the dependent variable? Be sure to label each axis properly.

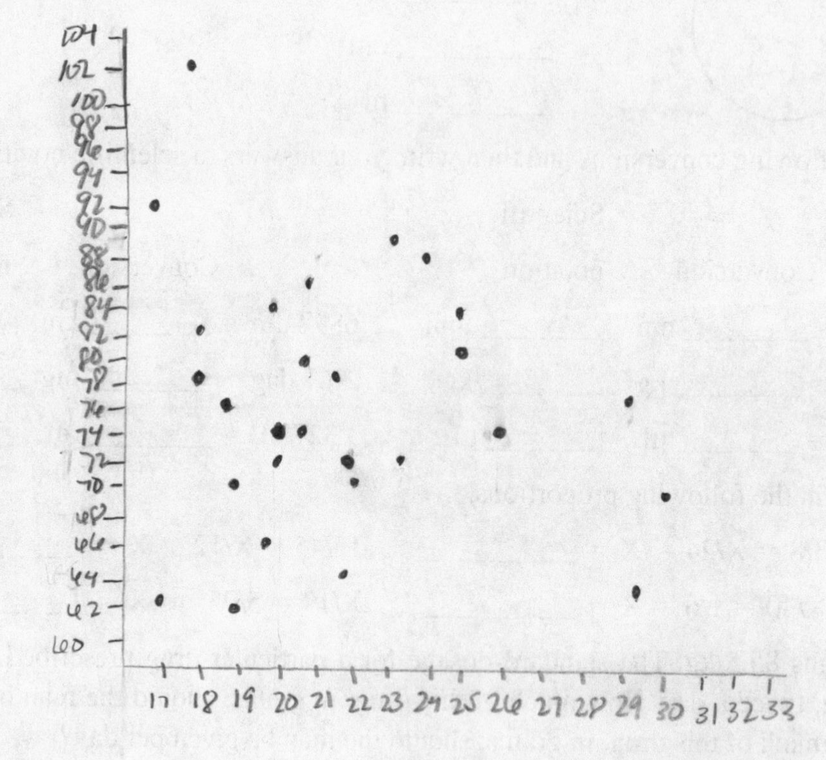

# Exercise 3

# Compound Light Microscope

## Objectives

- Be able to identify the structures of the microscope and explain their functions.
- Be able to calculate the *total magnification* when looking through the *ocular* and *objective lenses*.
- Be able to explain the following terms associated with microscope use: *magnification*, *field of view*, *depth of field*, and *parfocal*.
- Be able to calculate the actual size of an image based on its apparent magnified size.
- Be able to explain how changing the objective lens will alter the amount of light necessary for viewing the specimen.
- Be able to explain how changing the objective lens will alter the *field of view*.
- Be able to explain how changing the objective lens will alter the *depth of field*.
- Be able to prepare a wet mount slide.
- Be able to explain the procedure for the use of the *oil immersion lens*.
- Be able to explain proper use and care of the microscope.

## Lab Safety Reminders

- Handle your own saliva and cells.
- Dispose of hazardous materials properly.
- Disinfect your work area.

## Introduction

The microscope is a tool that allows us to see specimens or details of specimens that we could not see with the eyes only. In either case, it is the ability of the microscope to magnify the specimen and increase its apparent size that allows us to do this. A **compound microscope** (Figure 1) uses two or more lenses to magnify a specimen. The lenses you look directly through are called the **ocular lenses**. Because our microscopes have two ocular lenses, one for each eye, they are called **binocular microscopes**. Our microscopes also have a set of four **objective lenses**. The objective lenses are located on the nosepiece or revolving turret of the microscope. These lenses can be rotated into place so that you can observe your specimen using different powers of magnification. All lenses should be cleaned *only* with lens paper, never paper towels or other types of paper or cloths. If any lenses are extremely dirty, they may be cleaned with the lens cleaning solution available. To clean either an ocular or an objective lens, apply a drop of the lens cleaning solution to a piece of the lens paper and rub the lens. You may then want to rub the lens with a dry piece of lens paper.

The microscope should always be carried with two hands, one hand holding the arm that connects the head of the microscope to the body of the microscope and one hand under the base of the microscope. When storing the microscope, the lowest power objective lens (the scanning lens) should be in place, and the mechanical stage should be at its lowest position. The light should be turned off before unplugging the microscope from the power source. The light cord should be wound around the microscope and secured in place for storage.

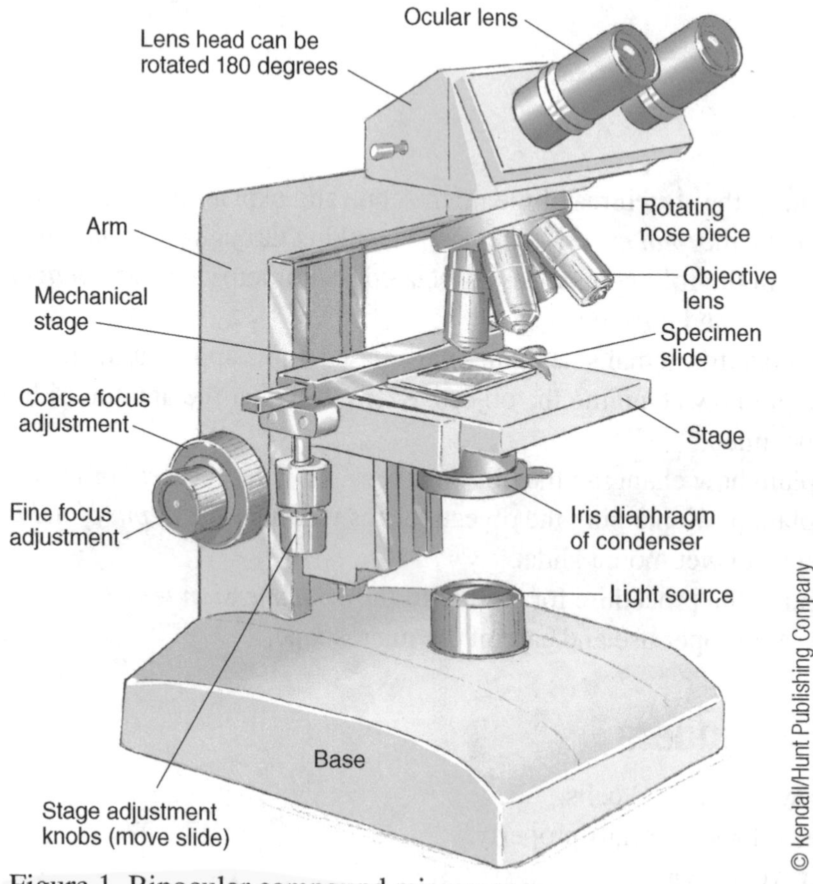




Figure 1 Binocular compound microscope

be able to identify

## Magnification

The ocular lenses on our microscopes have a **magnifying power** of 10 or 10X. That means they magnify the specimen to 10 times its true size. The objective lenses are each labeled with their power. The smallest lens or the **scanning lens** is labeled with a 4X. That means it magnifies your specimen four times. The **low-power** lens has a magnification power of 10X. The **high-power** lens has a magnification power of 40X. The **oil immersion** lens has a magnification power of 100X. To calculate the total magnification power of the microscope, you must multiply the power of the ocular lens and the power of the objective lens in use.

If you know the apparent size of the specimen you are looking at, you can calculate the actual size of the specimen if you also know the magnification you are using to look at the specimen. Because the magnification of a specimen basically multiplies its size, the actual size can be found by dividing the apparent size of the specimen by the total magnification being used to observe it. For example, if your specimen appears to be 3 mm in length when looking through the ocular lens and the scanning lens, you would use the following equation to calculate the actual size of your specimen: 3 mm / 40X.

Although it would seem that greater magnification is always better, you do not always want to use the highest power lens. There are several characteristics of the compound light microscope that change when you change magnification. These should be taken into consideration when deciding what power of magnification to use. We will look at these characteristics as we work through the lab and become more familiar with the light microscope.

## Light

The light source on your microscope is adjustable so you can control the amount of light being produced. You can also control the amount of light reaching your specimen by adjusting the **iris diaphragm** lever located at the base of the stage (see Figure 1). By adjusting the diaphragm lever, you are increasing or decreasing the diameter of the opening that allows light to pass through the slide and your specimen.

The **condenser**, located below the diaphragm (see Figure 1), can also be adjusted. The purpose of the condenser is to change the focal point of the light. You will want to adjust the condenser, by moving it up or down, so the light is focused on your specimen.

The oil immersion lens provides the greatest magnification of all the objective lenses. However, there are some restrictions with using this lens. This lens has a very small diameter and because light rays scatter as they pass through air, not much light actually gets into the lens. This makes the specimen difficult to see. By placing a very small drop of immersion oil between the specimen and the lens, fewer light rays are scattered, and more light is directed into the lens. Your lab instructor will demonstrate the procedure for using the oil immersion lens. Important points: Use oil only with the oil immersion lens. The oil *must* be cleaned off the lens. This is done by wiping the lens with lens paper until no more oil is observed on the lens paper. The lens cleaning solution may then be used in a final cleaning of the lens. The oil must also be cleaned off the slide. The oil may be blotted off the slide with a paper towel or wipe. Then clean the slide with lens paper.

General rules of use for light levels are as follows: the lower the power of your objective lens and the wider your field of view, the less light you will need. As you increase magnification, you will need to increase light, and you will also need to adjust your condenser to better focus the light on the specimen.

- **Light Activity**
  Turn the microscope on, and using the scanning lens, observe the changes in light as you adjust the diaphragm and the condenser.

## Finding Your Specimen

Place a prepared slide of three crossed threads in the slide clip on the mechanical stage (see Figure 1). Adjust the position of your slide with the mechanical stage so that the specimen is directly below the objective lens that is currently in use. When looking for your specimen, always start with the scanning lens in place. You may want to adjust the intensity of the light when you first look through the microscope. Adjust the light so that you have plenty of light, but it is not painful to your eyes. When you look through the microscope your specimen will probably not be initially visible. Move the slide using the mechanical stage until you can see something, even if it is out of focus. You may even be finding the edge of the coverslip. That is okay, because it will give you an object to focus on.

## Focusing

When you begin looking at a specimen, it should be with the scanning lens in place. When you have located your specimen on the slide, you can then focus on the specimen so that it comes into sharp focus. You may use the **coarse focus** (see Figure 1) with the scanning lens to initially locate your specimen. Once you have focused with the coarse focus, the **fine focus** (see Figure 1) should be used to make minor adjustments to fine-tune the focusing. Once you have found and focused on your specimen, move your specimen to the center of your field of view. You may then rotate the low-power lens into place. If you centered the specimen before changing the objective lens, it should be visible but may not be in perfect focus. You may use the coarse focus with the low-power lens, but because your microscopes are **parfocal**, you should need to use only the fine focus. Again, once you have located, centered, and focused on your specimen with the low-power lens, you may rotate the high-power lens into place. You may use only the fine focus to adjust the focus with this lens in place.

Rules to remember: Start with the lowest magnification and work your way up. You may use the coarse focus only with the scanning and low-power lenses. You may use the fine focus with all lenses.

## Field of View

The **field of view** is the diameter of the area you can see when looking through the microscope.

- **Field of View Activity**
  - With the scanning lens in place, place a small metric ruler on the stage of the microscope.
  - While looking through the microscope, find the millimeter lines on the ruler.
  - Center the ruler across the field of view and focus so that you can count the millimeter lines.
  - Measure the field of view at its widest diameter.
  - Record the diameter below.
  - Change to the low-power lens and repeat the procedure.
  - Change to the high-power lens and repeat the procedure.

Scanning field of view ____5____ mm

Low-power field of view ____2____ mm

High-power field of view ____1____ mm

## Depth of Field

The depth of field is the depth or thickness of the specimen you can focus on with a particular lens.

- **Depth of Field Activity**
  - Obtain a prepared slide of three crossed threads.
  - Using the scanning lens, find the area where the three threads cross.
  - How many threads can you see in focus at one time? Record the number below.
  - Change to the low-power lens and repeat the procedure.
  - Change to the high-power lens and repeat the procedure.

    Number of threads in focus: Scanning lens ___3___

    Low-power lens ___3___

    High-power lens ___1___

## Optical Inversion

Because you are looking through a lens, the specimen will have been optically inverted.

- **Optical Inversion Activity**
  - Obtain a prepared slide of the letter *e*.
  - Place the slide on the stage so that the letter *e* is in the normal orientation.
  - With the scanning lens in place, find the specimen.
  - How does the letter *e* appear when looking through the microscope? Draw the letter *e* as you see it through the microscope. ___e___
  - Move the stage so that the slide moves away from you. What direction did the letter *e* appear to move when looking through the microscope?

    ___up___
  - Move the stage so that the slide moves toward you. What direction did the letter *e* appear to move when looking through the microscope? ___down___
  - Move the stage so that the slide moves to your right. What direction did the letter *e* appear to move when looking through the microscope?

    ___right___
  - Move the stage so that the slide moves to your left. What direction did the letter *e* appear to move when looking through the microscope? ___left___

## Making a Slide

Figure 2 illustrates the steps for preparing a wet mount slide. Your instructor will also demonstrate the procedure. You will make a slide of cheek cells. The cheek cells are simple squamous epithelium and should appear light blue when stained. The nucleus will be stained a darker blue than the cytoplasm of a cell as will any bacteria you may have also collected.

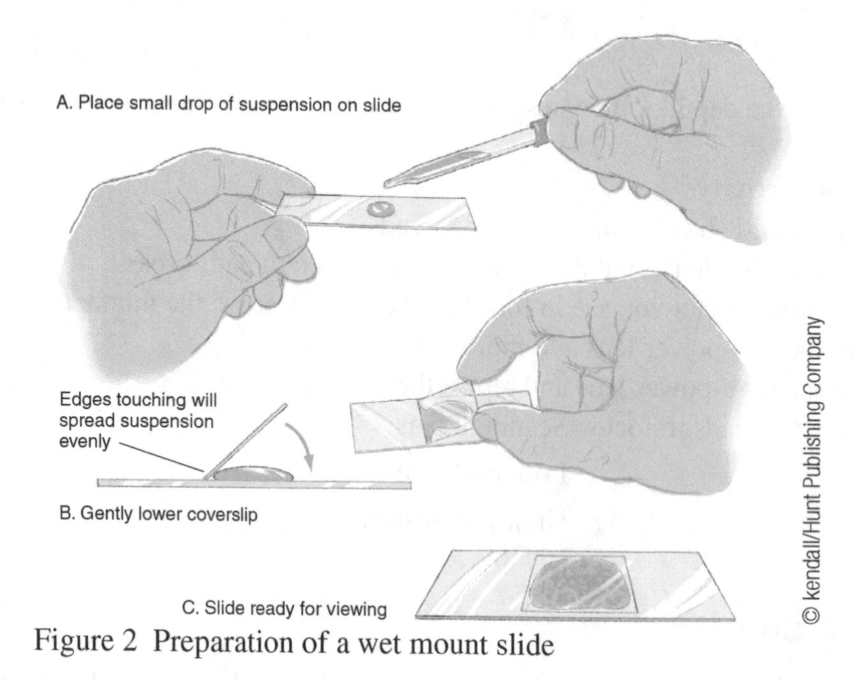
A. Place small drop of suspension on slide
Edges touching will spread suspension evenly
B. Gently lower coverslip
C. Slide ready for viewing
© kendall/Hunt Publishing Company

Figure 2 Preparation of a wet mount slide

**Figure 3 Buccal smear showing a squamous epithelial cell. Note the more darkly stained nucleus in the cell and the darkly stained bacterial cells on and near the epithelial cell.**

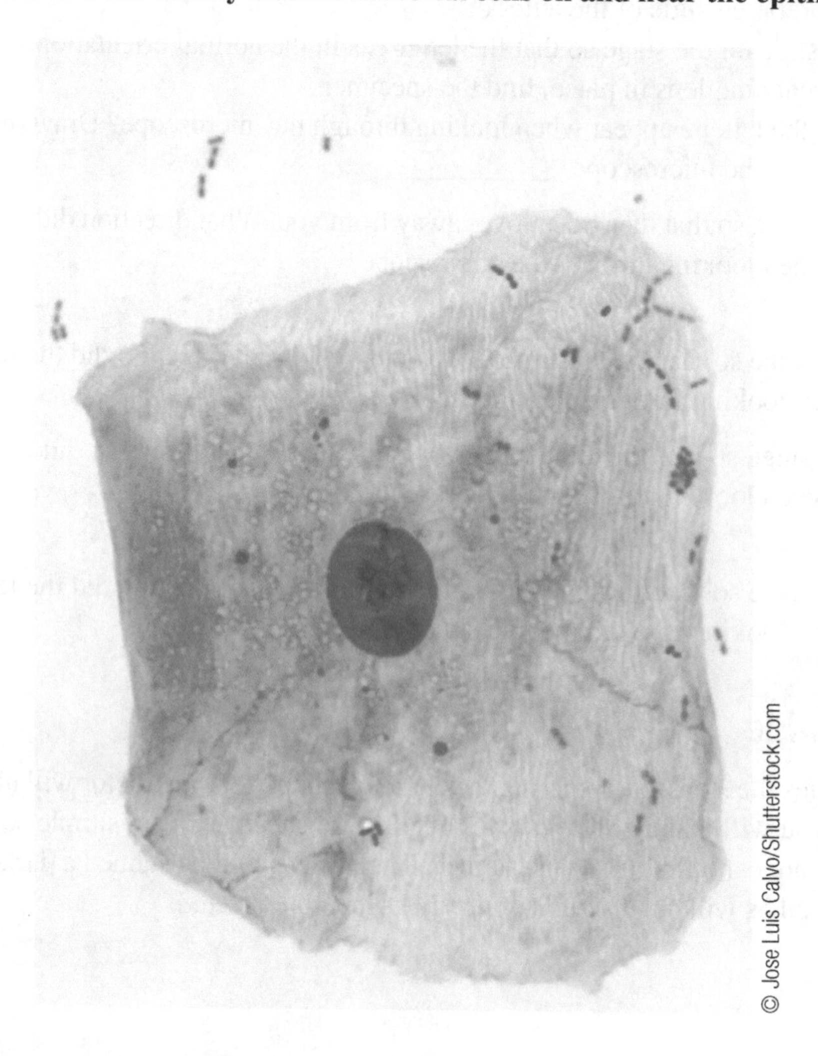
© Jose Luis Calvo/Shutterstock.com

- **Slide Preparation Activity**
  - Obtain a toothpick, a clean microscope slide and coverslip, and a dropper of dilute methylene blue stain.
  - Gently scrape the inside of your cheek with the toothpick and spread the scraping in the center of a clean microscope slide, making an even smear.
  - Next, add a small drop of dilute methylene blue to the smear. Do not touch the tip of the dropper to the smear; this will contaminate the bottle of stain. Caution, methylene blue will stain your skin as well as your clothes.
  - Place a coverslip over the specimen (see Figure 3).
  - Wipe off any excess stain with some absorbent paper.
  - With the scanning lens in place, find the specimen.
  - Make a sketch of the cheek cells in the space provided. Draw the cells so they are proportional to the field of view (as measured in the Field of View Activity).
  - Look at the cheek cells using all the remaining objective lenses, including the oil immersion lens. Make sketches of the cheek cells as observed with each of the lenses.
  - Observe the following characteristics of the microscope: parfocal, field of view, and depth of field and also observe the necessary changes in light as you use the different objective lenses for viewing the cheek cells.

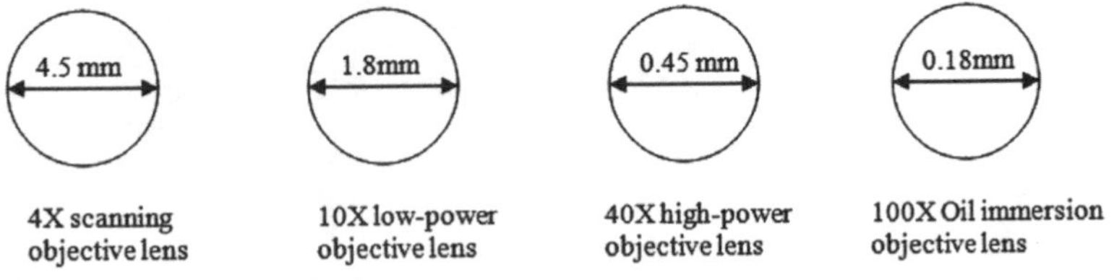


Figure 4 Field of view

Because you know the measurement of the field of view for the scanning, low-power, and high-power objective lenses (see Field of View Activity and Figure 4), you can estimate the diameter of the cheek cells. Return to either the low-power or the high-power objective lens and estimate the diameter of your cheek cells.

Diameter of cheek cells = __________ mm

## Care of the Microscope

### Cleaning the Microscope

When you finish using the microscope, the ocular and all objective lenses should be cleaned. Use ONLY the lens paper to clean the microscope lenses. Other types of paper can scratch the lenses. You may apply a drop of lens cleaning solution to a piece of lens paper and make one circular pass on the lens. Additional wiping of the lens with the same piece of lens paper may grind dirt or grit into the lens. If the lens is still dirty, obtain another piece of lens paper and repeat the procedure.

how to calculate actual size of organism?

3mm in length

$$\frac{\text{apparent size}}{\text{total magnification}}$$ = know lenses you are looking through

scanning lense = 40x
lower lense = 100x

$\frac{3}{40}$ = actual size

# Exercise 4

# Diffusion, Osmosis, Dialysis, and Tonicity

## Objectives

- Be able to distinguish between intracellular fluid and the extracellular fluids.
- Be able to define the terms *permeable* and *semi-* or *selectively permeable*.
- Be able to define and differentiate between the processes of *diffusion* and *osmosis*.
- Be able to explain *net diffusion* and *equilibrium*.
- Be able to explain *Fick's law* and the effects of each of the components of Fick's law on the rate of diffusion.
- Be able to explain the process of *dialysis* and *filtration*.

## Lab Safety Reminders

- Dispose of hazardous materials properly.
- Disinfect your work area.

## Hints for Success

You will be conducting multiple lab exercises simultaneously. Be sure to begin with the exercises that require a longer time to complete. The other labs can be started or conducted during incubation or waiting periods required for the longer exercises.

## Introduction

The human body consists of atoms organized into molecules and compounds. These are further organized into cells. Cells of a similar type are organized into tissues. Combinations of tissues form organs. A collection of organs forms an organ system. The organ systems then make up the body. In addition to the organized collection of cells, the body also contains a large amount of fluid. The fluid can be divided into two categories based on location. **Intracellular fluid** (ICF) is located within cells. **Extracellular fluid** (ECF) surrounds cells. The extracellular fluid can be divided into two categories: **interstitial fluid** and **plasma**. Approximately 60% of body weight is fluid (total body water, TBW). Of the total body water, two-thirds of the volume is intracellular fluid and one-third is extracellular fluid (ECF). Of the extracellular fluid, one fifth or 20% is plasma, and four fifths or 80% is interstitial fluid. Figure 1 represents the distribution of body fluids for a 70-kg man.

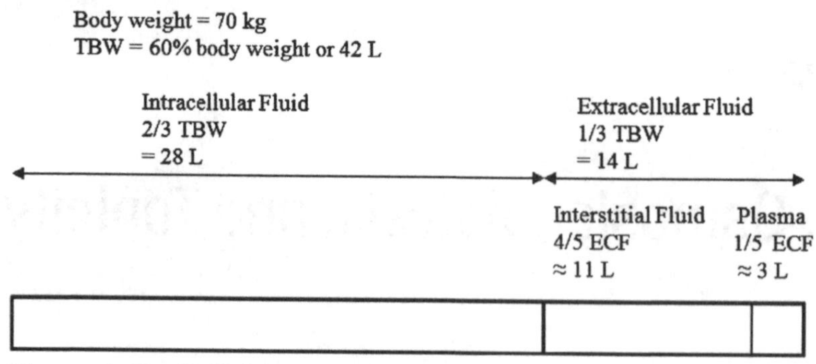


Figure 1 Body fluid compartments

Body fluids move among the compartments. For example, within the extracellular fluids, plasma and interstitial fluid are constantly exchanged at the level of the capillaries. The composition of these two fluids is very similar. Fluid is also exchanged between the interstitial fluid and the intracellular fluid. However, the composition of the intracellular fluid can vary widely among different types of cells and from the extracellular fluid. The movement of water and dissolved ions within and among the different water compartments can occur by several means, depending on the specific substance and movement.

## Diffusion, Osmosis, and Dialysis

The simplest movement within and between the compartments is by diffusion. **Diffusion** is the random movement of atoms or molecules from one area to another. Diffusion can occur in any state: gas, liquid, or solid. This random movement of atoms and molecules between two places occurs in both directions. However, the movement of atoms or molecules from the area of higher concentration to the area of lower concentration will exceed the movement of atoms or molecules from the area of lower concentration to the area of higher concentration. The difference in the amount of movement between the areas is the net diffusion. The term **net diffusion** refers to the overall movement of the atoms or molecules, from the area of higher concentration to the area of lower concentration. In the process of diffusion, net movement continues until there is an equal distribution of the atoms or molecules. In Figure 2, the solute (stars) can move in both directions, from area A to area B and from area B to area A. However, the concentration of the solute is higher in area B. As a result of the concentration gradient, the net diffusion of solute will be from B to A (from the area of high concentration to the area of low concentration). The net diffusion of the solute from B to A will continue until there is an equal distribution of solute (no concentration gradient). When this occurs, equilibrium has been reached, diffusion in both directions will be equal, and net diffusion will equal zero.

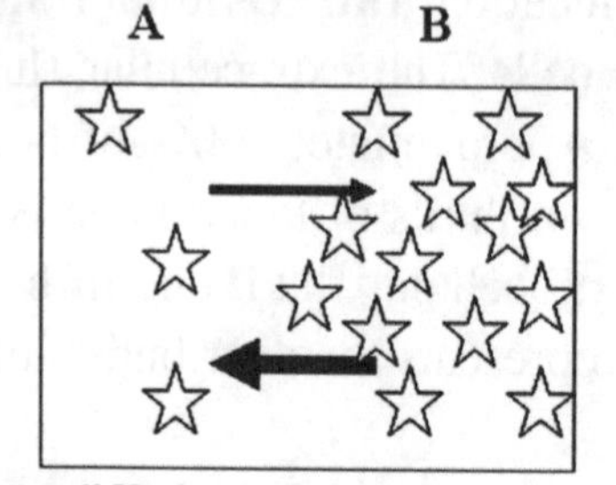


Net diffusion is from area B to area A

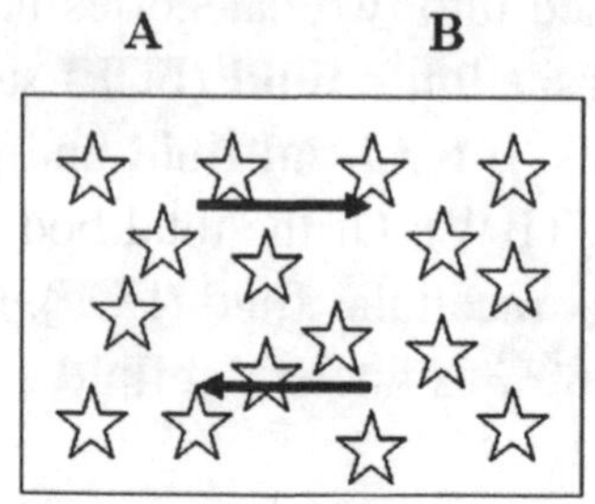


Equilibrium: net diffusion = 0

Figure 2 Net diffusion

One example of diffusion in the body is the net diffusion of oxygen from the air in the alveoli of the lungs to the plasma of the blood and then the majority diffuses into the red blood cells. That oxygen is transported through the body and then diffuses from the red blood cells to the plasma, then to the interstitial fluid and into the cells.

**Flux** is the rate of movement of a solute measured as amount of movement of the solute per time. Flux is influenced by several factors. **Fick's law** relates some of these factors. - don't need to focus on this

Flux is proportional to $(C_o - C_1)(A)(°C) / (D \cdot MW)$

where: $C_o - C_1$ = difference in concentration between side A and side B
A = area available for diffusion between side A and side B
°C = temperature
D = distance from side A to side B
MW = molecular weight of the diffusing substance

A **semi- or selectively permeable membrane** will allow some molecules to move freely through the membrane (freely permeable), whereas other molecules cannot pass through (impermeable) at all. The permeability of a membrane for other molecules will be somewhere between freely permeable and impermeable. In a cell, the permeability of the membrane will influence flux.

**Osmosis** differs from diffusion because it is specifically the movement of a solvent across a semi- or selectively permeable membrane. Although the solute cannot cross the membrane, the solvent can. Just as a solute would be expected to diffuse following its concentration gradient to reach equilibrium, so would the solvent. In the human body, the solvent is water, so osmosis generally refers to the movement of water across a selectively permeable membrane. Figure 3 illustrates the outcome of osmosis. In beaker A, the two sides of the beaker are separated by a selectively permeable membrane, and the solutions in each side of the beaker have different concentrations. Although the solute can't diffuse through the membrane, the water (solvent) can. Notice that in beaker B, so much water has moved through the membrane that the volume of water is now greater on one side of the membrane than on the other. The movement of water stops only when the concentration of the solution is equal on both sides of the selectively permeable membrane or when the forces opposing osmosis (example: hydrostatic pressure) equal the osmotic pressure.

The process of **dialysis** is a form of ultrafiltration of fluid and diffusion of solutes across a semi- or selectively permeable membrane. Dialysis tubing is a semi- or selectively permeable membrane usually made from cellulose or cellophane. Solvent can pass through the wall of the dialysis tubing following a pressure gradient and also following a concentration gradient. Solutes diffuse through the wall of the dialysis tubing following their gradients, but only if the solutes are small enough to fit through the pores in the semi- or selectively permeable membrane. Molecules too large to fit through the pores are not able to diffuse through the membrane. The size of the pores in the dialysis membrane determines which solutes can pass through. This is the basic concept behind the process used in kidney dialysis, although the actual process is much more complex. Although we normally think of kidney dialysis as a process used to remove excess fluid and solutes from the blood, dialysis can also be used to add solutes to the blood. One such solute is bicarbonate ($HCO_3^-$). The

Dissolve ≠ Diffusion

* Starch stayed in bag
* Salt leached through bag

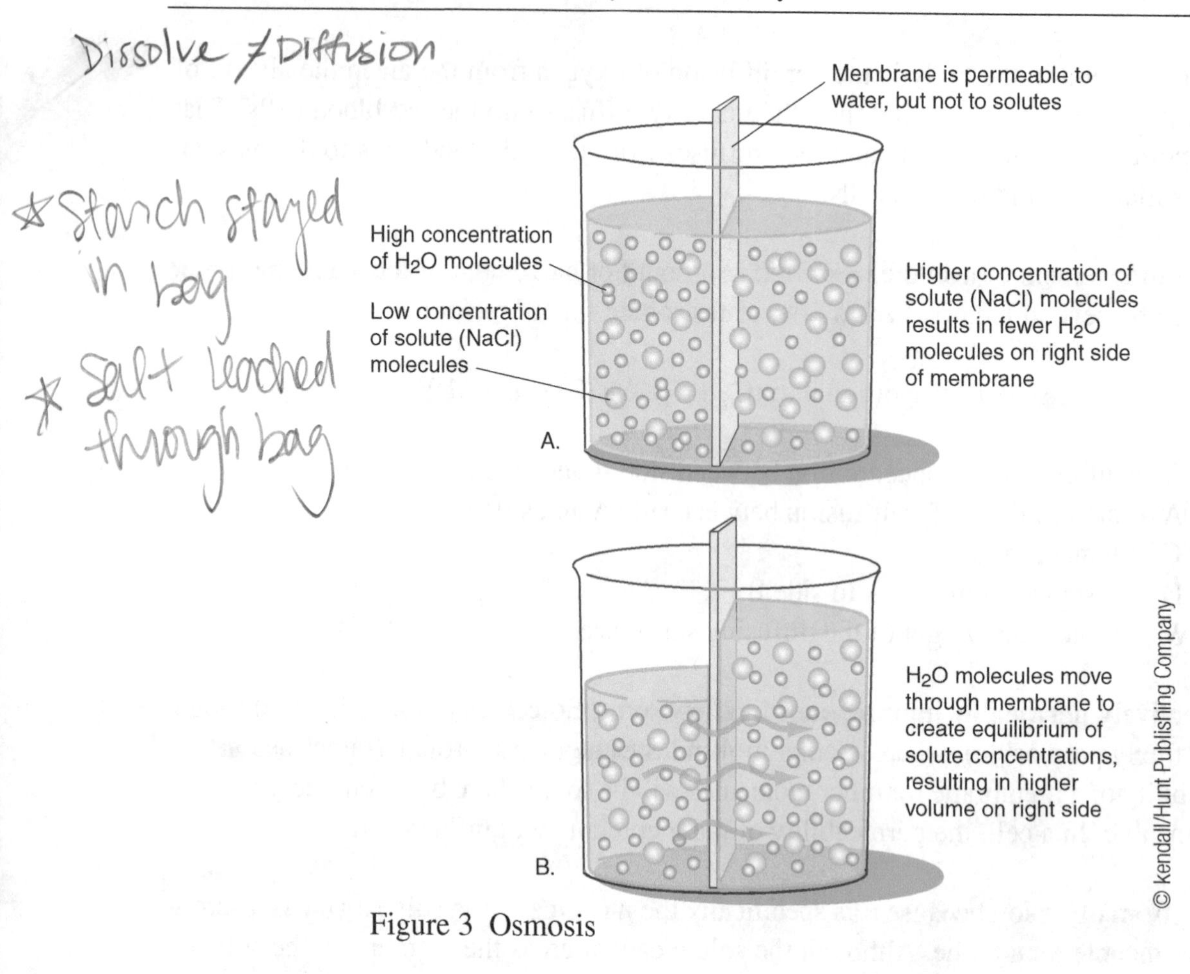




Figure 3 Osmosis

concentration gradient established in kidney dialysis causes bicarbonate to diffuse into the blood. In the blood, bicarbonate acts as a pH buffer to help neutralize metabolic acids. In the osmosis and dialysis lab activity, we will focus on the effects of the solvent and solute concentration gradients and the pore size of the selectively permeable dialysis membrane.

## Simple Diffusion Activity

- Obtain a 50 mL beaker and fill it approximately one fourth full of tap water.
- Sit the beaker on your lab bench in a place out of the way but still visible.
- Using a pair of forceps, place one small crystal of potassium permanganate ($KMnO_4$) in the bottom of the beaker.
- Make the indicated observations on the solution. For each observation, describe any changes occurring in the water.
- Dispose of the $KMnO_4$ solution in the large waste beaker provided.

Initial observation little change

30-minute observation getting bigger

45-minute observation changing in shape & size

60-minute observation dissolves in water

75-minute observation disperses into tiny particles in all of beaker

90-minute observation Water is purple color & crystal can't be seen to naked eye

## Flux Activity

- Obtain a wax pencil.
- Obtain one room temperature Petri plate containing 1.5% agar. There should be four small holes or wells in the agar.
- On the bottom half of the plate, use the wax pencil to label one of the wells "A", another well "B", another well "C", and the remaining well "D" (see Figure 4).

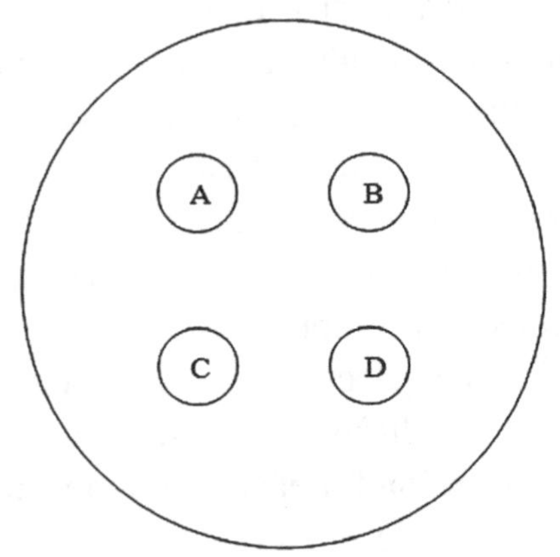


Figure 4 Flux 4 wells

- Well A: Using a dropper or pipette, carefully add 1 small drop of 0.05 M $KMnO_4$ to well A. Note the time you added the drop to the well.
- Well B: Using a dropper or pipette, carefully add 1 small drop of 0.01 M $KMnO_4$ to well B. Note the time you added the drop to the well.
- Well C: Using a dropper or pipette, carefully add 1 small drop of 0.05 M congo red to well C. Note the time you added the drop to the well.
- Well D: Using a dropper or pipette, carefully add 1 small drop of 0.01 M congo red to well D. Note the time you added the drop to the well.
- Using the wax pencil, put your initials on the lid of the Petri plate, and set the plate in a location on the lab bench where it will not be disturbed.
- Obtain one chilled Petri plate containing 1.5% agar. There should be two small holes or wells in the agar.
- On the bottom half of the plate, use the wax pencil to label one of the wells "E" and the other well "F" (see Figure 5).

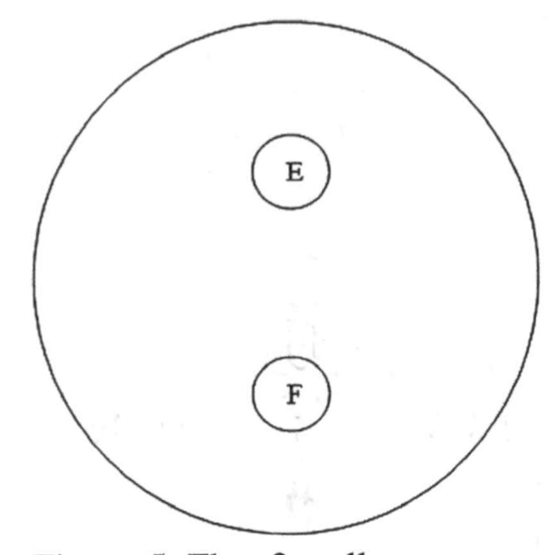


Figure 5 Flux 2 wells

- Well E: Using a dropper or pipette, carefully add 1 small drop of 0.05 M $KMnO_4$ to well E. Note the time you added the drop to the well.
- Well F: Using a dropper or pipette, carefully add 1 small drop of 0.05 M congo red to well F. Note the time you added the drop to the well.
- Using the wax pencil, put your initials on the lid of the Petri plate, and return the plate to your instructor so the plate can be returned to the refrigerator.
- Allow the solutions to diffuse for at least 1 hour.
- Divide each of the wells into four quadrants. Measure the distance (in mm) each solution has moved from the edge of the well to the outer edge of the color ring in each of the quadrants. Record the measurements and the time each was taken in the following table. Calculate the average of the four measurements obtained for each well and record the average in the following table. Please note that the solution has actually moved through a volume or three-dimensional space rather than simply linearly. However, we will not calculate the exact volume of the area of diffusion. Because we can assume the agar is uniformly or nearly uniformly thick, distance or linear diffusion will serve as an adequate estimate of flux.
- Calculate the flux or rate of diffusion for each solution using the average distance traveled. Record the flux in the following table.

Flux = (C₀-Cᵢ)(A)(°C)/(D·MW)

| Well | Solution | Temp. (°C) | Distance diffused (mm) | Avg. distance diffused (mm) | Time elapsed (min) (end time minus begin time) | Flux (mm/min) |
|---|---|---|---|---|---|---|
| A | 0.05 M $KMnO_4$ 158 g/mol, MW | 22°C | 14<br>6<br>7<br>7 | 8.5 | 60 Mins | 8.5 |
| B | 0.01 M $KMnO_4$ 158 g/mol, MW | 22°C | 7<br>10<br>5<br>3 | 6.25 | 60 Mins | 6.25 |
| C | 0.05 M congo red 697 g/mol, MW | 22°C | 4<br>4<br>7<br>7 | 5.5 | 60 mins | 5.5 |
| D | 0.01 M congo red 697 g/mol, MW | 22°C | 3<br>6<br>3<br>7 | 4.75 | 60 mins | 4.75 |

| | | | | | | |
|---|---|---|---|---|---|---|
| E | 0.05 M $KMnO_4$<br>158 g/mol, MW | 15°c | 6 | 5.5 | 60mins | 5.5 |
| | | | 5 | | | |
| | | | 6 | | | |
| | | | 5 | | | |
| F | 0.05 M congo red<br>697 g/mol, MW | 15°c | 3 | 2.5 | 60mins | 2.5 |
| | | | 2 | | | |
| | | | 2 | | | |
| | | | 3 | | | |

Which wells and solutions should be compared to determine only the effect of concentration on flux?
Well A and Well F ?
Well C and Well D

Which wells and solutions should be compared to determine only the effect of temperature on flux?
Well A and Well E
Well C and Well F

Which wells and solutions should be compared to determine only the effect of molecular weight on flux?
Well A and Well C
Well B and Well D
Well E and Well F

Which solution had the highest flux? A
Which solution had the lowest flux? F

## Osmosis and Dialysis Activity

- Obtain a flat-base stand, burette clamp, thistle tube, and 500 mL beaker. Assemble these pieces using the example on the instructor's bench.
- Obtain a wax pencil.
- Fill the 500 mL beaker approximately half full with deionized water (not tap water) from the faucet.
- In a small beaker combine approximately 25 mL of 10% NaCl solution and 25 mL of 1% starch solution.
- Obtain an 8-inch section of wetted dialysis tubing and two pieces of string or one piece of string and a dialysis clamp.
- Fold one end of the dialysis tubing back on itself and tie or clamp the tubing closed in this doubled-over section. Note that the knot must be tied tightly to prevent leaks.
- Open the other end of the dialysis tubing by rubbing it back and forth between your thumb and forefinger. If it is difficult to open, re-wet the dialysis tubing and try rubbing it again.
- Fill the dialysis tubing half full with the NaCl and starch solution.
- Place the dialysis tubing on the stem-end of the thistle tube and tie the dialysis tubing onto the stem of the thistle tube. The knot must be tied tightly to prevent leaks.

- Using a pipette, add more of the NaCl and starch solution to the dialysis tubing by slowly dropping the solution into the thistle tube. Air bubbles can be removed by gently squeezing the dialysis tubing and forcing the air out of the thistle tube. Fill the dialysis tubing until some of the NaCl and starch solution extends a short distance into the stem of the thistle tube. The solution must not extend far up into the stem of the thistle tube or you will not be able to measure changes in fluid level.
- When the dialysis tubing is filled, place the tubing into the 500 mL beaker of deionized water. Be sure there is enough water to immerse the dialysis tubing. You may need to add more deionized water.
- Using the wax pencil, mark the level of the solution in the stem of the thistle tube and note the start time.
- Measure the distance the solution traveled from the origin in 15-minute increments. Record your data in the following table.

| Time elapsed (min) | Distance traveled (mm) |
|---|---|
| 0 | 0 (origin) |
| 15 | 63mm |
| 30 | 12 mm |
| 45 | 4 mm |
| 60 | 34 mm |

? don't have actual measurements

- Obtain two test tubes and a test tube rack.
- Fill one test tube half full with 10% NaCl solution. Add three drops of 3% silver nitrate to the test tube. The silver nitrate reacts with chloride and forms a white precipitate. This serves as a positive test for the presence of chloride ions.
- Fill the second test tube one quarter full with 1% starch solution. Add three drops of Lugol's iodine (or IKI). Iodine causes starch to turn black or blue-black. This serves as a positive test for the presence of starch.
- Discard the test solutions and thoroughly rinse the test tubes.
- After collecting the 60-minute data from the dialysis experiment, fill one test tube half full with water from the 500 mL beaker used in the experiment. Add three drops of silver nitrate to the test tube and record your result below.

  Was $Cl^-$ present in the beaker water? Yes

  From your results, how do the size of a chloride ion and the size of the pores in the dialysis tubing compare?

  The $Cl^-$ is smaller than dialysis pores.

- After collecting the 60-minute measurement, fill the second test tube one quarter full with water from the 500-mL beaker used in the experiment. Add three drops of Lugol's iodine to the test tube and record your result below.

Was starch present in the beaker water? Yes

From your results, how do the size of a starch molecule and the size of the pores in the dialysis tubing compare?

Starch molecules are smaller than the pores

## Tonicity - know definitions

**Tonicity** refers to a solution and the effect of the solution on the volume of a cell. The relative differences in concentration of nonpenetrating solutes in the intracellular fluid versus the extracellular fluid will have an effect on the movement of water (solvent) into or out of the cell. This movement of water will have an effect on the volume of the cell. When a cell is placed in an **isotonic** solution, the volume of the cell remains the same. The concentrations of nonpenetrating solutes in the two solutions (the intracellular fluid and the extracellular fluid) are the same. When a cell is placed in a **hypertonic** solution, the volume of the cell decreases. The concentration of nonpenetrating solutes in the extracellular fluid is greater than in the intracellular fluid, and water diffuses out of the cell causing the cell to shrink. When a cell is placed in a **hypotonic** solution, the volume of the cell will increase. The concentration of nonpenetrating solutes in the extracellular fluid is less than in the intracellular fluid, and water diffuses into the cell causing the cell to swell.

Red blood cells can be used to measure the effects of tonicity on a cell. Although osmosis occurs in all cells when placed in solutions of different concentrations, most cells can restrict the amount of water gain or loss and thus cell swelling or shrinking. Red blood cells, however, cannot regulate the movement of water. As a result, when a red blood cell is placed in a hypertonic solution, it will lose water and shrivel or **crenate**. If a red blood cell is placed in a hypotonic solution, it will gain water and burst or **lyse**. The term **hemolysis** refers specifically to red blood cell lysing.

- **Tonicity Activity**
  - Obtain a pair of gloves and a bleach wipe.
  - Obtain a wax pencil.
  - Obtain a microscope.
  - Obtain three clean microscope slides. With the wax pencil, label one slide "A," one slide "B," and one slide "C."
  - Place a **very small drop** of blood on each of the slides. This may be done by simply touching the tip of the blood pipette to the slide.
  - On slide A, add a drop of distilled water to the blood and place a coverslip over the sample. Do not touch the blood with the dropper.
  - Observe the red blood cells on slide A under the microscope, and sketch the appearance of a few of the red blood cells in the space provided.
  - On slide B, add a drop of 0.85% saline solution to the blood and place a coverslip over the sample. Do not touch the blood with the dropper.
  - Observe the red blood cells on slide B under the microscope, and sketch the appearance of a few of the red blood cells in the space provided.

- On slide C, add a drop of 5% saline solution to the blood and place a coverslip over the sample. Do not touch the blood with the dropper.
- Observe the red blood cells on slide C under the microscope, and sketch the appearance of a few of the red blood cells in the space provided.
- Discard all blood contaminated material, including the gloves, in a biohazards box.
- Disinfect your lab space and microscope if necessary.

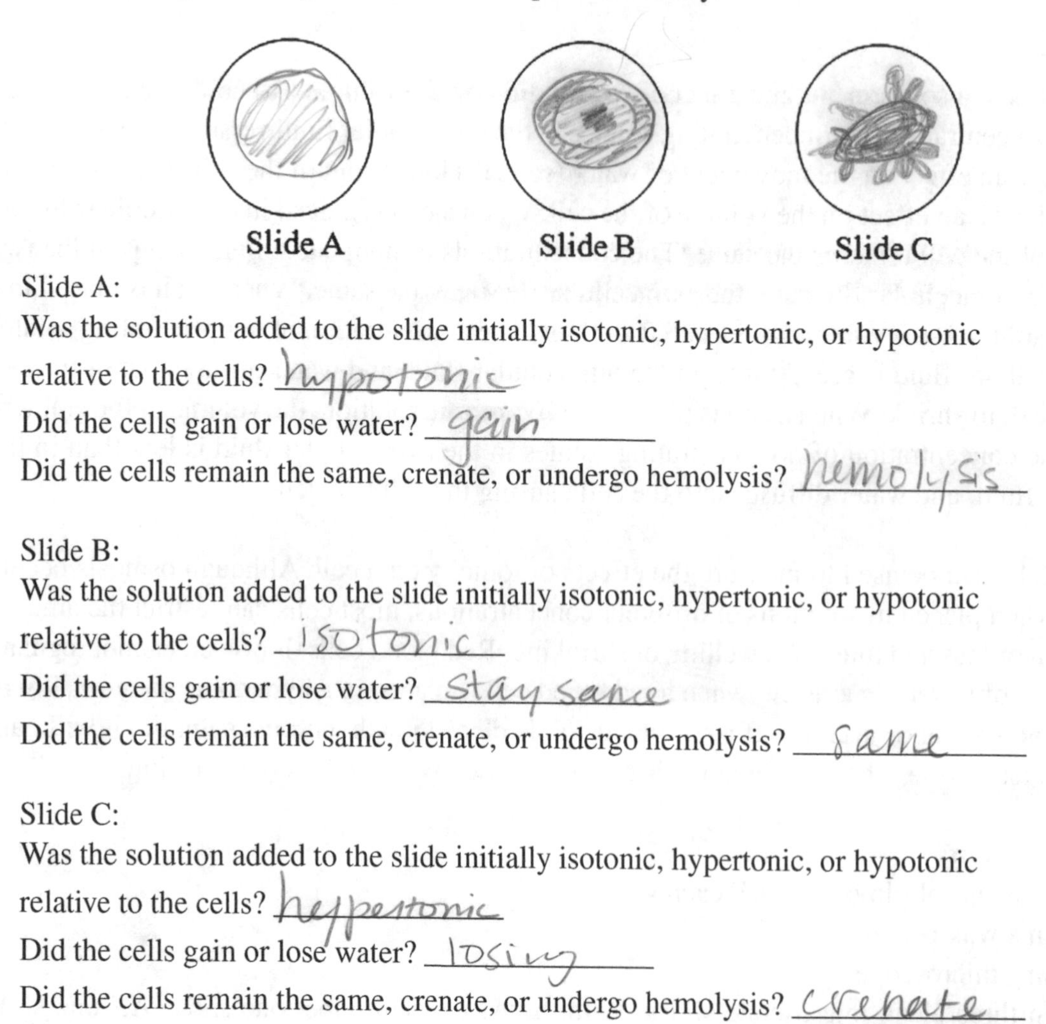


Slide A:
Was the solution added to the slide initially isotonic, hypertonic, or hypotonic relative to the cells? hypotonic
Did the cells gain or lose water? gain
Did the cells remain the same, crenate, or undergo hemolysis? hemolysis

Slide B:
Was the solution added to the slide initially isotonic, hypertonic, or hypotonic relative to the cells? isotonic
Did the cells gain or lose water? stay same
Did the cells remain the same, crenate, or undergo hemolysis? same

Slide C:
Was the solution added to the slide initially isotonic, hypertonic, or hypotonic relative to the cells? hypertonic
Did the cells gain or lose water? losing
Did the cells remain the same, crenate, or undergo hemolysis? crenate

# Exercise 5

# The iWorx Student Lab Data Collection and Analysis System: Tutorial

## Objectives

- Become familiar with the operation of the iWorx data acquisition system.

## Lab Safety Reminders

- Use caution with any electrical equipment.

## Hints for Success

All members of your lab group should feel comfortable operating the iWorx system.

**Never unplug any sensors from the iWorx box!** If you unplug a sensor during an experiment, you will lose all data. You will need to turn the iWorx box off, plug the sensor in and start the lab from the beginning.

On the iWorx recording screens, it is helpful to remember that the blue bars are for data that are being recorded. The green bars are for data that is being calculated based upon the data you have collected.

## Introduction

The iWorx system is a system that will allow you to collect real-time data on a subject in your lab group. This tutorial lab is designed to allow you to become familiar with the basic working of the iWorx system. It is important that you become fairly comfortable with how the system works. This is important because you will need to assess your data and determine if the data you collected is appropriate or a good representation of the desired outcome. If your data does not appear as expected, you may need to adjust your data collection procedure. Check to be sure you have followed the instructions correctly. Some of the exercises require some initial setup with regard to selection of units, addition of temperature information, and some calibration time. The data you collect today is not important. Future labs will require more careful collection of data since you will use this data to investigate and demonstrate the physiologic principles being covered in lecture.

## Starting the iWorx Software

1. Be sure that the iWorx box is turned on. A small green light will be lit when the power is on. If you need to turn the power on, the on/off switch is on the back of the iWorx box.
2. Click on the **LabScribe3 shortcut** on the computer's desktop to open the program.
   You should see an information box that says "Hardware Found". Click **OK**; this will put you on the Main Window page. If the hardware is not found, check to be sure the iWorx box is turned on.
3. On the Main Window, pull down the **Settings** menu. Scroll to **Tutorial,** then select the **Tutorial-Pulse** settings file from the "Tutorial" list.
   Note: If "Tutorial" is empty, then complete the following instructions. On the Main Window, pull down the "Settings" menu and select "Load Group." Scroll down to locate the folder that contains the settings group, "IPLMv6Complete.iwxgrp." Select this group and click "Open." At this point, you should be able to then complete step 3.
4. Instructions for the Tutorial Setup will automatically open. These instructions are not necessary. Close this page.
5. LabScribe will appear on the computer screen as configured by the Tutorial settings file.
6. For this lab, you do not need any information from the following files. But as a reminder, once the settings file has been loaded, clicking the **Experiment** button on the toolbar will allow you to access any of the following documents:
   - Introduction—introductory material for iWorx lab exercises
   - Appendix—may have additional information and resources
   - Background—may have some background information regarding the lab topic
   - Labs—the lab exercise as written by iWorx
   - Setup (opens automatically)—instructions for setting up this particular iWorx lab. You already closed this out.
7. If you choose to save your data, it should be saved in the **iWorx Saved Files** folder on the **Desktop**. You may create your own folder within the iWorx Saved Files folder.

## Pulse Sensor Setup

The pulse plethysmograph has already been connected to the iWorx box. Do not unplug any of these cables. Doing so will cause you to lose your data. For adequate electrical signals, the subject should remove all jewelry from the wrists. Place the plethysmograph (see Figure 1) on the volar surface (fingerprint surface) of the distal phalange (segment) of the subject's middle finger or thumb and wrap the Velcro strap around the end of the finger to attach the unit firmly in place. Attachment

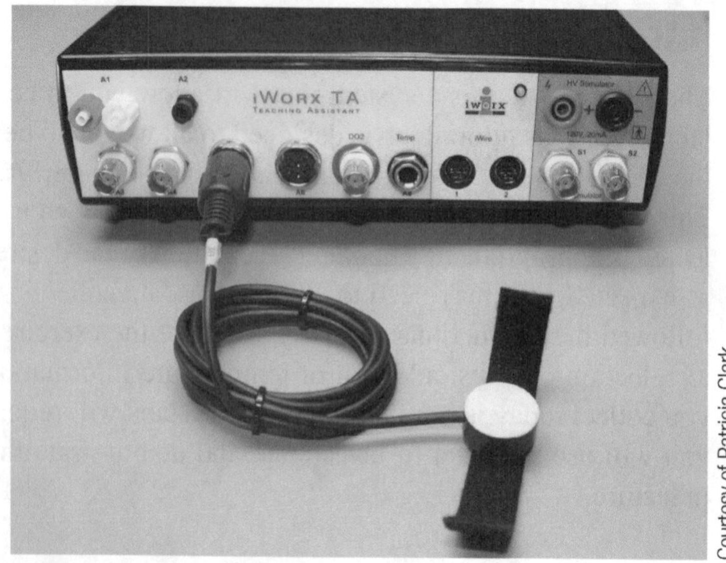


Figure 1 Equipment

should be firm enough that the unit cannot be removed by shaking or simply sliding it off the finger. However, be careful it is not wrapped so tightly that blood flow is cut off to the fingertip. **Important: do not squeeze the unit; this will damage the sensor.**

## Lab Exercises

Your exercises will be presented in the following order:

General instructions for the specific exercise
Specific instructions for the subject for the specific exercise
Specific instructions for the tester for the specific exercise
Examples of what the data should look like for the specific exercise
Instructions for saving data for the specific exercise
Instructions for data analysis for the entire lab

- **Exercise 1: Obtaining a Recording**
  - Read through all instructions before beginning the testing.
  - **Subject Instructions**
    - The subject should sit quietly with both hands in his or her lap, palms facing upward. There should be no extraneous muscle activity.
  - **Tester Instructions**
    - You will conduct a short test run to be sure your data is being collected and recorded properly.
    - When you are ready to begin your test run, click the "Record" button in the upper right corner of your screen. Instruct the subject that the testing has begun and to remain still.
    - Click the "Autoscale" buttons in the upper left corner of the "Pulse," and the "Heart Rate" channels. The data recording should look similar to Figure 2. If the signal in the "Pulse" channel is upside down, click on the downward arrow at the front of the "Pulse" channel and select "Invert" from the dropdown menu.
    - After 10 to 20 seconds, click "Stop."

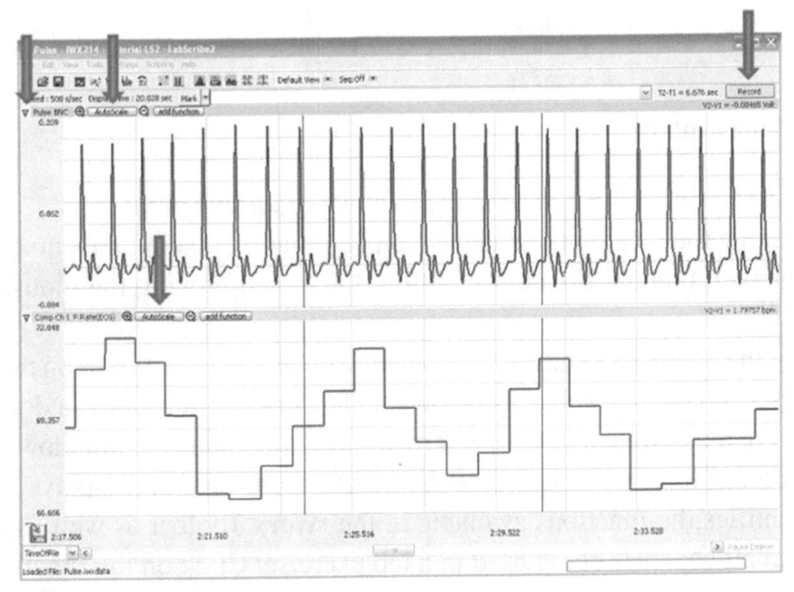

Figure 2 Sample data

## Exercise 2: Becoming familiar with LabScribe

- To utilize the data you have just recorded, it is necessary to become comfortable with the functions available in LabScribe. The majority of the functions you will use in the labs conducted are available on the Main Window of each particular lab. Each of these labs has had the specific configurations set and saved within the file settings.
- Figure 3 identifies the functions available for a particular channel. In this case, the "Pulse" channel is illustrated. Click on the functions as you investigate each of the following:
  - Channel Menu: dropdown menu for altering channel appearance, hiding or minimizing the channel, changing channel name, units, or Y-axis scale
  - Channel Title
  - Mode/Function: dropdown menu for altering input used and computed functions
  - Zoom In: doubles the height of the display but not the amplitude of the signal
  - AutoScale: maximizes the visibility of the signal by increasing or decreasing the height of the signal to within 5% of both the upper and lower limits of the Y-axis
  - Zoom Out: halves the height of the display but not the amplitude of the signal
  - Add Function: dropdown menu for adding computed functions to the analysis of data in the channel
  - Values or Difference between values at the cursors: calculates the difference between the second cursor and the first cursor for the Y-axis values, or if only one cursor is shown, it gives the Y-axis value at that cursor

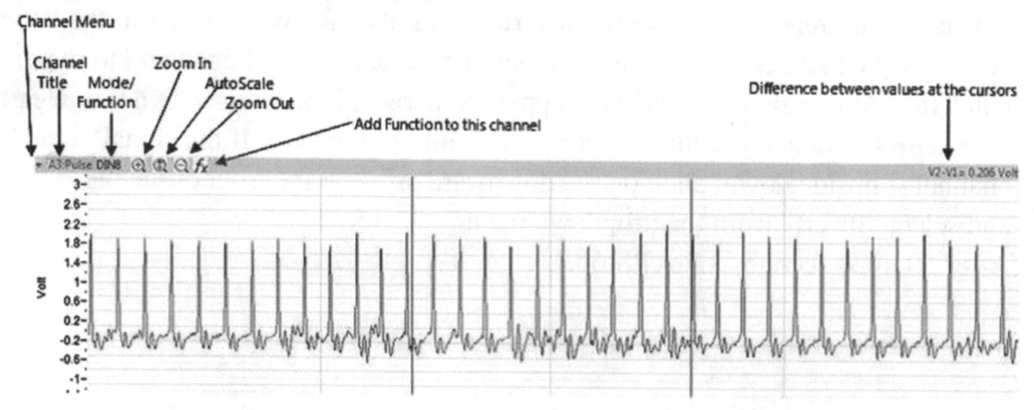


Figure 3 Main window

- Additional alterations can be made to the channel recording area:
  - Repositioning the data on the channel can be done by placing the cursor in the data area of the channel, holding the mouse button down, and moving the mouse (and thus data) up or down.
  - Adjusting the height of a channel can be done by placing the cursor on the border between the channel and the one above or below it. The cursor will become a double arrow cursor when it is in the correct location, and by holding the mouse button down and moving the mouse you can increase or decrease the height of the channel display.
- Figure 4 identifies the functions available in the iWorx Toolbar as well as data information that relates to all channels being used in a lab exercise. Click on the functions as you investigate each of the following:

- General information:
  - Sampling Rate/Speed: the number of samples taken per second or at the rate set within the "Preferences" in the "Edit" menu of the experiment
  - Screen Time/Display Time: length of time visible on the Main Window. This can be altered in the "Preferences" in the "Edit" menu of the experiment or by using the time display tools in the iWorx toolbar.
  - Mark: clicking on the "Mark" button allows you to place a mark at the current point of the data as the data is being recorded. A text box will appear that will allow you to name the mark being inserted. A mark may also be inserted into the data as it is being recorded by clicking the "Enter" key on the keyboard.
  - Mark Menu: this is opened by clicking on the arrow at the right of the "Mark Box." A dropdown menu will appear that allows you to go to a list of "Marks," "Add to Mark Presets," "Delete from Mark Presets," or "Reset Location of Displayed Marks."
  - Mark Box: preset marks can be activated and comments can be added to preset marks. The preset marks are activated by clicking on the down arrow at the right of the "Mark Box." The added comments can be helpful in the identification of testing conditions associated with the data collection at a particular mark.
  - Time difference between cursors/T2–T1: calculates the time difference between the second cursor and the first cursor, or if only one cursor is shown, it gives the time at that cursor
- iWorx Toolbar:
  - File tools:
    - New File: opens a new data file with the same program settings
    - Open File: opens a data file that had been saved
    - Save File: saves a data file
  - LabScribe3 tools: only those we will be using in our iWorx lab exercises or that you may find useful in troubleshooting or additional analysis of the data will be discussed.
    - Main: opens the Main Window. This is the window data is recorded in. You can toggle between the Main Window and the Analysis Window to maximize use of the data by adjusting data display in the Main Window.
    - Analysis: opens the Analysis Window. This is the window that allows you to make additional measurements on the data. You can toggle between the Main Window and the Analysis Window to maximize use of the data by adjusting data display in the Main Window.
    - X-Y View: allows you to display the data in graphical form. You will select the data you want to use as the dependent variable (X-axis) and the independent variable (Y-axis).
    - Journal: although you are provided with the data tables necessary for each lab, you may find it beneficial to utilize the "Journal." In the "Journal" you can record notes and data that you may want to transfer to a word processor program.
    - Mark: the pencil icon is an alternative way to get to the list of marks generated during the data recording.
    - Half display: reduces the time displayed on the screen to half of that currently displayed. As a result, the length of the recorded data displayed is reduced to half of that currently displayed.

- Zoom display: alters the display between the two cursors so that the data that was between the two cursors now extends across the entire width of the screen. It is important to note that the cursors do not move even though the data points have moved.
- Double display: increases the time displayed on the screen to twice that currently displayed. As a result, the length of the recorded data displayed is increased to twice that currently displayed.
- Two cursors: places two cursors on the screen so that differences between data points can be measured. It is also necessary to have two cursors on the screen when you manipulate the length of time and data displayed on the screen.
- One cursor: places one cursor on the screen so that the data at a particular point can be measured.
- Views: returns you to the default view of the exercise
- Record: allows you to start recording data. This icon will switch to a "Stop" icon during recording. Clicking "Stop" will halt data collection and the icon will return to the "Rec" icon. At this point, data collection can be resumed by once again clicking "Rec."

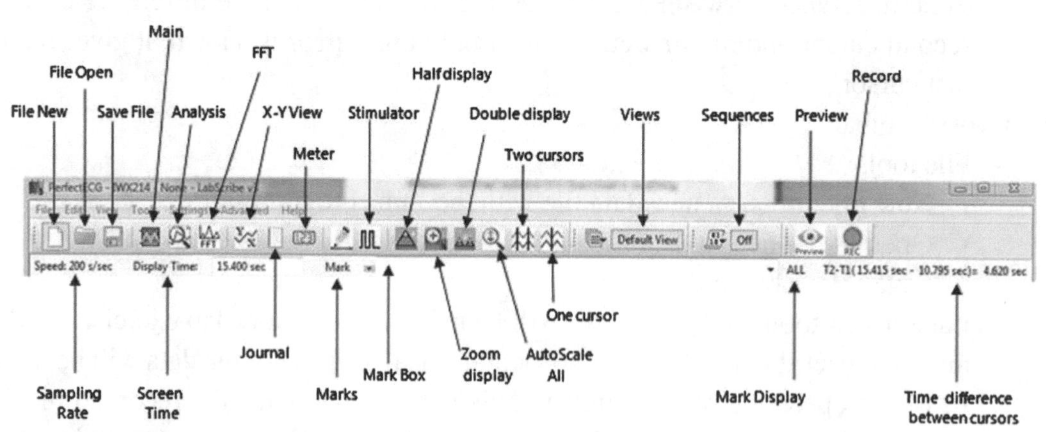


Figure 4 iWorx toolbar

## Exercise 3: Using Marks

- Read through all instructions before beginning the testing. You will conduct a short test run to practice using the "Marks" tools.
- **Subject Instructions**
  - The subject should sit quietly with both hands in his or her lap, palms facing upward. There should be no extraneous muscle activity.
- **Tester Instructions**
  - To make a preset mark:
    - Click in the "Mark Box" to activate the cursor and type the phrase "Pulse 1" in the box.
    - Pull down the "Marks" menu and select "Add to Mark Presets."
    - Click on the down arrow at the right of the "Mark Box" to confirm that your mark has been added to the list of preset marks.
    - When you are ready to begin your test run, click the "Record" button in the upper right corner of your screen. Instruct the subject that the testing has begun and to remain still.

  - Click the "Autoscale" buttons in the upper left corner of the "Pulse," and the "Heart Rate" channels. The data recording should look similar to the following figure. If the signal in the "Pulse" channel is upside down, click on the downward arrow at the front of the "Pulse" channel and select "Invert" from the dropdown menu.
  - Pull down the preset mark list (the down arrow to the right of the "Mark Box") and select the preset mark you created ("subject name—preset mark"). The mark name should appear in the "Mark Box." Click the "Mark" button or "Enter" on the keyboard to insert the mark in your data recording.
  - Click "Stop" to suspend the data recording while you read the next set of instructions.
- To add a mark to the data during recording:
  - When you are ready to begin your test run, click the "Record" button in the upper right corner of your screen. Instruct the subject that the testing has begun and to remain still.
  - Click the "Autoscale" buttons in the upper left corner of the "Pulse," and the "Heart Rate" channels. The data recording should look similar to the following figure. If the signal in the "Pulse" channel is upside down, click on the downward arrow at the front of the "Pulse" channel and select "Invert" from the dropdown menu.
  - Move the cursor to the "Mark" text box to the right of the "Marks" button and click to activate the text box. Type the phrase "Pulse 2" in the "Mark" text box. The mark will be inserted and labeled when you click on the "Mark" button or click "Enter" on the keyboard.
  - After allowing the recording to run an additional 10 to 20 seconds, click "Stop."
  - Click "Stop" to suspend the data recording while you read the next set of instructions.
- To add a mark to a data file after the data has been recorded:
  - Find an area of data in the last 10 to 20 seconds of recorded data.
  - Click the "One Cursor" icon and using the mouse, position the cursor on the data you want to mark.
  - Move the cursor to the "Mark" text box to the right of the "Marks" button and click to activate the text box. Type the phrase "Pulse 3" in the "Mark" text box. The mark will be inserted and labeled when you click on the "Mark" button or click "Enter" on the keyboard.

- You may choose to save your data at this point or you may continue on with the experiment or data analysis. To save your data, on the Main Window, either

  1. click on the **File** menu. Scroll to **Save As**, then save your data; it should be saved in the **iWorx Saved Files** folder on the **Desktop**. You may create your own folder within the iWorx Saved Files folder. Give your file a name and designate the file type as .iwxdata, then click **Save**.

  Or

  2. click on the **Save File** icon in the iWorx toolbar, then save your data; it should be saved in the **iWorx Saved Files** folder on the **Desktop**. You may create your own folder within the iWorx Saved Files folder. Give your file a name and designate the file type as .iwxdata, then click **Save**.

- **Data Analysis Instructions**

  Figure 5, a diagram of icons in the iWorx toolbar may be helpful.

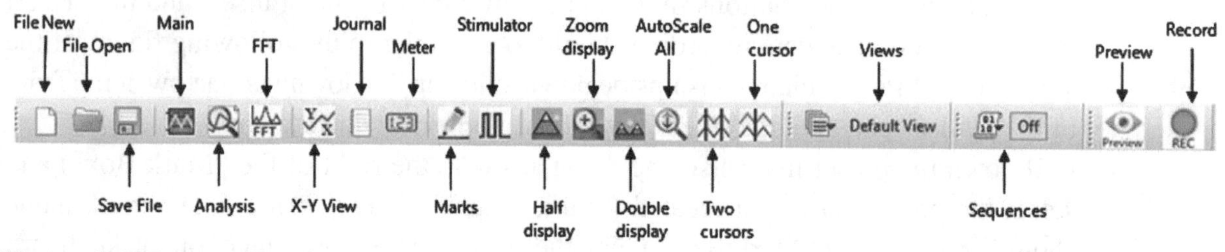


Figure 5 iWorx toolbar icons

- You can move through your data in the Main Window by using the scroll bar at the bottom of the Main Window screen, either by clicking on the scroll bar and holding down the mouse button and sliding the bar to the right or left, or by clicking on the arrows to the right or left of the scroll bar.
- You can also move to a specific section of data by selecting the appropriate mark used to identify that section of data. Go to the beginning of your recordings. Click on the **Marks** icon to bring up the Marks Dialogue box with a table of the marks created in your lab (Figure 6). Once you have the table of marks, click on the number at the beginning of the row for the mark that identifies the beginning of the experimental data segment you are going to analyze. This will highlight that mark. Next, click on the **Go To Mark** button near the bottom of the Marks Dialogue box. Data at that mark will be displayed on your screen.

Marks Dialog

| | Time | Mark | Channel | Show |
|---|---|---|---|---|
| 1 | 12:36:54.591 | Pulse 1 | All | ✓ |
| 2 | 12:36:55.471 | Pulse 2 | All | ✓ |
| 3 | 12:36:56.333 | Pulse 3 | All | ✓ |

Delete | Go To Mark | Export | OK | Cancel

Figure 6 Marks dialogue box

- You can also alter the amount of time displayed on the screen by using the "Half display" or "Double display" icons.
- Some measurement can also be made in the Main Window.
- Be sure that two cursors are present on your screen. If necessary, click on the **Two Cursors** icon on the iWorx toolbar.
- Click on a cursor and drag it to the peak of one of the pulse waves in the **Pulse channel.** Click on the second cursor and drag it to the peak of the adjacent pulse wave (Figure 7).
- The difference in time between the two peaks is listed in the upper right section of the toolbar in the "T2–T1" data box.

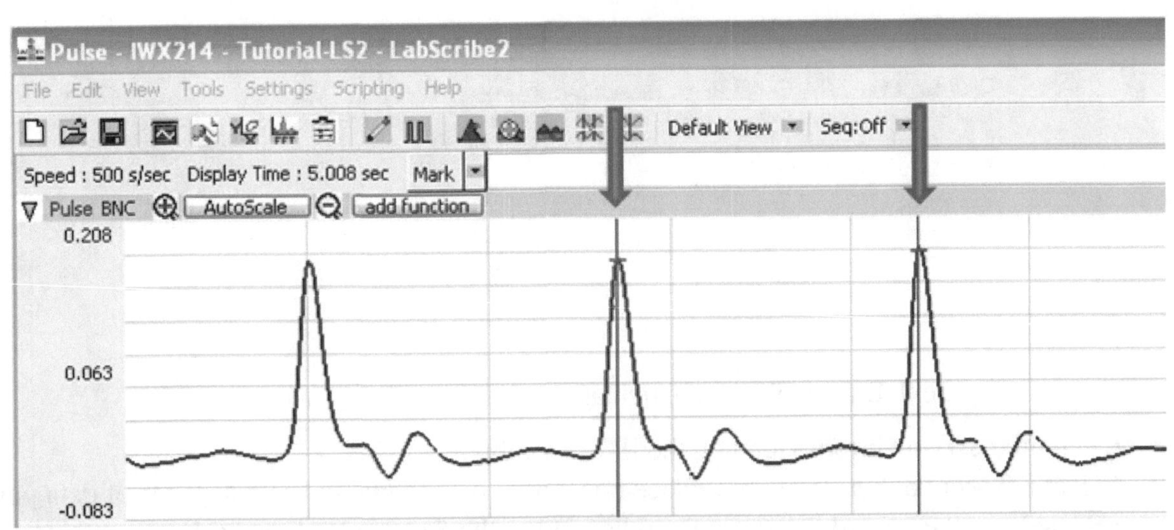


Figure 7 Data analysis of the pulse wave duration

- Click on a cursor and drag it to the baseline of one of the pulse waves in the **Pulse channel**. Click on the second cursor and drag it to the peak of the same pulse wave (Figure 8).
- The difference in amplitude between the baseline and the peak is listed in the upper right section of the toolbar "V2–V1" data box.

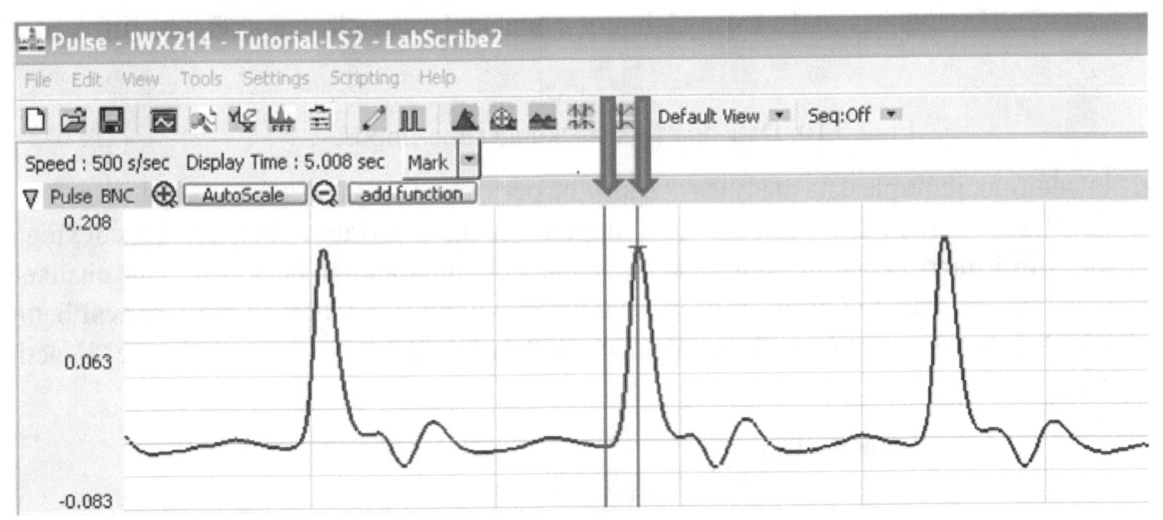


Figure 8 Data analysis of the pulse wave amplitude

- Click on the **One Cursor** icon on the iWorx toolbar.
- Move the cursor to a top plateau in the **Heart Rate channel** (Figure 9). In the figure the **Pulse channel** has been minimized; the data is still present but not visible on the screen. Your screen will have both channels visible unless you minimize the **Pulse channel**. Note that the heading bar of the **Heart Rate channel** is in green. This indicates that the data in this channel have been calculated from the data in a channel containing the directly measured data, in this case the **Pulse channel** (blue heading). A single value for time and a single value for amplitude will recorded for the point in the data indicated by the cursor. The data are listed in the upper right section of the toolbar in the time and amplitude boxes, respectively.

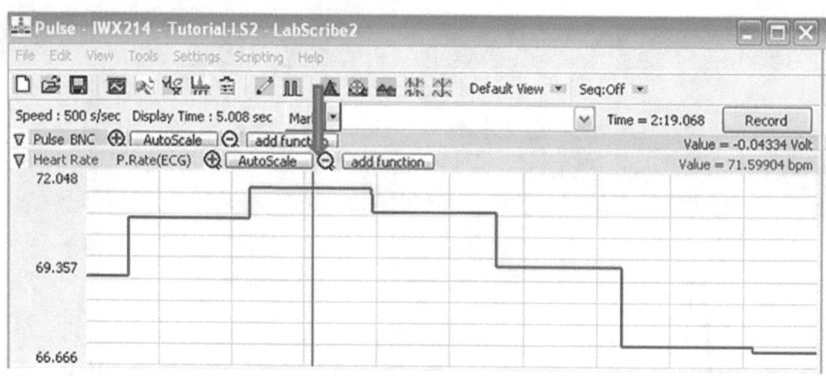


Figure 9 Data analysis of calculated heart rate

- Click on the **Two Cursors** icon on the iWorx toolbar.
- Select a 5-second section of data. Place one cursor on one end of the 5 seconds of data you selected and the second cursor on the other end of the 5 seconds of data selected.
- Click on the **Zoom between Cursors** icon on the iWorx toolbar.
- Click on the **Analysis** icon in the iWorx toolbar.
- Most of the same functions are available in the Analysis Window as in the Main Window. The channels can be superimposed on each other to have a more directly observe-possible relationships between the data in the channels. This can be done by clicking on the **Stack/ Superimpose** button in the lower left corner of the Analysis Window. Stacked channels can be unstacked by clicking on the **Unstack** button found in the same location (Figure 10).

Figure 10 Data analysis stack and superimpose

- In addition, multiple data analyses can also be performed. In the Analysis window, a "Function Table" will be displayed above the uppermost channel (Pulse channel). By clicking on the **Add Function** button, additional mathematical functions can be added to the channels.
- Click on the **Add Function**. A dropdown menu will appear. Click on the **General** button. Another dropdown menu will appear. Click on the desired function to add it to the "Function Table" (Figure 11).

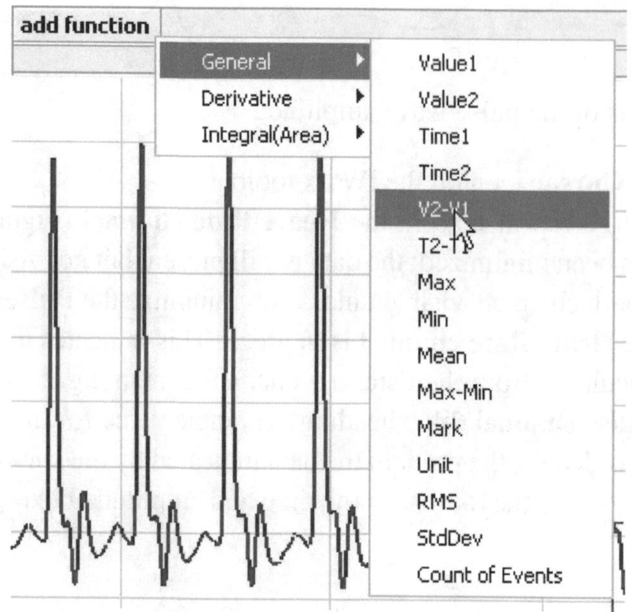


Figure 11 Data analysis add function dropdown menu

- Following the procedure for adding a function, click on **Value 1**. Value 1 should now be present in the "Function Table."
- Repeat the procedure to add **Value 2**, **V2–V1**, and **T2–T1**. Value 2, V2–V1, and T2–T1 should now be present in the "Function Table." The values for each of these functions should now be displayed in the top margin of each channel (Figure 12). The "Value 1" and "Value 2" functions measure the amplitudes at the cursors. The "T2–T1" and "V2–V1" functions measure the difference between cursor 2 and cursor 1 for time and amplitude, respectively.
- The values for each of these functions should now be displayed in the top margin of each channel (see the following figure). The "Value 1" and "Value 2" functions measure the amplitudes at the cursors. The "T2–T1" and "V2–V1" functions measure the difference between cursor 2 and cursor 1 for time and amplitude, respectively.

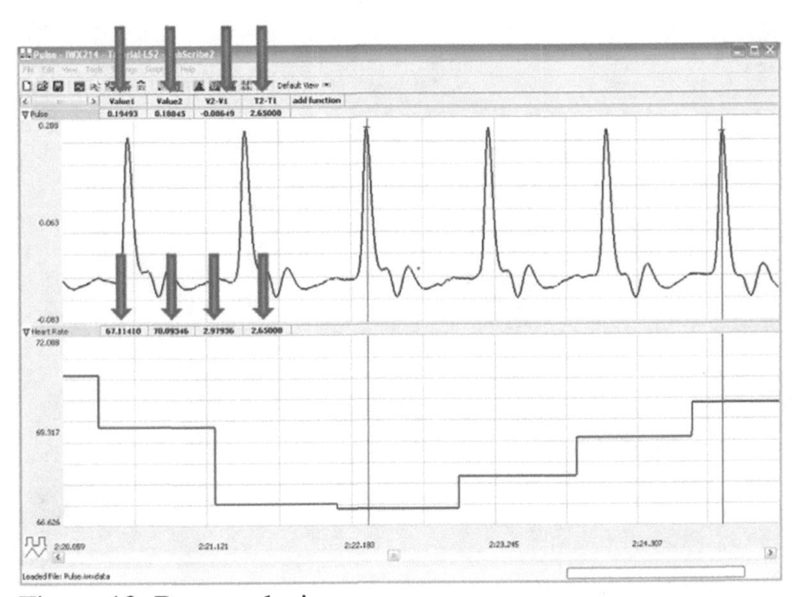

Figure 12 Data analysis

- Using the techniques learned in collecting data in the Main Window, collect the same type of data in the Analysis Window.
- The difference in time between the peaks of two adjacent pulse waves from the "T2–T1" data box.
- The amplitudes of the two adjacent pulse waves.
- The difference in amplitude between the baseline and the peak of one pulse wave from the "V2–V1" data box.
- The amplitude of the baseline and the amplitude of the peak of the same pulse wave.

Somatic Reflex arcs:

① receptor cell - detect stimulus
[receptor on surface of skin/muscle]

② Afferent neuron - transmits sensory info to integration center

③ Integration center - asses the info + direct response
[brainstem / spinal cord / CNS)

④ Efferent Neuron - transmits info to effector cells

⑤ Effector Cell - gives response via motor neuron
[muscle fiber / motor neuron]

# Exercise 6

## iWorx Reflex Response Times

### Objectives

- Become familiar with a basic reflex arc.
- Become more familiar with the operation of the iWorx system and data collection.
- Be able to identify the components of a simple reflex arc.
- Be able to differentiate between a *spinal reflex* and a reflex that utilizes integration in the brain.
- Be able to define and differentiate between *monosynaptic* and *multisynaptic reflexes*.
- Be able to discuss the integration necessary in a reflex that requires conscious responses.
- Be able to explain the *Stroop effect* and the impact of interference of stimuli on associative tasks.

### Lab Safety Reminders

- Use caution with any electrical equipment.

### Hints for Success

All members of your lab group should feel comfortable operating the iWorx system.

**Never unplug any sensors from the iWorx box!** If you unplug a sensor during an experiment, you will lose all data. You will need to turn the iWorx box off, plug the sensor in and start the lab from the beginning. On the iWorx recording screens, it is helpful to remember that the blue bars are for data that are being recorded. The green bars are for data that is being calculated based upon the data you have collected.

### Introduction

You will learn more about reflex arcs when you perform the lab, Somatic Reflex Arcs, with specific examples of somatic reflexes that utilize the nervous system. As a basic introduction, a nervous system reflex arc requires a receptor capable of detecting a stimulus, an afferent neuron that transmits the sensory information to an integration center, an efferent neuron that transmits motor information to an effector cell which in turn gives a response to the initial stimulus.

For spinal reflexes, the integration center is the spinal cord, and the brain is not necessary for integration. Some spinal reflexes are simple monosynaptic reflexes that involve only two neurons, an afferent neuron that synapses with an efferent neuron in the spinal cord.

Other spinal reflexes are multisynaptic reflexes that include one or more interneurons that lie completely within the central nervous system (CNS), between the afferent neuron and the efferent neuron.

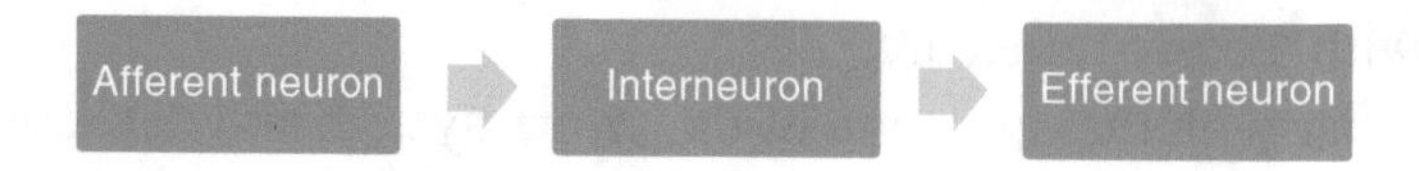


The degree of integration is often reflected in the number of interneurons and neural networks involved in the completion of the reflex.

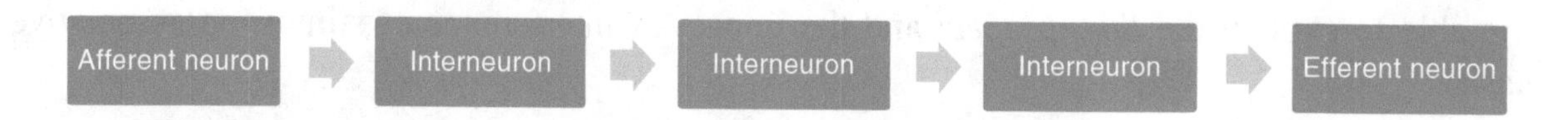


Being a spinal reflex does not necessarily mean that information is not sent to the brain, but it does mean the immediate reflex is controlled by the spinal cord. Sending information from a spinal reflex to the brain allows perception and possibly additional responses to the stimulus. For other CNS reflexes, the integration center is the brain. In these reflexes the spinal cord or cranial nerves carry the sensory information to the brain, and they carry the efferent information from the brain to the PNS.

The reflexes being used in this lab are integrated in the brain and involve complicated pathways and integration. This lab allows you to make a comparison of the time required to complete a reflex arc using visual stimuli versus auditory stimuli. You also investigate the effect of priming and prediction on auditory stimulation and reflex time. In the final experiments, you will examine the *Stroop effect* and the impact of stimulus interference on the ability to complete a task effectively. The Stroop effect is the result of research conducted by J.R. Stroop. This exercise adds the effect of conscious processing of visual information. In testing the Stroop effect, the subject is given stimuli with varying degrees of interference, distraction or confliction in the stimulus presented. Subjects will be presented color words printed in different colors in a variety of combinations. The hypothesis is that the greater the degree of interference, the longer it takes for the subject to respond correctly.

## Starting the iWorx Software

1. Be sure that the iWorx box is turned on. A small green light will be lit when the power is on. If you need to turn the power on, the on/off switch is on the back of the iWorx box.

2. Click on the **LabScribe3 shortcut** on the computer's desktop to open the program.
   You should see an information box that says "Hardware Found". Click **OK**; this will put you on the Main Window page. If the hardware is not found, check to be sure the iWorx box is turned on.
3. On the Main Window, pull down the **Settings** menu. Scroll to **Human Nerve,** then select the **Auditory-VisualReflexes** settings file from the "Human Nerve" list.
   *Note: If "Human Nerve" is empty, then complete the following instructions. On the Main Window, pull down the "Settings" menu and select "Load Group." Scroll down to locate the folder that contains the settings group, "IPLMv6Complete.iwxgrp." Select this group and click "Open." At this point, you should be able to then complete step 3.*
4. Instructions for the Auditory-VisualReflexes Setup will automatically open. These instructions are not necessary. Close this page by clicking on the close button in the upper right corner of this document.
5. LabScribe will appear on the computer screen as configured by the Auditory-VisualReflexes settings file.
6. For this lab, you do not need any information from the following files. But as a reminder, once the settings file has been loaded, clicking the **Experiment** button on the toolbar will allow you to access any of the following documents:
   - Introduction—introductory material for iWorx Human Nerve lab exercises
   - Appendix—may have additional information and resources
   - Background—may have some background information regarding the lab topic
   - Labs—the lab exercise as written by iWorx
   - Setup (opens automatically)—instructions for setting up this particular iWorx lab. You already closed this out.
7. If you choose to save your data, it should be saved in the **iWorx Saved Files** folder on the **Desktop**. You may create your own folder within the iWorx Saved Files folder.

## Event Marker Setup

The event marker (Figure 1) has already been connected to the iWorx box. One student will serve as the tester and will hold the event marker out of sight of the subject. The tester should push the event marker button quietly so the subject does not use the sound of the button as a cue.

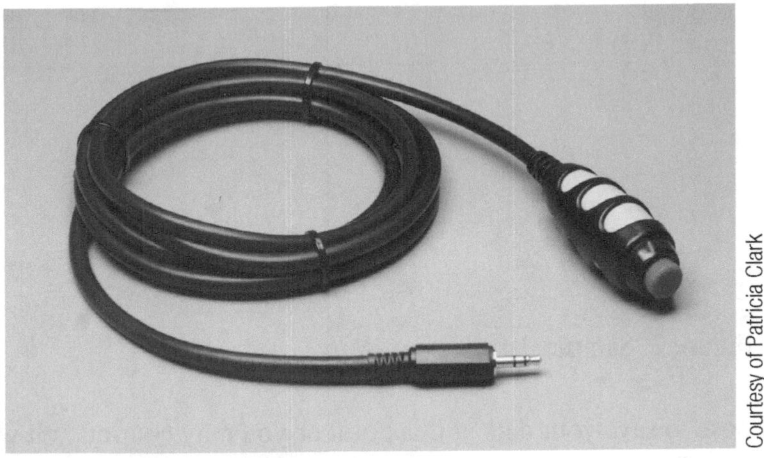

Courtesy of Patricia Clark

Figure 1 Event marker

## Lab Exercises

### Exercise 1: Simple Visual Cues and Reaction Time

- Read through all instructions before beginning the testing.
- To label the data for exercise 1, move the cursor to the "Mark" text box to the right of the "Mark" button and click to activate the text box. Type the subject's name followed by "visual cues" in the "Mark" text box. The mark will be inserted and labeled the first time the subject responds during the testing period.
- **Subject Instructions**
  - The subject should sit in a chair facing the computer with his or her hand positioned so the subject can push the "Enter" key as quickly as possible.
  - The subject should be watching for a vertical line (the visual cue) to appear on the right side of the computer screen. The subject should push the "Enter" key as soon as the visual cue is detected.
- **Tester Instructions**
  - A visual cue will be delivered when the tester pushes the button of the event marker. The tester should deliver 10 visual cues to the subject. The cues should be delivered between 5 and 10 seconds apart, and the time between cues should vary.
  - When you are ready to begin, click the "Record" button in the upper right corner of your screen. Instruct the subject that the testing has begun and to click the "Enter" key whenever the subject detects a new visual cue on the right side of the screen.
  - After the delivery and detection of the tenth visual cue, click the "Stop" button to quit recording.
- Your data should look similar to the data in Figure 2 where three visual signals were given by the tester, each followed by the subject's response. The additional two lines are the cursors.

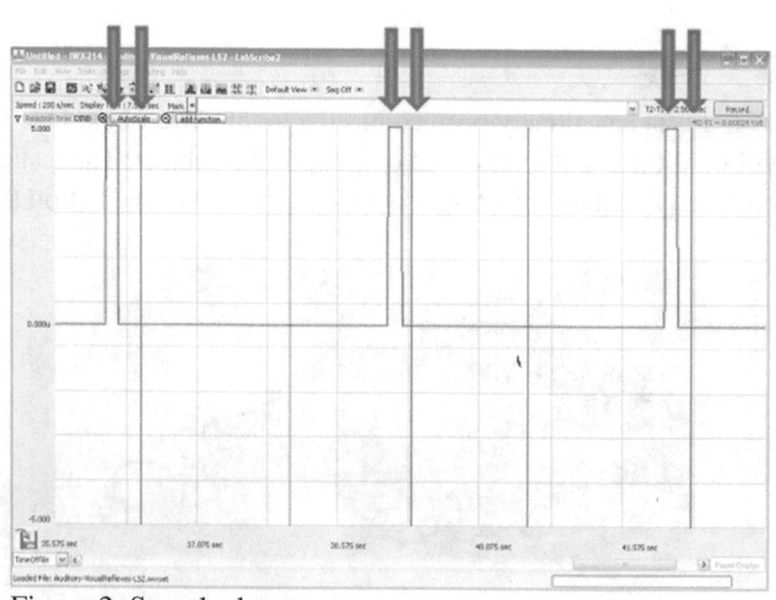

Figure 2 Sample data

- You may choose to save your data at this point or you may continue on with the experiment or data analysis. To save your data, on the Main Window, either

1. click on the **File** menu. Scroll to **Save As**, then save your data; it should be saved in the **iWorx Saved Files** folder on the **Desktop**. You may create your own folder within the iWorx Saved Files folder. Give your file a name and designate the file type as .iwxdata, then click **Save**.

Or

2. click on the **Save File** icon in the iWorx toolbar, then save your data; it should be saved in the **iWorx Saved Files** folder on the **Desktop**. You may create your own folder within the iWorx Saved Files folder. Give your file a name and designate the file type as .iwxdata, then click **Save**.

## Exercise 2: Simple Auditory Cues and Reaction Time

- Read through all instructions before beginning the testing.
- To label the data for exercise 2, move the cursor to the "Mark" text box to the right of the "Mark" button and click to activate the text box. Type the subject's name followed by "auditory cues" in the "Mark" text box. The mark will be inserted and labeled the first time the subject responds during the testing period.
- Cover the computer screen with a piece of paper so the subject cannot see the visual signal and use it as a cue.
- **Subject Instructions**
  - The subject should sit in a chair facing the computer with his or her hand positioned so the subject can push the "Enter" key as quickly as possible.
  - The subject should be listening for a sound. You may choose to use a tap on the table or you may snap the fingers, anything that the subject will recognize as an auditory cue. Familiarize the subject with the auditory cue and use the same sound as the auditory cue throughout the exercise. The subject should push the "Enter" key as soon as the auditory cue is detected.
- **Tester Instructions**
  - An auditory cue will be delivered **at the same time that the tester pushes the button of the event marker**. Be sure that the presentation of the signal is outside the visual field of the subject; the subject must not see the audible cue being given. The tester should deliver 10 auditory cues to the subject. The cues should be delivered between 5 and 10 seconds apart, and the time between cues should vary.
  - When you are ready to begin, click the "Record" button in the upper right corner of your screen. Instruct the subject that the testing has begun and to click the "Enter" key whenever the subject detects a new auditory cue.
  - After the delivery and detection of the tenth auditory cue, click the "Stop" button to quit recording.
- Your data should look similar to the data you obtained using the visual signals.
- You may choose to save your data at this point or you may continue on with the experiment or data analysis. If you have not yet saved your data, follow the instructions for saving data under exercise 1. If you have previously saved your data, click on the **Save File** icon in the iWorx toolbar.

## Exercise 3: Prompted Auditory Cues and Reaction Time

- Read through all instructions before beginning the testing.
- To label the data for exercise 3, move the cursor to the "Mark" text box to the right of the "Mark" button and click to activate the text box. Type the subject's name followed by "prompted auditory cues" in the "Mark" text box. The mark will be inserted and labeled the first time the subject responds during the testing period.
- Cover the computer screen with a piece of paper so the subject cannot see the visual signal and use it as a cue.
- **Subject Instructions**
    - The subject should sit in a chair facing the computer with his or her hand positioned so the subject can push the "Enter" key as quickly as possible.
    - The subject should be listening for a verbal prompt before each auditory cue. The subject should push the "Enter" key as soon as the auditory cue is detected, **not when the prompt is heard**.
- **Tester Instructions**
    - A verbal prompt will be given before the auditory cue is given. It is best to use a one syllable word as the verbal prompt and use the same verbal prompt each time the auditory cue is given. Familiarize the subject with the verbal prompt. Use the same auditory cue as used in exercise 2. **The verbal prompt will be given immediately before the auditory cue, and the auditory cue will be delivered at the same time that the tester pushes the button of the event marker** (as in exercise 2). Be sure that the presentation of the signal is outside the visual field of the subject; the subject must not see the audible cue being given. The tester should deliver 10 prompted auditory cues to the subject. The cues should be delivered between 5 and 10 seconds apart, and the time between cues should vary.
    - When you are ready to begin, click the "Record" button in the upper right corner of your screen. Instruct the subject that the testing has begun and to click the "Enter" key whenever the subject detects a new auditory cue.
    - After the delivery and detection of the tenth prompted auditory cue, click the "Stop" button to quit recording.
- Your data should look similar to the data you obtained using the visual signals.
- You may choose to save your data at this point or you may continue on with the experiment or data analysis. If you have not yet saved your data, follow the instructions for saving data under exercise 1. If you have previously saved your data, click on the **Save File** icon in the iWorx toolbar.

## Exercise 4: Predictable Auditory Cues and Reaction Time

- Read through all instructions before beginning the testing.
- To label the data for exercise 4, move the cursor to the "Mark" text box to the right of the "Mark" button and click to activate the text box. Type the subject's name followed by "predictable auditory cues" in the "Mark" text box. The mark will be inserted and labeled the first time the subject responds during the testing period.
- Cover the computer screen with a piece of paper so the subject cannot see the visual signal and use it as a cue.

- **Subject Instructions**
  - The subject should sit in a chair facing the computer with his or her hand positioned so the subject can push the "Enter" key as quickly as possible.
  - The subject should be listening for an auditory cue, but this time the auditory cues will be delivered at regular time intervals. The subject should push the "Enter" key as soon as the auditory cue is detected.
- **Tester Instructions**
  - An auditory cue will be given at 5-second intervals. Use the same auditory cue as used in exercise 2. **The auditory cue will be delivered at the same time that the tester pushes the button of the event marker** (as in exercise 2). Be sure that the presentation of the signal is outside the visual field of the subject; the subject must not see the audible cue being given. The tester should deliver 10 auditory cues to the subject. The cue should be delivered every 5 seconds.
  - When you are ready to begin, click the "Record" button in the upper right corner of your screen. Instruct the subject that the testing has begun and to click the "Enter" key whenever the subject detects a new auditory cue.
  - After the delivery and detection of the tenth auditory cue, click the "Stop" button to quit recording.
- Your data should look similar to the data you obtained using the visual signals.
- You may choose to save your data at this point or you may continue on with the data analysis. If you have not yet saved your data, follow the instructions for saving data under exercise 1. If you have previously saved your data, click on the **Save File** icon in the iWorx toolbar.

## Exercise 5: Conscious Processing of Visual Information and Reaction Time, Part A

- Read through all instructions before beginning the testing.
- Obtain a stack of cue cards labeled as "Exercise 5."
- To label the data for exercise 5, move the cursor to the "Mark" text box to the right of the "Mark" button and click to activate the text box. Type the subject's name followed by "reading word" in the "Mark" text box. The mark will be inserted and labeled the first time the subject responds during the testing period.
- **Subject Instructions**
  - The subject should sit in a chair facing the tester with his or her hand positioned so the subject can push the "Enter" key as quickly as possible.
  - The subject will be shown a series of words that he or she will be asked to read. The subject should push the "Enter" key as soon as the word is read. If the word is incorrectly read, the tester will leave the word in place and the subject should attempt to read the word and push the "Enter" key again.
- **Tester Instructions**
  - A card will be presented to the subject that has the name of a color printed on it. However, the word is printed in a color different than the color identified by the word. The subject is to **read the word**. Note that the word is also printed, in black, on the back of the card so the tester can identify when the subject has correctly read the word. **The visual cue will be delivered at the same time that the tester pushes the button of the event marker** (as in exercise 2). If the subject does not read the word correctly, leave the card in place

  until the subject successfully reads the word. The tester should deliver 5 cues to the subject. The timing of the cue will vary depending upon the number of attempts the subject makes before correctly reading the word.
  - When you are ready to begin, click the "Record" button in the upper right corner of your screen. Instruct the subject that the testing has begun and to click the "Enter" key whenever the subject detects a new auditory cue.
  - After the delivery and detection of the fifth visual cue, click the "Stop" button to quit recording.
- Your data should look similar to the data you obtained using the visual signals except that there may be more than one mark per presentation of the cue.
- You may choose to save your data at this point or you may continue on with the data analysis. If you have not yet saved your data, follow the instructions for saving data under exercise 1. If you have previously saved your data, click on the **Save File** icon in the iWorx toolbar.

## Exercise 6: Conscious Processing of Visual Information and Reaction Time, Part B

- Read through all instructions before beginning the testing.
- Obtain a stack of cue cards labeled as "Exercise 6."
- To label the data for exercise 6, move the cursor to the "Mark" text box to the right of the "Mark" button and click to activate the text box. Type the subject's name followed by "reading color" in the "Mark" text box. The mark will be inserted and labeled the first time the subject responds during the testing period.
- **Subject Instructions**
  - The subject should sit in a chair facing the tester with his or her hand positioned so the subject can push the "Enter" key as quickly as possible.
  - The subject will be shown a series of words that he or she will be asked to read. The subject should push the "Enter" key as soon as the color of the word is identified. If the color is incorrectly identified, the tester will leave the word in place and the subject should attempt to read the word and push the "Enter" key again.
- **Tester Instructions**
  - A card will be presented to the subject that has the name of a color printed on it. However, the word is printed in a color different than the color identified by the word. The subject is to **identify the color**. **The visual cue will be delivered at the same time that the tester pushes the button of the event marker** (as in exercise 2). If the subject does not identify the color correctly, leave the card in place until the subject successfully identifies the color. The tester should deliver 5 cues to the subject. The timing of the cue will vary depending upon the number of attempts the subject makes before correctly identifying the color.
  - When you are ready to begin, click the "Record" button in the upper right corner of your screen. Instruct the subject that the testing has begun and to click the "Enter" key whenever the subject detects a new auditory cue.
  - After the delivery and detection of the fifth visual cue, click the "Stop" button to quit recording.
- Your data should look similar to the data you obtained using the visual signals except that there may be more than one mark per presentation of the cue.
- You may choose to save your data at this point or you may continue on with the data analysis. If you have not yet saved your data, follow the instructions for saving data under exercise 1. If you have previously saved your data, click on the **Save File** icon in the iWorx toolbar.

## Data Analysis Instructions

Figure 3, a diagram of icons in the iWorx toolbar may be helpful.

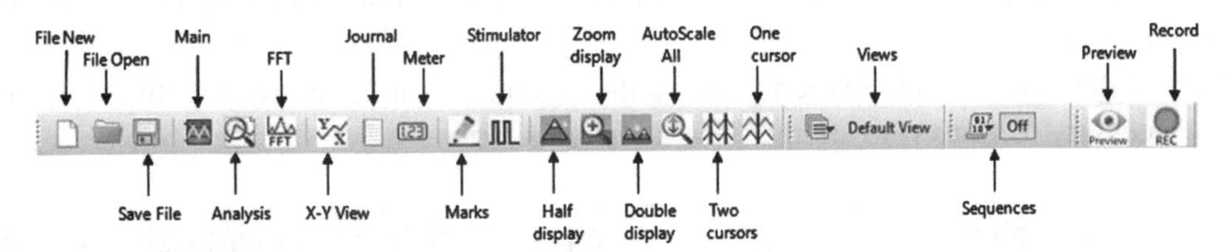


Figure 3 Data analysis Instructions

- Click on the **Analysis** icon in the iWorx toolbar.
- Go to the beginning of your recordings. Click on the **Marks** icon to bring up the Marks Dialogue box with a table of the marks created in your lab. Once you have the table of marks, click on the number at the beginning of the row for the mark that identifies the beginning of the data segment you are going to analyze. This will highlight that mark. Next, click on the **Go To Mark** button near the bottom of the Marks Dialogue box. Data at that mark will be displayed on your screen.
- Be sure that two cursors are present on your screen. If necessary, click the **Two Cursors** icon on the iWorx toolbar.
- Drag and place one cursor at the beginning of the cue. Drag and place the second cursor on the mark made by the subject. If necessary, you can adjust the amount of time displayed on the graph by clicking on the time display icons on the iWorx toolbar. See Figure 4.

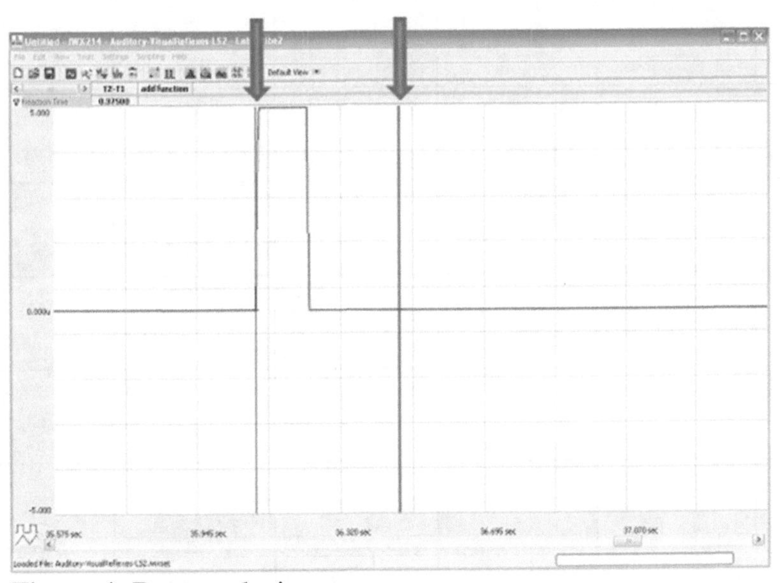

Figure 4 Data analysis

- The **T2–T1** function is displayed above the graph of the data. Just below this is the actual time difference between the time at the beginning of the cue and the time the subject made the mark. This time difference is the subject's reaction time.
- Record the reaction time in the following data table. Be sure to notice the time units. It may be necessary to convert to another unit in the data table.
- Scroll to the next cue and response and repeat the data analysis process. Repeat this process until all ten visual cue response times have been recorded in the following data table. Repeat this process for data analysis for exercises 2, 3, and 4.

- Data analysis for exercises 5 and 6 follows the same procedures except there may be more than one mark per presentation of the visual cue. Drag and place one cursor at the beginning of the cue. Drag and place the second cursor on the final mark made by the subject for the presentation of that cue.
- Record the reaction time in the following data table. Be sure to notice the time units. It may be necessary to convert to another unit in the data table.

| | **Simple Visual Cues T2 – T1 (sec)** | **Simple Auditory Cues T2 – T1 (sec)** | **Prompted Auditory Cues T2 – T1 (sec)** | **Predicted Auditory Cues T2 – T1 (sec)** | **Reading Word Cues T2 – T1 (sec)** | **Reading Color Cues T2 – T1 (sec)** |
|---|---|---|---|---|---|---|
| * Trial 1 | .666 | .354 | .316 | .224 | .269 | .64 |
| Trial 2 | .336 | .488 | .391 | .411 | .662 | .577 |
| Trial 3 | .391 | .252 | .331 | .321 | .550 | 1.033 |
| Trial 4 | .405 | .119 | .101 | .082 | .544 | .654 |
| Trial 5 | .595 | .333 | .401 | .193 | .464 | .513 |
| Trial 6 | .440 | .230 | .270 | .100 | .420 | .550 |
| Trial 7 | 1.558 | .459 | .628 | .442 | .694 | 1.976 |
| Trial 8 | .299 | .140 | .235 | .170 | .503 | .712 |
| Trial 9 | | | | | | |
| Trial 10 | | | | | | |
| Mean Reaction Time | .5855 | .2958 | .334 | .243 | .5133 | .8319 |

- After dropping the longest and shortest reaction times in each exercise, calculate the mean reaction time for each exercise. Record these values in the table and your report sheet.
- Collect the mean reaction time from all subjects for each of the exercises and calculate the means. Record these values in your report sheet.

# Exercise 7

# Somatic Reflex Arcs

## Objectives

- Be able to identify the components of a reflex arc.
- Be able to define *visceral* or *autonomic reflex.*
- Be able to define *somatic reflex.*
- Be able to define and differentiate between *monosynaptic* and *multisynaptic reflexes.*
- Be able to define and differentiate between *ipsilateral* and *contralateral reflexes.*
- Be able to define and differentiate between *agonist* and *antagonist muscle actions.*
- Be able to define and differentiate between *alpha* and *gamma motor neurons.*
- Be able to explain the function of extrafusal muscle fibers.
- Be able to identify and explain the function of the components of a muscle spindle: intrafusal fibers and muscle stretch receptors.
- Be able to define the term *reciprocal innervation.*
- Be able to define and differentiate between *upper* and *lower motor neurons.*
- Be able to define and differentiate between *spastic* and *flaccid paralysis.*
- Be able to explain the *patellar reflex.*
- Be able to explain the *Achilles tendon reflex.*
- Be able to explain the *bicep reflex.*
- Be able to explain the *plantar reflex.*
- Be able to explain the *corneal reflex.*
- Be able to explain the *ciliospinal reflex.*

## Lab Safety Reminders

- Use caution when performing any of the tests, especially those involving the eyes.

## Introduction

Reflexes are actions we often think of as occurring automatically and involuntarily. This concept is the result of several factors. Reflex actions we are aware of usually occur very quickly, and we are often aware that they have occurred only after completing the action. We also tend to think of reflexes as actions involving our skeletal muscles. In reality, almost all the functions that occur in our bodies involve some sort of reflex. A reflex uses a **reflex arc.** The simplest form of a reflex arc involves each of the following: receptor, afferent neuron, integration (usually in the central nervous system, CNS), efferent neuron, and an effector.

The **receptor** is a specialized sensory cell or the modified dendrite of a sensory neuron that detects some type of stimulus. The **afferent or sensory neuron** carries the information regarding the stimulus to the integrating center. **Integration** assesses the information and directs the response. When integration occurs in the CNS (most reflexes), it may occur in the spinal cord or in the brain, depending on the specific reflex. The **efferent or motor neuron** carries information regarding the directed response from the CNS to the effector cells. The **effector cells** respond as a result of the instructions sent via the motor neuron. A **spinal reflex** generally allows a rapid, automatic response to a stimulus. In a spinal reflex, the information only needs to be carried to the spinal cord for integration. The motor response is immediately sent out to the effector cells via the efferent neurons. The information may also be sent to the brain for cognitive analysis of and response to the stimulus. The cognitive analysis generally occurs after the reflex has been completed. CNS-integrated reflexes other than spinal reflexes require integration in the brainstem or other areas of the brain. **Visceral** or **autonomic reflexes** utilize smooth muscle, a gland, or cardiac muscle as the effector. A **somatic reflex** has a skeletal muscle cell as the effector. The following diagram illustrates a simple somatic system spinal reflex arc that would be expected to result in the withdrawal of the arm from the painful stimulus.

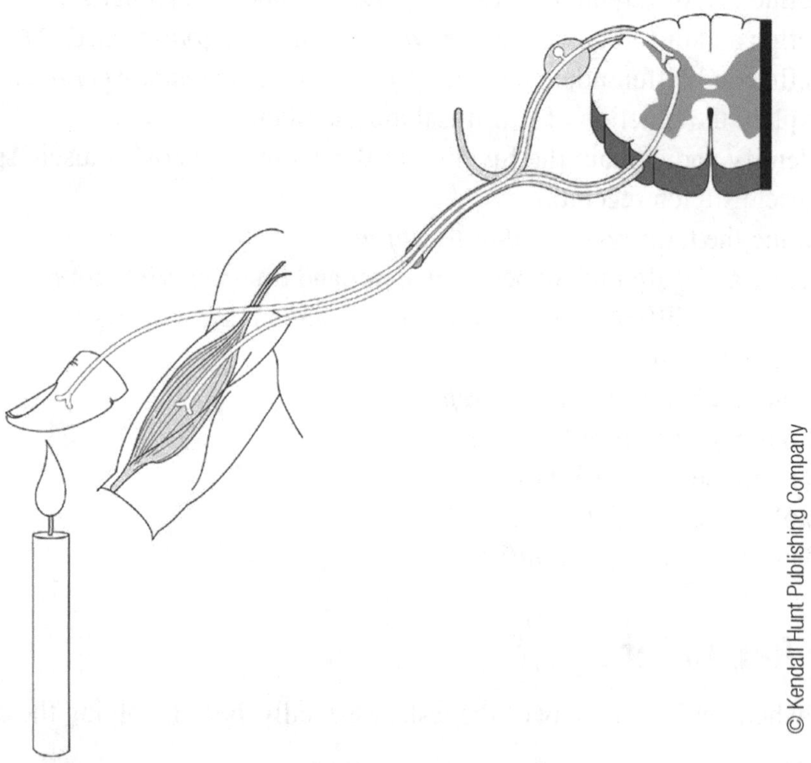


Figure 1 A simple monosynaptic spinal reflex

All somatic reflex arcs require one sensory and one motor neuron. In addition, some reflex arcs also include one or more interneurons. Somatic reflex arcs can be categorized based on the number of synapses in the reflex arc. A **monosynaptic reflex** has only one synapse, between the sensory neuron axon and the motor neuron dendrite, both located within the CNS (Figure 1). There are no interneurons or neurons totally contained within the CNS.

Afferent sensory neuron → Somatic motor neuron

A **multisynaptic** reflex requires more than one synapse within the CNS and thus more than the two neurons (the sensory and the motor neurons) that make up a monosynaptic reflex arc. The additional neurons in the multisynaptic reflex arc are interneurons located totally within the CNS.

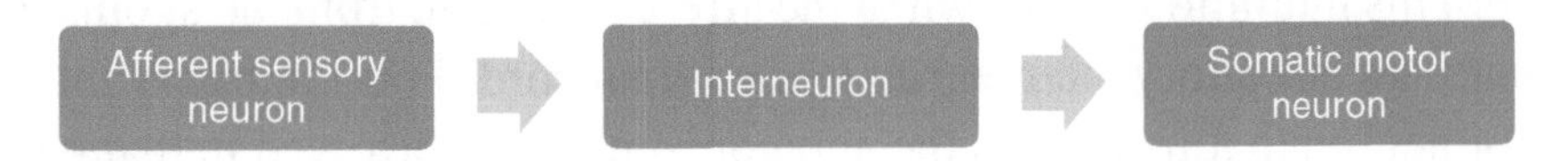


Somatic reflex arcs can also be categorized based on the side of stimulation and the side of response. In an **ipsilateral** reflex, the stimulus and the response are on the same side of the body. Stimulation on the right side would lead to a response on the right side. Depending upon the specific reflex, an ipsilateral reflex may be monosynaptic or multisynaptic. In a **contralateral** reflex, the stimulus and the response are on opposite sides of the body. Stimulation on the right side would lead to a response on the left side. All contralateral reflexes are multisynaptic. Some reflexes may be **consensual** and produce a **bilateral** response that occurs on both sides of the body.

## Reflexes Involving Skeletal Muscles

Somatic reflexes generally stimulate **agonist** muscle action and inhibit **antagonist** muscle action. **Alpha motor neurons** are the neurons that stimulate the extrafusal fibers of a skeletal muscle. At a neuromuscular junction, a skeletal muscle myofiber can only be stimulated by the alpha motor neuron and not inhibited. To inhibit a skeletal muscle myofiber, the alpha motor neuron that stimulates the myofiber must be inhibited. Inhibiting the alpha motor neuron hyperpolarizes the neuron, and it cannot send an action potential to the neuromuscular junction. In this way, the agonist muscle can be stimulated to contract while the antagonist muscle relaxes.

There are different types of reflexes involving skeletal muscles as the effectors. A **stretch reflex** detects stretch in a skeletal muscle and results in the contraction of the muscle to overcome the stretch. Skeletal muscles are composed of two types of fibers, extrafusal fibers and intrafusal fibers. **Extrafusal fibers** are innervated with alpha motor neurons and are responsible for the development of force by skeletal muscles. **Intrafusal fibers** are modified skeletal muscle fibers located in **muscle spindles** and are used in the detection of muscle stretch rather than to develop muscle force. The central region of the intrafusal fibers lack actin and myosin and are wrapped with sensory endings that act as **stretch receptors**. These sensory endings detect the intrafusal fiber stretch that occurs as a result of the application of a load to that skeletal muscle. As the intrafusal fibers are stretched, the stretch receptors send action potentials to the CNS. Integration occurs in the CNS as the sensory neuron synapses with the alpha motor neuron (monosynaptic reflex) and an action potential is immediately sent out of the CNS to the extrafusal fibers of the agonist muscle to stimulate concentric contraction and development of force to overcome the applied load. An action potential is also sent out of the CNS via a gamma motor neuron to the intrafusal fibers of the agonist muscle to cause them to shorten concurrently. The shortening of the intrafusal fibers occurs at the ends of the intrafusal fibers and results in the stretch of their central region. This resulting stretch of the central sensory region of the intrafusal fibers maintains the original length of the sensory region rather than the normal shortening that occurs with concentric muscle contraction. The maintenance of the length of the sensory region allows the sensory receptors to continue to monitor the stretch of the muscle even as the overall length of the muscle shortens.

While observing the stimulation and contraction of the agonist muscles, parallel processing occurs that results in the inhibition of the antagonist muscles. This **reciprocal inhibition** is important because it allows the relaxation of the antagonist muscle so the agonist muscles can better contract and develop force. This inhibition occurs when the afferent neuron, from the agonist muscle stretch receptors, stimulates an inhibitory interneuron in the spinal cord. This inhibitory interneuron inhibits the alpha motor neuron that innervates the antagonist muscles involved in that particular stretch reflex. Because the reciprocal inhibition requires an afferent neuron, an interneuron, and an efferent neuron, it is an example of a multisynaptic reflex.

Skeletal muscle reflexes can be used as diagnostic tools to identify nerve damage as well as the location of damage in the spinal cord. **Upper motor neurons** are located in the motor cortex and descending pathways, and they synapse with other neurons. Upper motor neurons are located completely within the CNS. These neurons carry action potentials to the lower motor neurons that then exit the CNS. When upper motor neurons are damaged, the muscles under their control can no longer perform movements directed by the cerebral motor cortex. The muscles will maintain tonus (a spinal reflex) and can still be stimulated by other spinal reflexes. The person will thus experience **spastic paralysis** in which the muscles can only be stimulated to move reflexively. **Lower motor neurons** are the true motor neurons that exit the spinal cord and synapse directly with muscle fibers and stimulate their contraction. When alpha lower motor neurons are damaged, the extrafusal fibers of the motor units under the control of those alpha motor units can no longer be stimulated, will lose tonus, and will exhibit **flaccid paralysis**. When gamma lower motor neurons are damaged, the intrafusal fibers of the motor units under the control of those gamma motor units can no longer be stimulated and will no longer maintain the sensitivity of the muscle spindle to stretch.

## Deep Tendon Reflexes

The stretch reflexes conducted in this lab are a subset of the deep tendon reflexes. The deep tendon reflexes include the biceps brachii, triceps brachii, brachioradialis, patellar, and Achilles tendon reflexes. The body position of the subject is important for proper implementation of the tests. During testing the limbs should be relaxed and in symmetrical positions to reduce any impact of position on reflex amplitude. In practice, both sides of the body should be tested with the contralateral test being performed immediately after testing the reflex on the first side so any differences in response can be observed and compared. The following scale is often used to assess the subject's responses.

- 0: absent reflex
- 0.5+: a reflex that is elicited only with reinforcement
- 1+: low normal or diminished
- 2+: normal
- 3+: brisker or more reflexive than normal
- 4+: very brisk, hyper-reflexive, non-sustained clonus (repetitive vibratory movements in the stimulated muscle)
- 5+: sustained clonus

Normal responses include scores of 1+, 2+, and 3+ unless there is asymmetry between sides of the body or there is a large difference between responses in the arms and responses in the legs. Abnormal responses include lack of response (0), a response that requires reinforcement (0.5+), and excessive response (4+ and 5+), as well as asymmetry and large differences between arms and legs.

- **Stretch Reflex Activity: Biceps Brachii Tendon Reflex**
  - This tests the musculocutaneous nerve, C5–C6 (cervical) level.
  - Obtain a reflex hammer.
  - Have the subject sit in a chair at the edge of the lab table with one arm fully relaxed but extended on the table or on their leg.
  - Locate the subject's biceps brachii tendon by pressing in the antecubital fossa with your thumb.
  - While pressing on the bicep tendon with your thumb, **strike your thumb** with the hammer. This will cause the bicep tendon and the bicep muscle to be stretched.
  - Observe the biceps brachii muscle for contraction rather than just the potential movement in the forearm.
  - What was your subject's response in the agonist muscles?
  - What was your subject's response in the antagonist muscles?
  - Were the responses ipsilateral, contralateral, or bilateral?
  - Expected Results
    The biceps brachii muscle should be stimulated to contract, but the contraction is usually only a twitch rather than a large flexion of the forearm. The triceps muscle should be inhibited.

- **Stretch Reflex Activity: Patellar Reflex**
  - This tests the femoral nerve, L2–L4 (lumbar) level.
  - Obtain a reflex hammer.
  - Have the subject sit on the edge of the lab table with legs relaxed and hanging over the edge of the table.
  - Locate the subject's patella and the patellar ligament that connects the patella to the tibia.
  - Using the broad edge of the hammer, strike the patellar ligament. This will cause the tendon of the quadriceps and the quadriceps femoris muscles to be stretched. Observe the muscles for a contraction rather than the movement of the lower leg. The vastus medialis is the last muscle to lose the reflex.

- What was your subject's response in the agonist muscles?

- What was your subject's response in the antagonist muscles?

- Were the responses ipsilateral, contralateral, or bilateral?

- Expected Results
  The quadriceps femoris muscle (knee extensor) should be stimulated to contract, and the biceps femoris muscle (knee flexor) should be inhibited, resulting in the extension of the foreleg.

**Stretch Reflex Activity: Achilles Tendon Reflex**

- This tests the tibial nerve, S1 (sacral) level.
- Obtain a reflex hammer.
- Have the subject sit on the edge of the lab table with legs relaxed and hanging over the edge of the table. Shoes and socks should be off.
- Locate the subject's Achilles or calcaneal tendon.
- Hold the toes/sole of the subject's foot in your fingertips and put a small amount of tension on the sole of the foot.
- Using the broad edge of the hammer, strike the Achilles tendon. This will cause the tendon and the gastrocnemius muscle to be stretched.
- Observe the muscle for contraction rather than just the potential movement in the foot.
- What was your subject's response in the agonist muscles?

- What was your subject's response in the antagonist muscles?

- Were the responses ipsilateral, contralateral, or bilateral?

- Expected Results
  The gastrocnemius muscle should be stimulated to contract resulting in the plantar flexion of the foot. At the same time the muscles responsible for dorsiflexion of the foot should be inhibited.

## Other Diagnostic Reflexes

Other reflexes can be used as diagnostic tools to test the various spinal and cranial nerves. Some of these reflexes involve skeletal muscles, and some of them involve visceral muscles rather than skeletal muscles.

- **Cutaneous Reflex Activity: Plantar Reflex and the Babinski Sign**
  - This tests the corticospinal or pyramidal tracts and also the tibial and fibular nerves, L4–S1 (lumbar, sacral) level.
  - Obtain a reflex hammer.
  - Have the subject lie on the lab table with legs uncrossed and relaxed. Shoes and socks should be off.
  - Press on the plantar surface of the heel to hold the foot in place.
  - Use the handle end of the hammer to stimulate the subject.
  - Run the tip of the handle up the lateral edge of the sole of the foot from the heel to the ball of the foot and across the ball of the foot from the little toe to the big toe. This should be one continuous movement.
  - What was your subject's response?
  - Was the response ipsilateral, contralateral, or bilateral?
  - Expected Results
    A normal response results in the adduction and plantar flexion of the toes. An abnormal response is a positive **Babinski sign** and results in the extension of the big toe and the abduction or fanning of the other toes. This is also called a complete Babinski sign. The response is an incomplete Babinski sign if the big toe dorsiflexes and the other toes plantar flex. It is normal to see a positive Babinski sign in an infant (12 to18 months) because their neural system is not completely developed.

- **Corneal Reflex Activity**
  - This tests the trigeminal nerve for sensory function and the facial nerve for motor function.
  - Obtain a small piece of cotton.
  - Have the subject sit so that he/she is facing you.
  - With a fine wisp of cotton and approaching the eye from the lateral side, gently touch the outer corner of the cornea of the right eye.
  - Repeat but stimulate the outer corner of the left eye.
  - What was the response of your subject when the right cornea was stimulated?

- What was the response of your subject when the left cornea was stimulated?

- Were the responses ipsilateral, contralateral, or bilateral?

- Expected Results
  When one eye is stimulated, the subject should blink the lids of both eyes.

## Ciliospinal Reflex Activity

- This tests the trigeminal nerve for sensory function and the upper thoracic sympathetic motor neurons (T1 through T3, lateral horn) and the ascending cervical sympathetic chain.
- Obtain a pin.
- Have the subject sit facing straight ahead.
- Using the pin, gently scratch the right side of the dorsal surface of the neck. The scratch should be strong enough to be uncomfortable but not strong enough to hurt or puncture the subject.
- Observe both pupils of the subject as the neck is being stimulated.
- Repeat but stimulate the subject on the left side of the neck.
- What was the response of your subject when the right side of the neck was stimulated?

- What was the response of your subject when the left side of the neck was stimulated?

- Were the responses ipsilateral, contralateral, or bilateral?

- Expected Results
  The pupil of eye on the same side of the body as the stimulation should dilate.

Bilateral = Both sides
Contralateral = opposite sides
Ipsilateral = Same side

# Exercise 8

# Somatic Senses

## Objectives

- Be able to define the term *sensory receptor*.
- Be able to define the term *adequate stimulus*.
- Be able to define the *law of specific nerve energies*.
- Be able to define and distinguish between *first-*, *second-*, and *third-order neurons*.
- Be able to define and explain *decussation*.
- Be able to trace the sensory pathway of the posterior column.
- Be able to trace the sensory pathway of the anterior spinothalamic pathway.
- Be able to trace the sensory pathway of the lateral spinothalamic pathway.
- Be able to identify the somatic senses carried in each of these three pathways.
- Be able to define the term *anesthesia* and explain the anesthesia experienced by a person who has experienced damage to the spinal cord.
- Be able to define the term *threshold stimulus*.
- Be able to define the terms *adaptation* and *afterimage* and explain the difference between *tonic receptors* and *phasic receptor*.
- Be able to define the term *two-point discrimination* and explain the effect of receptor density and *primary* and *secondary receptor fields* on two-point discrimination.
- Be able to define and explain the term *shifting of physiologic zero*.
- Be able to define the term *referred pain* and explain how it can occur.
- Be able to define the term *phantom limb pain* and explain how it can occur.

## Lab Safety Reminders

- Use caution when performing any of the tests, especially those near the eyes.

## Introduction

The sensory system allows us to monitor both the external and internal environment and thus respond to changes in our environments. The first step in this process is the detection of the stimulus. The detection of stimuli is accomplished by **sensory receptors**. These receptors may be specialized cells that can pass the information on to the afferent neuron, or they may be neurons with specialized dendrites able to detect the stimulus directly. Pain and temperature receptors are free nerve endings. **Exteroceptors** are used to detect changes in the external environment. They include the cutaneous receptors as well as the special senses of taste, olfaction, vision, and hearing. **Interoceptors** are used to detect changes in the internal environment. They include the visceroreceptors (internal pain receptors, baroreceptors, and osmoreceptors) and proprioceptors. Each type of receptor is more sensitive to a particular type of stimulus. This particular stimulus is the **adequate stimulus**. The threshold necessary for the generation of an action potential should be lower for the adequate stimulus than for any other type of stimulus. When a particular type of receptor responds to multiple types of stimuli, all the stimuli are perceived in the same way. As an example, when you close your eye and put pressure on the eye, the stimulation is pressure, a mechanical force. But the pressure applied to the rods and cones, the receptors in the retina, is perceived as light. This illustrates the **law of specific nerve energies**. The sensory pathway carries the information from a sensory receptor to a specific area in the brain where it is interpreted. The interpretation of the stimulus is determined by the final destination in the brain.

The **somatic** senses, in contrast to special senses, are senses with receptors located across the body rather than concentrated in specialized sensory organs. After the detection of a stimulus, the next step in the process of somatic sensory perception is **transduction** or conversion of the stimulus from a physical stimulus to an action potential. **Sensory pathways** carry somatic sensory information regarding stimuli from the location of stimulation to the sensory cortex in the central nervous system. The nonspecific ascending pathways and the specific ascending pathway include three neurons: the first-, second-, and third-order neurons. The **first-order neuron** is the afferent neuron that when stimulated carries the action potential from the sensory receptor in the peripheral nervous system into the central nervous system (CNS) and synapses with the second-order neuron. The **second-order neuron** is contained within the CNS and when stimulated carries an action potential to the thalamus, where it synapses with the third-order neuron. The **third-order neuron** is also contained within the CNS and when stimulated carries an action potential to the somatosensory cortex. The perception or integration of a stimulus in the somatosensory cortex is contralateral to the sensory receptor that detected the stimulus. This occurs because the second-order neuron **decussates** or crosses from one side of the CNS to the other side. The second-order neuron decussates immediately after the synapse between the first-order neuron and the second-order neuron.

Six pathways carry somatic sensory information to the somatosensory cortex. The right and left **posterior columns** (specific ascending pathways) carry information regarding discriminatory touch, vibration, conscious muscle sense, and joint movement (proprioception). The first-order neuron extends from the receptor into the CNS to the medulla, where it synapses with the second-order neuron and where the second-order neuron decussates. The right and left **lateral spinothalamic tracts** (nonspecific ascending tracts) carry information regarding pain and temperature. The first-order neuron extends from the receptor to the spinal cord, where it synapses with the second-order neuron and where the second-order neuron decussates. The right and left **anterior spinothalamic tracts** (nonspecific ascending tracts) carry information regarding light touch and light pressure. The first-order neuron extends from the receptor to the spinal cord, where it synapses with the second-order neuron and where the second-order neuron decussates.

The consequence of decussation in the specific and the nonspecific ascending pathways is evident in anesthesia. Although we generally think of anesthesia as a medication or drug, **anesthesia** in general terms refers to the inability to detect a stimulus. Consider the effect of damaging the spinal cord. Damaging the right half of the spinal cord would result in the contralateral loss of senses carried by the right anterior and right lateral spinothalamic pathways and ipsilateral loss of senses carried by the right posterior column. Note: Arms are not affected if the spinal cord damage is below the level of T1 or T2.

## Threshold Stimulus

The lowest strength of stimulus required to stimulate the initiation of an action potential is considered the **threshold stimulus**. Stimuli below this strength will not be detected.

- **Threshold Stimulus Activity**
  - Obtain a porcelain sample plate, a wax pencil, the sugar and salt solutions, and a dropper.
  - Obtain a glass of distilled water.
  - Label the depressions of the plate with the following concentrations: 1:1,000, 1:800, 1:500, 1:300, 1:200, and 1:100 (see Figure 1).
  - Place approximately 1 mL of each of the sugar solutions into the appropriate depression of the sample plate.
  - The subject should rinse his or her mouth with the distilled water.
  - Place one drop of the weakest sugar solution (1:1000) on the midline of the tip of the subject's tongue.
  - Have the subject rinse his or her mouth with the distilled water. Repeat the process with the next higher concentration of the sugar solution. Continue the process until the subject can detect a sweet taste.
  - Repeat the entire process with the salt solution.

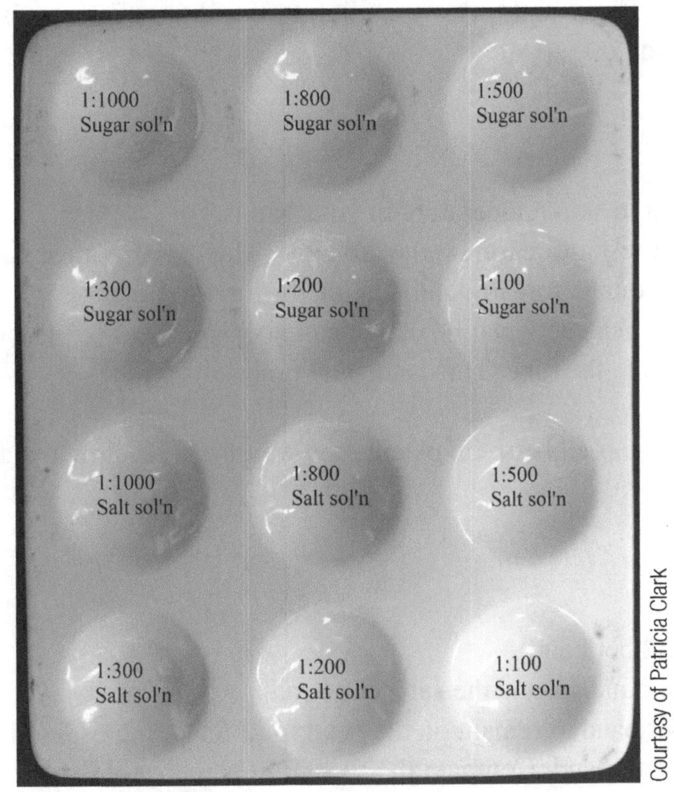


Courtesy of Patricia Clark

Figure 1 Spot plate w solutions

- What was the weakest concentration of sugar solution the subject could taste?

1:800

- What was the weakest concentration of salt solution the subject could taste?

1:100

- Was the subject more sensitive to sugar or salt?

## Adaptation and Afterimage

When continually stimulated with an appropriate stimulus that exceeds the threshold for stimulation, some sensory receptors will initially generate action potentials but then reduce the rate or even stop generating action potentials after a period of time. This is an example of sensory **adaptation.** Some receptors adapt quickly (light touch), some receptors adapt slowly (taste), and some do not adapt at all (skin nociceptors and muscle stretch receptors). **Tonic receptors** have slow adaptation characterized by initial rapid firing followed by a reduced but constant or tonic rate of firing even though the same intensity of stimulation is present. **Phasic receptors** have rapid adaptation characterized by initial rapid firing followed by no firing even though the same intensity of stimulation is present. Phasic receptors that adapt quickly to a stimulus will sometimes again send action potentials when the stimulus is removed. The perceived sensation after removal of the stimulus is called the **afterimage**. The receptor is responding to the change in stimulation that occurs when the stimulus is removed.

### Adaptation Activity

- Obtain a weight.
- Warm the weight to skin temperature in your hands.
- Have the subject rest a forearm on the lab table. The forearm must now remain motionless.
- Place the weight on the forearm and note the beginning time.
- Have the subject indicate when he or she can no longer feel or no longer notice the weight on the arm and note the time. This is the end time.

- How much time elapsed between placing the weight on the arm and adaptation?

### Afterimage Activity

- Obtain a large rubber band.
- Place the rubber band around the subject's head.
- Allow the rubber band to remain on the head for a few minutes, until adaptation occurs.
- Remove the rubber band. Compare the sensation perceived after the removal of the rubber band with the sensation perceived immediately after putting the rubber band on the head.

## Two-Point Discrimination

**Two-point discrimination** is the minimum distance between two points on the skin that are touched simultaneously, but the touches are detected as separate touches. The receptors and primary sensory neurons responsible for detecting somatic sensations are distributed unequally across the surface of the body. The number of receptors and primary sensory neurons in different areas of the body is reflected in a somatotopic map (see the somatotopic maps in your textbook). The larger the number of sensory receptors and primary sensory neurons, the larger the area of the somatosensory cortex designated to the sensory perception of that area of the body. The receptors and primary sensory neurons detect stimuli within their **primary receptor field**. The size of the primary receptor fields can vary among the first-order sensory neurons. Some have large receptor fields. In this case, simultaneous touches within the primary receptor field may be farther apart but still within the same receptor field. Although two stimuli were given, they will be detected as one touch. See Figure 2.

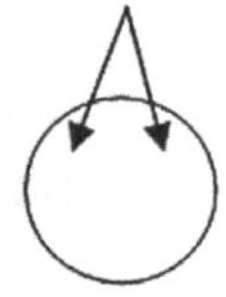

One primary receptor field
Simultaneous two point touch
One touch felt

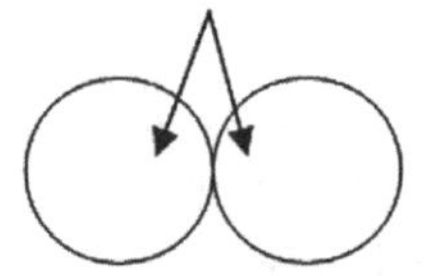

Two primary receptor fields
Simultaneous two point touch
Two touches felt

Figure 2 Simultaneous touch in 1 & 2 primary receptive fields

Convergence of multiple first-order sensory neurons on the same second-order neuron in the sensory pathway can also result in an increased distance in two-point discrimination. When multiple first-order sensory neurons converge on one second-order sensory neuron, the total area covered by the receptors and the converging first-order neurons comprises the **secondary receptor field**. In this case, simultaneous touches in different primary receptor fields but within the same secondary receptor field will be detected as one touch because the same second-order neuron is being stimulated. If first-order sensory neurons of adjacent primary receptor fields do not converge, simultaneous touches in adjacent primary receptor fields will be detected as separate touches. See Figure 3.

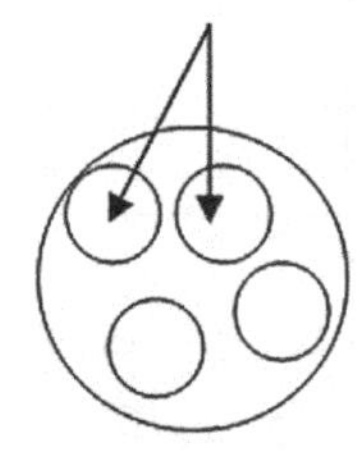

Four primary receptor fields
One secondary receptor field
Simultaneous two point touch
One touch felt

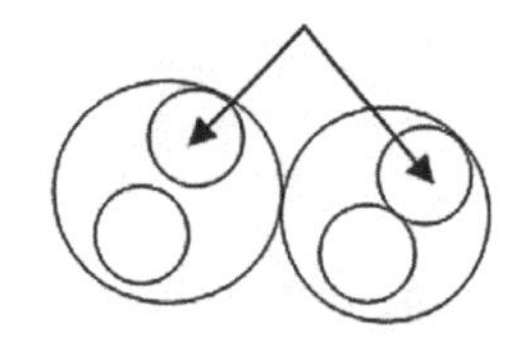

Four primary receptor fields
Two secondary receptor fields
Simultaneous two point touch
Two touches felt

Figure 3 Secondary receptive fields with simultaneous touch in primary receptive fields

The two-point discrimination test allows you to determine the size of receptor fields. Areas of the skin with a smaller two-point discrimination distance will have smaller primary receptor fields and be closer to a one first-order sensory neuron to one second-order sensory neuron ratio (1:1 ratio). Areas of the skin with a larger two-point discrimination distance will have larger primary receptor fields and more sensory neuron convergence.

### Two-Point Discrimination Activity

- Obtain a pair of adjustable calipers.
- Obtain two alcohol prep pads.
- Obtain a small metric ruler.
- Clean the tips of the adjustable calipers with an alcohol prep pad.
- Set the tips of the calipers as close together as possible.
- Have the subject sit with eyes closed and his or her hand, palm up, resting on the lab table.
- Simultaneously touch the tips of the calipers to the palm of the subject's hand. It is important that both tips touch the skin at the same time.
- The subject should indicate if he or she felt one touch or two.
- Increase the distance between the caliper tips and touch the subject again.
- You should periodically touch the subject with only one point so the subject is not anticipating giving an appropriate response.
- Continue to increase the distance between the two tips until the subject can correctly detect two points.
- When the subject can detect two points, measure the distance between the tips of the caliper and record the distance in the chart below. Repeat the process two more times and calculate the mean distance for two-point discrimination.
- Repeat the entire process for the other areas listed in the following table.
- Clean the tips of the adjustable calipers with an alcohol prep pad before returning them to their appropriate location.

| | Trial 1 distance (mm) | Trial 2 distance (mm) | Trial 3 distance (mm) | Mean distance (mm) |
|---|---|---|---|---|
| Palm of hand | >1 MM | 1 MM | 1.5 MM | 2 MM |
| Index fingertip | >1 MM | 1 MM | | |
| Back of neck | >1 MM | 1 MM | | |
| Cheek near lips | >1 MM | 1 MM | | |
| Upper arm | >1 MM | 1 MM | | |

- Check the somatotropic map in your textbook to see if your results are as expected based on the area of the somatic cortex occupied by the specific body region on the map.
- Although the fingertips are expected to be sensitive and have a small two-point discrimination, they do not always exhibit this. What factors might reduce the sensitivity of the fingertips and reduce the ability to detect touch? Hint: Have you burned your fingertips or developed calluses?

## Shifting of Physiologic Zero

**Shifting of physiologic zero** is a form of adaptation. In this case, the free nerve endings that detect temperature change are exposed to a particular temperature for a period of time. As the hand adapts or adjusts to the temperature, the baseline temperature or the zero point shifts to the temperature the hand has been exposed to. When the hand is then placed in a different temperature, the receptors respond to the change in temperature as the new temperature is compared to the temperature the hand was adapted to. As a result, the hand feels the new temperature as warm or cold as compared to the zero temperature.

- **Physiologic Zero Activity**
  - Obtain three beakers large enough to fit your hand in.
  - Fill the first beaker half full of warm water (about 40°C).
  - Fill the second beaker half full of cold water (about 20°C).
  - Fill the third beaker half full of lukewarm water (about 30°C).
  - Place the right hand in the warm water beaker (40°C) and at the same time place the left hand into the cold water beaker (20°C) (Figure 4).
  - Allow the hands to remain in their respective beakers for 1 to 2 minutes.
  - Remove the hands from these beakers and place both hands in the lukewarm water (30°C) and observe the sensations.

Figure 4 Beakers

- The right hand was originally placed in the warm water (40°C). What sensation was felt when the right hand was placed in the lukewarm water (30°C)?

  It didn't take long for the adjustment.
  It felt slightly colder.

- The left hand was originally placed in the cold water (20°C). What sensation was felt when the left hand was placed in the lukewarm water (30°C)?

  It was warmer, but took longer to adjust than warm water

## Sensory Interpretation

It is difficult to stimulate only one type of cutaneous sensory receptor at a time. When multiple somatic senses are stimulated in the same area of the body, they are integrated in the brain. Sensory interpretation of these integrated senses involves not only the somatosensory cortex but also the association areas.

■ **Interpretation Activity**

- Obtain a tight-fitting nitrile glove and put it on one hand.
- Obtain one beaker large enough to fit your hand in.
- Fill the beaker half full of lukewarm water (about 30°C).
- Place the gloved hand in the beaker of water.

- What sensation was felt in the submerged hand? If you did not notice any distinct sensation when the hand was submerged, what was your impression when you removed the glove from your hand?

## Referred Pain

**Nociceptors** are receptors for pain. Several different extreme stimuli may serve as the adequate stimulus for nociceptors. **Referred pain** is pain that originates in one part of the body but is "felt" in another part of the body. A common example of referred pain is the somatic pain felt in response to visceral pain. For example, the angina pectoris or pain felt in the left pectoral region and the left arm actually originates in the heart. The first-order neurons that carry information from these two areas (heart and pectoral/arm region) synapse on the same second-order neurons (convergence) and share the same sensory pathway to the brain. As a result, the sensory information is carried to the same region of the somatosensory cortex. In the case of angina pectoris, the information is carried to the area of the somatosensory cortex that detects sensations in the left pectoral region and the left arm. Referred pain can be used as an indicator for problems in the deep viscera of the thoracic and abdominal cavities just as angina pectoris may serve as an indicator of ischemia to the heart muscle. **Phantom limb pain** is another example of referred pain. In this case, the severed nerve fibers are the source of stimulation. The remaining portions of the sensory pathway can carry these erroneous signals to the somatosensory cortex where they may be perceived as pain, itching, or any other sensation felt in the amputated limb even though the actual sensory receptors are no longer present.

■ **Referred Pain Activity**

- Obtain a reflex hammer.
- Have the subject sit in a chair at the edge of the lab table with one arm extended but bent at the elbow.
- Locate the subject's ulnar nerve in the groove between the median epicondyle of the humerus and the olecranon process of the ulna.
- Use the hammer to tap the ulnar nerve.

- Where did the subject feel pain or a tingling sensation? How does this compare to the location of stimulation?

- Expected Results
  Although you are stimulating sensory neurons of the ulnar nerve proximally, the pain associated with the stimulation is felt or perceived to have originated not only in the elbow region but also in the hands.

Ch. 7

upper Motor neuron
[CNS or Motor cortex]

↓

lower Motor neuron
[Alpha motor neuron
Gamma motor neuron]

↓

Effector cell [muscle]

if damaged,
"Spastic paralysis"

if damaged,
"faccid paralysis"

# Exercise 9

# Cranial Nerve Assessments

## Objectives

- Be able to identify each cranial nerve by number and name.
- Be able to identify each cranial nerve as a sensory, motor, or mixed nerve.
- Be able to identify the functions of each cranial nerve.

## Lab Safety Reminders

- Use caution when performing any of the tests, especially those near the eyes.

## Introduction

Cranial nerves are a part of the peripheral nervous system (PNS). The cranial nerves do not enter the central nervous system (CNS) via the spinal cord as spinal nerves do. Most enter the CNS directly through the brainstem (olfactory nerve is the exception). Most of the functions of the cranial nerves are in the head or cervical region. The exception is the vagus nerve that innervates viscera in the thoracic and abdominal cavities. Another difference between cranial nerves and spinal nerves is that the cranial nerves do not decussate. The right cranial nerves function on the right side of the body, and the left cranial nerves function on the left side of the body. One exception is in the optic pathway. A portion of the right and left optic nerves cross to the opposite side of the brain at the optic chiasma. The axons of the sensory neurons that carry the information from the temporal fields of view cross to the opposite side of the brain. The axons of the sensory neurons that carry the information from the nasal fields of view do not cross but remain ipsilateral. See the figures in your text or on the lab posters as references.

Use the following general table as a guide for learning the cranial nerves. You should notice that for the reflexes controlled by cranial nerves, one nerve may be responsible for the sensory (afferent) component whereas another nerve may be responsible for the motor (efferent) component of the reflex. Note: Not all reflexes are listed.

| **Cranial Nerve** | | | **Overview of Major Functions** | |
|---|---|---|---|---|
| # | Name | Sensory (S)<br>Motor (M)<br>Mixed (B) | Sensory Function | Motor Function |
| I | Olfactory | S | Smell | |
| II | Optic | S | Vision<br>Accommodation reflex<br>Pupillary light reflex | |
| III | Oculomotor | M | | Iris:<br>pupillary light reflex<br>Ciliary body:<br>accommodation reflex<br>Upper eyelid<br>Extraocular eye muscles:<br>superior rectus<br>medial rectus<br>inferior rectus<br>inferior oblique |
| IV | Trochlear | M | | Extraocular eye muscle:<br>superior oblique |
| V | Trigeminal<br>facial sensation | B | Facial sensations:<br>ophthalmic branch<br>maxillary branch<br>mandibular branch | Masseter muscle<br>Temporalis muscle<br>Tympanum muscle |
| VI | Abducens | M | | Extraocular eye muscle:<br>lateral rectus |
| VII | Facial<br>facial movement | B | Soft palate<br>Taste:<br>anterior 2/3 of the tongue | Glands:<br>lacrimal,<br>nasal,<br>submandibular,<br>sublingual<br><br>Facial expression:<br>temporal branch<br>zygomatic branch<br>buccal branch<br>mandibular branch<br>cervical branch |
| VIII | Vestibulocochlear or Auditory | S | Equilibrium<br>Hearing | |

| Cranial Nerve | | | Overview of Major Functions | |
|---|---|---|---|---|
| IX | Glossopharyngeal | B | Palate<br>Taste:<br>posterior 1/3 of the tongue<br>Pharynx and tonsils<br>Carotid sinus and carotid body | Pharyngeal muscles<br>Parotid gland |
| X | Vagus | B | Pharyngeal taste<br>Aortic body and aortic arch<br>Atria<br>Diaphragm<br>Esophagus<br>Respiratory tract<br>Abdominal organs | Heart<br>Smooth muscles and glands of the respiratory and digestive tracts |
| XI | Accessory or Spinal Accessory | M | | Soft palate<br>Pharynx<br>Larynx<br>Sternocleidomastoid<br>Trapezius |
| XII | Hypoglossal | M | | Tongue movement |

## Cranial Nerve I, Olfactory Nerve

The sensory receptors for olfaction are located in the nasal epithelium. In order for an odorant to be detected, it must be dissolved and come in contact with the appropriate receptors in the nasal epithelium. The axons of these receptors join to form the olfactory nerve. Loss of the olfactory receptors or damage to the neural pathway for olfaction can result in **anosmia**, the loss of the sense of smell. Some of the factors that can reduce the ability of an individual's olfactory receptors to detect smells are smoking, cocaine use, inflammation, and age. Tumors that apply pressure to the olfactory bulbs and tracts can also reduce an individual's ability to smell.

- **Cranial Nerve I (Olfactory) Activity**
  - Obtain small bottles of each of the following common odors: lemon, peppermint, vanilla, cinnamon, and wintergren.
  - Have the subject close the eyes and block the left nostril. Pass one of the sample bottles, opened, a small distance under the subject's right nostril.
  - Ask the subject to smell the sample.
  - Based upon the subject's responses, mark the appropriate boxes in the following table.
  - Repeat the procedure with the right nostril blocked, but randomize the presentation of the scents.

| Scent | Ability to Smell (+ or −) | | Ability to Identify (+ or −) | |
|---|---|---|---|---|
| | Right nostril | Left nostril | Right nostril | Left nostril |
| Lemon | + | + | + | + |
| Peppermint | + | + | + | + |
| Vanilla | + | + | + | + |
| Cinnamon | + | + | + | + |
| Wintergreen | − | − | − | − |

- Does the subject suffer from anosmia?

Subject cannot smell wintergreen.
Cranial nerve I may have slight anosmia.

## Cranial Nerve II, Optic Nerve

Although we normally think of the optic sensory receptors and nerve as being responsible for visual images, they have some additional functions. The receptors and optic nerve are also responsible for providing sensory information regarding the amount of light entering the eye through the pupil. This sensory information is used to regulate the contraction of the muscles of the iris and the resulting dilation and constriction of the pupil. The receptors and optic nerve also provide sensory information regarding whether or not an object is in focus. This sensory information is used to regulate the ciliary muscle in the process of **accommodation**, changing the focal point between far and close.

- **Cranial Nerve II (Optic) Activity**
  - Optic nerve activities will be performed in another lab.

## Cranial Nerve III, Oculomotor Nerve; Nerve IV, Trochlear Nerve; Nerve VI, Abducens Nerve

The oculomotor, trochlear, and abducens nerves contain the efferent neurons that innervate the six extraocular or extrinsic eye muscles of each eye. The oculomotor nerve innervates the superior, inferior, and medial rectus muscles as well as the inferior oblique muscle. It also innervates the pupillary constrictor muscle of the iris, the ciliary muscle, and the elevator muscles of the upper eyelid (the dilator muscle of the iris is innervated by the sympathetic system via the superior cervical ganglion). Differences in the size of pupils may be normal (slight differences) or may be the result of a variety of diseases or drugs. The trochlear nerve innervates the superior oblique muscle. The abducens nerve innervates the lateral rectus muscle.

There are **six cardinal directions of gaze**, dextroelevation (up/right), dextroversion (right), dextrodepression (down/right), levoelevation (up/left), levoversion (left), and levodepression (down/left). These gazes test all extraocular muscles and thus the three cranial nerves associated with them.

The six cardinal directions of gaze and the expected direction of gaze as a result of the contraction of the appropriate extraocular muscle are:
SR—superior rectus, IR—inferior rectus, MR—medial rectus, LR—lateral rectus, IO—inferior oblique, SO—superior oblique (See Figure 1)

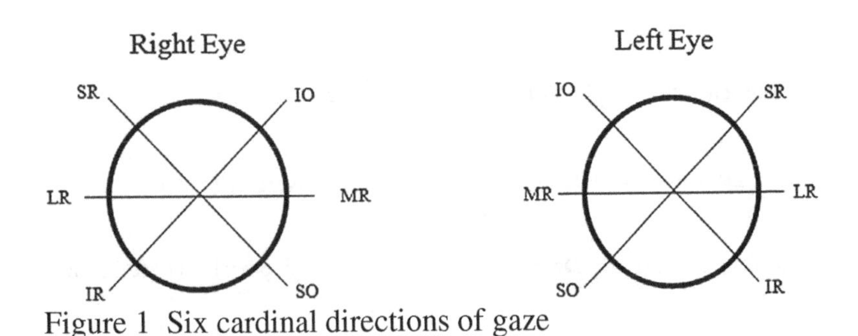


Figure 1 Six cardinal directions of gaze

In addition to the six cardinal gazes, a gaze test may also include infraversion (down), supraversion (up), and convergence (crossed). Notice that for all eye movements other than convergence, the movements of the extraocular muscles in the two eyes are yoked, meaning that both eyes work together to move in the same direction at the same time.

- **Cranial Nerve III (Oculomotor), IV (Trochlear), and VI (Abducens) Activity**
  - Observe the normal, relaxed position of the subject's upper eyelids. Indicate the results in the following table.
  - Obtain a small flashlight.
  - Observe the size of the subject's pupils. They should be the same size.
  - Briefly shine the flashlight in the subject's right eye and observe the pupillary response in both eyes. The eyes should exhibit a consensual or bilateral reflex (both ipsilateral and contralateral).
  - Test the pupillary response when shining the light in the subject's left eye. Indicate the responses for both eyes in the following table.
  - To test the subject's extraocular muscles, the tester should begin by holding a pencil and positioning their hand with the tip of the pencil pointing up and held directly in front of the subject's eyes and a slight distance from the subject's face. Ask the subject to track the movement of the pencil.

    From the beginning position, move the hand with the pencil in each of the six cardinal directions of gaze as well as the three additional directions of gaze (see Figure 2) and indicate the subject's response in the following table.

    Six cardinal directions of gaze

    Move the hand: Right/Up—Dextroelevation
    Right—Dextroversion
    Right/Down—Dextrodepression
    Left/Up—Levoelevation
    Left—Levoversion
    Left/Down—Levodepression

Three additional directions of gaze

Move the hand: Down (midline)—Infraversion
Up (midline)—Supraversion
Level of the nose (midline)—Convergence

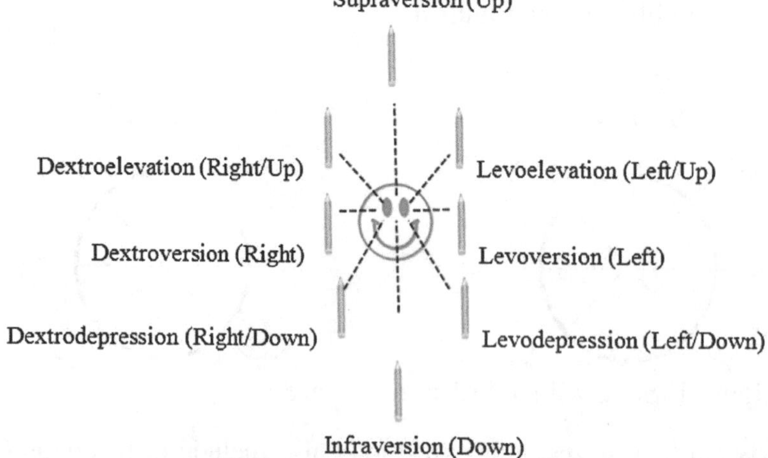


Figure 2 Gaze testing positions

| Action | Nerve and Muscle | Muscle Function | Right Eye Response (+ or –) | Left Eye Response (+ or –) |
|---|---|---|---|---|
| Upper eyelid position | III: Elevator muscles | Normal: Upper lids cover 1/3 of iris | + | + |
| | | | | |
| Comparison of pupil size | III: Pupillary muscles | Pupils should be nearly the same size | + | + |
| Pupil response to light | III: Pupillary muscles | Excess light results in constriction | + | + |
| | | | | |
| Dextroversion | Right VI: Lateral rectus (LR) | Abduction<br>Gaze shifts laterally | + | + |
| | Left III: Medial rectus (MR) | Adduction<br>Gaze shifts medially | + | + |
| Levoversion | Right III: Medial rectus (MR) | Adduction<br>Gaze shifts medially | + | + |
| | Left VI: Lateral rectus (LR) | Abduction<br>Gaze shifts laterally | + | + |

| Action | Nerve and Muscle | Muscle Function | Right Eye Response (+ or –) | Left Eye Response (+ or –) |
|---|---|---|---|---|
| Dextroelevation | Right III: Superior rectus (SR) | Elevation, maximal on lateral gaze<br>Gaze shifts up and laterally | + | + |
| | Left III: Inferior oblique (IO) | Elevation, maximal on medial gaze<br>Gaze rotates up and medially | + | + |
| Levoelevation | Right III: Inferior oblique (IO) | Elevation, maximal on medial gaze<br>Gaze rotates up and medially | + | + |
| | Left III: Superior rectus (SR) | Elevation, maximal on lateral gaze<br>Gaze shifts up and laterally | + | + |
| Dextrodepression | Right III: Inferior rectus (IR) | Depression, max. on lateral gaze<br>Gaze shifts down and laterally | + | + |
| | Left IV: Superior oblique (SO) | Depression, max. on medial gaze<br>Gaze rotates down and medially | + | + |
| Levodepression | Right IV: Superior oblique (SO) | Depression, max. on medial gaze<br>Gaze rotates down and medially | + | + |
| | Left III: Inferior rectus (IR) | Depression, max. on lateral gaze<br>Gaze shifts down and laterally | + | + |

| Action | Nerve and Muscle | Muscle Function | Right Eye Response (+ or –) | Left Eye Response (+ or –) |
|---|---|---|---|---|
| Infraversion | Right III: Inferior rectus (IR) | Depression, max. on lateral gaze<br>Gaze shifts down and laterally | + | + |
| | Right IV: Superior oblique (SO) | Depression, max. on medial gaze<br>Gaze rotates down and medially | + | + |
| | Left IV: Superior oblique (SO) | Depression, max. on medial gaze<br>Gaze rotates down and medially | + | + |
| | Left III: Inferior rectus (IR) | Depression, max. on lateral gaze<br>Gaze shifts down and laterally | + | + |
| Supraversion | Right III: Superior rectus (SR) | Elevation, maximal on lateral gaze<br>Gaze shifts up and laterally | + | + |
| | Right III: Inferior oblique (IO) | Elevation, maximal on medial gaze<br>Gaze rotates up and medially | + | + |
| | Left III: Inferior oblique (IO) | Elevation, maximal on medial gaze<br>Gaze rotates up and medially | + | + |
| | Left III: Superior rectus (SR) | Elevation, maximal on lateral gaze<br>Gaze shifts up and laterally | + | + |
| Convergence | Right III: Medial rectus (MR) | Adduction<br>Gaze shifts medially | + | + |
| | Left III: Medial rectus (MR) | Adduction<br>Gaze shifts medially | + | + |

- Did the subject exhibit any abnormal responses in the eye muscles?

- Ask the subject to gaze straight ahead. Check for placement of the pupils. Pupils that are unintentionally oriented medially result in crossed eyes or **strabismus**. This may occur in both eyes or in only one eye.
- Does the subject demonstrate strabismus?

## Cranial Nerve V, Trigeminal Nerve

The sensory components of the trigeminal nerve detect cutaneous stimulation of the face and the cornea. The motor components of this nerve innervate the muscles of mastication.

You have already performed the corneal blink reflex in another lab. You will test the cutaneous touch receptors of the face and the stimulation of the temporalis muscle in this lab.

- **Cranial Nerve V (Trigeminal) Activity**
  - Obtain a cotton ball, a pin, and alcohol swabs.
  - Ask the subject to close his or her eyes.
  - Using the cotton ball, lightly touch the subject in the areas listed in the following table. Indicate the subject's ability to detect the stimulus.
  - Clean the pin with an alcohol swab.
  - Using the pin, lightly prick the subject in the areas listed in the following table. Indicate the subject's ability to detect the stimulus.
  - Clean the pin with an alcohol swab and return to the supply area or discard in the biohazards/sharps container provided.

| Location of Stimulation | Response to Light Touch (+ or −) | | Response to Pain (+ or −) | |
|---|---|---|---|---|
| | Right side | Left side | Right side | Left side |
| Forehead | + | + | + | + |
| Zygomatic area | + | + | + | + |
| Upper lip | + | + | + | + |
| Cheek | + | + | + | + |
| Chin | + | + | + | + |

- Place your hands on the skin covering the subject's temporalis muscle.
- Ask the subject to clench the jaw.
- Could you feel the contraction of the temporalis muscle as the subject clenched the jaw?

## Cranial Nerve VII, Facial Nerve

The primary sensory component of the facial nerve is used for gustation or taste (anterior two-third of the tongue). You have already performed a taste test in a previous lab. The motor function of this nerve is for facial expressions. Bell's palsy results in the paralysis of the facial nerve, usually on one side of the face. There is no one cause of Bell's palsy, but the problem is usually the result of inflammation or compression of the facial nerve.

- **Cranial Nerve VII (Facial) Activity**
  - Ask the subject to perform the following movements. Movements should be equal on both sides of the face.
  - Indicate the subject's ability to perform the movement in the following table.

| Movement | Ability to Perform Movement (+ or −) | | Movement Equal Bilaterally (+ or −) | |
|---|---|---|---|---|
| | Right side | Left side | Right side | Left side |
| Raise eyebrows | + | + | + | + |
| Smile | + | + | + | + |
| Show teeth | + | + | + | + |
| Frown | + | + | + | + |
| Close eyes | + | + | + | + |

- Did the subject exhibit any abnormalities in function of the facial nerve including uneven movement between the right and left sides of the face?

## Cranial Nerve VIII, Vestibulocochlear or Auditory Nerve

The sensory function of vestibulochoclear nerve is divided between two organs and senses. The vestibular branch of the nerve carries the sensory information from the vestibular apparatus regarding balance, equilibrium, acceleration and deceleration, and body position relative to gravity. The cochlear branch of the nerve carries the sensory information from the cochlea regarding hearing.

- **Cranial Nerve VIII (Vestibulocochlear or Auditory) Activity**
  - Auditory nerve activities will be performed in a separate lab.

## Cranial Nerve IX, Glossopharyngeal Nerve; Nerve X, Vagus Nerve

The glossopharyngeal and the vagus nerves are both mixed function nerves with a wide variety of functions. The glossopharyngeal nerve carries taste information from gustatory receptors in the posterior one-third of the tongue. The vagus nerve carries taste information from gustatory receptors in the pharynx and epiglottis. Other sensory receptors innervated by these cranial nerves detect changes in blood chemistry (partial pressures of carbon dioxide and oxygen, pH). Baroreceptors in the aortic body also detect blood pressure. We will be observing the motor functions of these nerves involving the muscles of the throat region.

- **Cranial Nerve IX (Glossopharyngeal) and X (Vagus) Activity**
  - Obtain a long cotton swab.
  - Ask the subject to open the mouth wide and say "aah."
  - Observe the position of the soft palate and uvula. They should be even, and the uvula should be midsaggital. yes
  - Ask the subject to swallow and watch for any difficulties. No diffiulties
  - Ask the subject to again open the mouth wide. Can do
  - Test for the gag reflex by gently touching the back of the throat with the cotton swab. Absence of a gag reflex is abnormal, although some people learn to control the reflex. Cant do
  - Did the subject exhibit any abnormalities?

Tonsils are in the way, so subject couldnt experience these examples. Nerves IX + 10 may have sligh effect on this.

## Cranial Nerve XI, Spinal Accessory Nerve

The motor function of the spinal accessory nerve regulates the action of some of the throat muscles as well as the sternocleidomastoid and the trapezius muscles. These muscles assist in the movement of the head and the shoulders. The sternocleidomastoid turns the head to the opposite side, and the trapezius muscles elevate the shoulders.

- **Cranial Nerve XI (Spinal Accessory) Activity**
  - With one hand on the subject's left cheek and the other hand on the right sternocleidomastoid muscle, ask the subject to turn his or her face against your hand.
  - Repeat on the other side.
  - The right and left sternocleidomastoids should contract with the same strength. Twitches should not be felt during contraction.

- Place your hands on both of the subject's shoulders and press down. Ask the subject to shrug the shoulders.
- The strength of both trapezius muscles should be roughly equal.

Yes

- Did the subject exhibit any abnormalities?

No

## Cranial Nerve XII, Hypoglossal Nerve

The hypoglossal nerve innervates the muscles of the tongue. The right and left hypoglossal nerves innervate the genioglossus muscles on the right and left sides of the tongue respectively. The genioglossus muscles of the tongue originate on the medial surface of the mandible, near the midline. They insert on the lateral sides of the body of the tongue and on the hyoid bone. The action of the genioglossus muscles is to depress and protract the tongue. In other words, these muscles help you stick your tongue out and depress it when you say "aah" during an examination of your pharynx. This muscle is also extremely important in infant nursing as it creates a "trough" that directs milk to the pharynx. Damage to the hypoglossal nerve will result in damage to a lower motor neuron and an ipsilateral effect. When the right hypoglossal nerve has infranuclear damage, damage distal to the nucleus and damage to a lower motor neuron, the tongue will deviate to the right or ipsilateral, because the right genioglossus muscle becomes flaccid and cannot contract and extend the tongue, but the left muscle retains its tonus and moves the left side of the tongue forward (protrusion of the tongue). As a result the tongue bends toward the flaccid side. When the left hypoglossal nerve is damaged, the tongue will deviate to the left or ipsilateral, because the left muscle becomes flaccid and cannot contract, but the right genioglossus muscle retains its ability to contract and the right side of the tongue protrudes (see Figure 3). A loss of tongue function can also lead to dysarthria or dysarticulation, the slurring of speech. The production of some letter sounds is more affected than others.

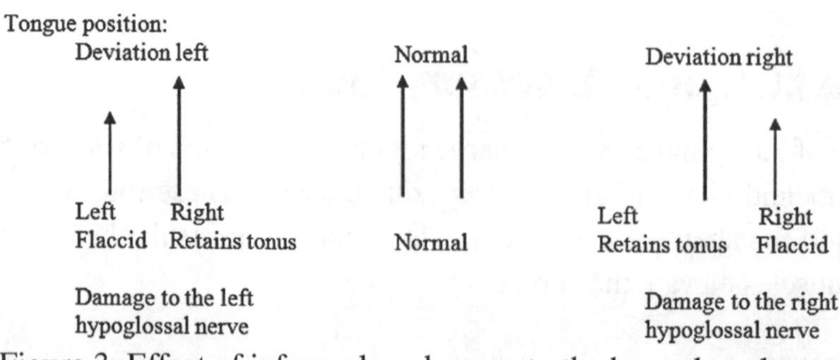


Figure 3 Effect of infranuclear damage to the hypoglossal nerve

It is important to note that the majority of the fibers in the hypoglossal nerve decussate in the corticobulbar tracts. As a result, damage to the upper motor neurons associated with movement of the genioglossus muscles will result in deviation of the tongue contralateral to the damaged upper motor neurons.

## Cranial Nerve XII (Hypoglossal) Activity

- Ask the subject to stick out the tongue.
- Does the tongue deviate right or left?
- Indicate the subject's ability to perform the test in the following table.

- Ask the individual to poke the inside of their right cheek with their tongue. At the same time apply counter pressure on the outside of the cheek. Can the person resist the pressure?
- Repeat the test on the left side of the cheek.
- Indicate the subject's ability to perform the test in the table.

- Ask the person to make the sound of one or two of the following letters: L, T, D, N, R.
- Indicate the subject's ability to perform the test in the following table.

| Test | Right Side Affected (+) | Left Side Affected (+) | Normal (+) |
|---|---|---|---|
| Tongue position | | | + |
| Tongue strength | | | + |
| Speech clarity | | | + |

- Did the subject exhibit any abnormalities in function of the hypoglossal nerve?

# Exercise 10

# Vision

## Objectives

- Be able to identify the structures of the eye associated with vision.
- Be able to explain the function of each of the structures of vision.
- Be able to distinguish between the vision as a result of the activation of rods and as a result of the activation of cones.
- Be able to determine the areas of vision lost as a result of lesions in the optic nerves, optic chiasma, and optic tracts.
- Be able to define the terms *visual acuity*, *emmetropia*, *myopia*, *hyperopia*, and *presbyopia*.
- Be able to associate the conditions of emmetropia, myopia, and hyperopia with visual acuity.
- Be able to explain the terms *OD (oculus dexter)* and *OS (oculus sinister)*.
- Be able to interpret the *prescriptive number* for a *divergent corrective lens* for myopia and a *convergent corrective lens* for hyperopia.
- Be able to define the term *astigmatism* and explain the effect it has on vision.
- Be able to define the term *visual field* and explain the components of a person's complete visual field and how this results in binocular vision.
- Be able to define the term *blind spot* and relate it to the optic disc and location in the visual field.
- Be able to explain how the brain and eyes compensate for the blind spot.
- Be able to define the term *accommodation*.
- Be able to explain the changes in the eye that allow for accommodation.
- Be able to define the term *near point of accommodation*.
- Be able to explain the changes in the eye that result in presbyopia.
- Be able to explain the use of an ophthalmoscope.

## Introduction

The sense of vision involves refraction of light and focusing of light on the **retina** and the detection of light waves by the receptors. Two types of receptors are present in the retina. The **rods** detect light and are responsible for black-and-white vision. The rods are very useful in dim light, but they provide less visual acuity. The **cones** detect specific wavelengths of light and are responsible for color vision. Cones require bright light but provide high visual acuity. The **optic nerves** (cranial nerve II) carry the impulses from the retina to the central nervous system (CNS). The remainder of the optic pathway carries the impulses to the visual cortex of the brain. Unlike most cranial nerves, all the information carried by the right optic nerve does not travel to the right side of the brain, and all the information carried by the left optic nerve does not travel to the left side of the brain. The optic nerves are responsible for conducting the impulse from the retina to the optic chiasma. At the **optic chiasma**, the fibers that carry sensory information from the temporal fields of vision cross to the opposite side of the brain, but the fibers that carry sensory information from the nasal fields of vision stay on the originating side. As a result, the **optic tracts** that carry the sensory information on to the thalamus carry information from the ipsilateral nasal field of vision and the contralateral temporal field of vision. It is important to note that the nasal field of vision is projected on the lateral side of the retina, and the temporal field of vision is projected on the medial side of the retina (see Figure 1 and the appropriate figures in your text or on the lab posters as references).

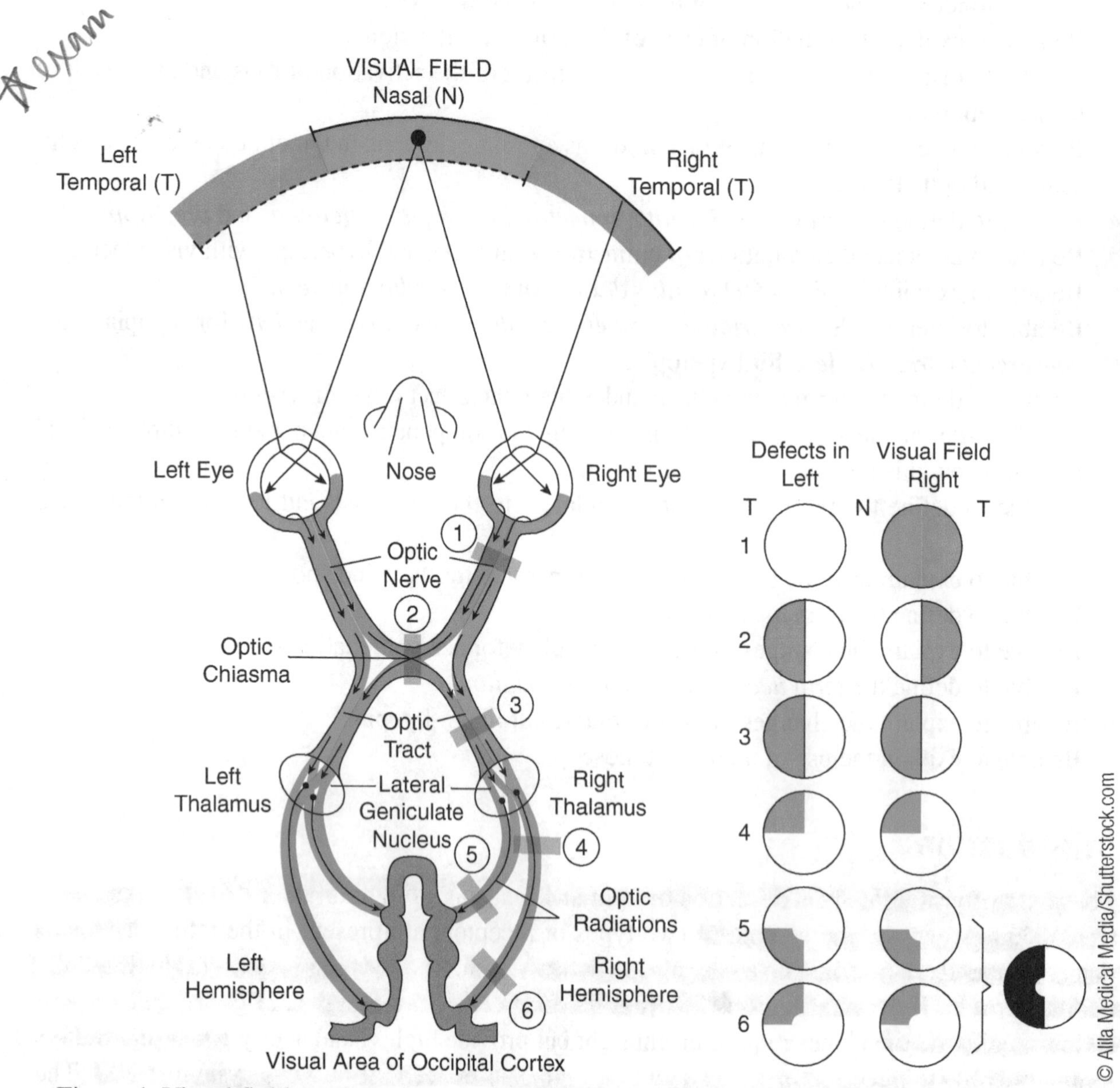


Figure 1 Visual fields and sensory visual pathways

## Vision Tests

Vision tests allow an individual to determine how well his or her eyes are functioning compared to normal. The tests you will perform in this lab are very basic tests, but many of them are the same tests that would be performed with more sophisticated equipment in an optometrist's office.

- **Visual Acuity Activity**

  The **Snellen chart** (see Figure 2) is used to test **visual acuity** or the ability to focus incoming light on the retina. A person with normal vision, **emmetropia**, has a visual acuity of 1 (20/20 vision). Visual acuity (V) is calculated as distance from the Snellen chart (d) divided by the distance at which the letter should be read (D): V= d/D. For normal vision, d = 20 feet and D = 20 feet, therefore 20/20 = 1 (the metric equivalent is 6/6, d = 6 meters and D = 6 meters). When the focal point of the incoming light is behind the retina because the eyeball is too short, the person has **hyperopia** or hypermetropia. The person may be referred to as farsighted because he or she can focus on objects that are farther away but cannot focus on objects that are close. The visual acuity of an individual with hyperopia may be greater than 1 (20/10 = 2.0). When the focal point of the incoming light is in front of the retina because the eyeball is too long, the person has **myopia**. The person may be referred to as nearsighted because he or she can focus on objects that are close but cannot focus on objects that are farther away. The visual acuity of a person with myopia is less than 1 (20/200 = 0.1).

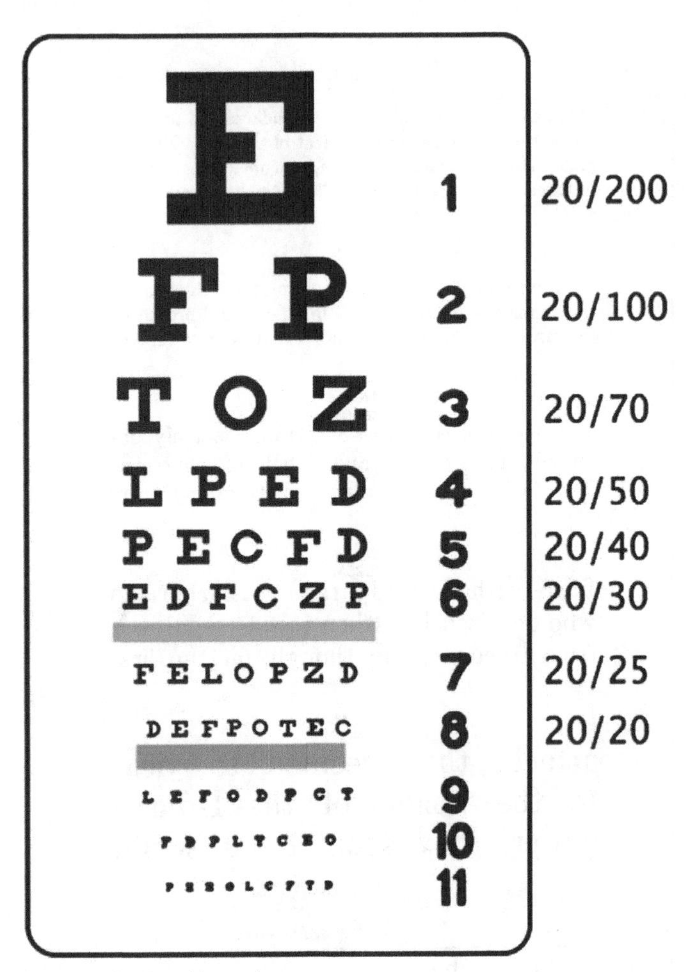


Figure 2 Snellen chart

- Stand 20 feet from the Snellen chart.
- Look at the Snellen chart with one eye. Close or use your cupped hand to cover the other eye but do not press on the eye (an occluder is the tool often used to cover the eye for the Snellen test).
- Read the smallest line possible. Record the ratio for the line and calculate your visual acuity (V). The top number of the ratio is always 20 (d). The lower number is D.
- Repeat for the other eye.
- If you wear glasses or contacts, you may want to determine your visual acuity both with and without your corrective aids.

- Ratio for the right eye. _____ Calculated visual acuity. _____
- Ratio for the left eye. _____ Calculated visual acuity. _____

Near visual acuity, testing for hyperopia, is done with a Jaeger eye chart (see Figure 3). This chart is similar to the Snellen test in that it has samples of print of different size or font. The Jaeger eye chart is held 12 to 14 inches from a person's eye. The larger the print must be in order to read it, the less near visual acuity the person has.

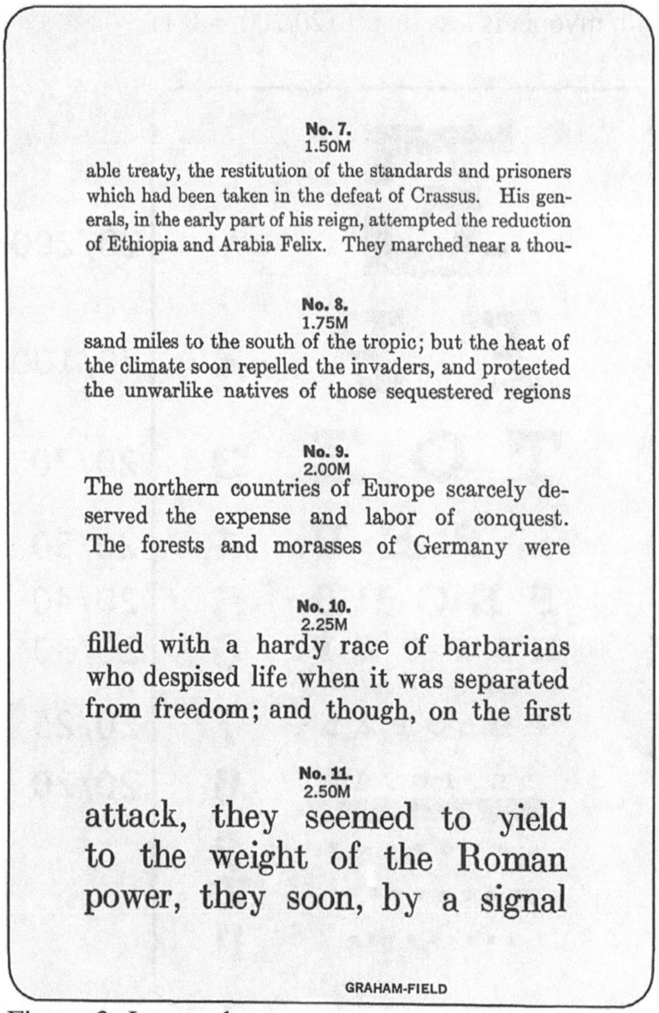

**No. 7.**
1.50M

able treaty, the restitution of the standards and prisoners which had been taken in the defeat of Crassus. His generals, in the early part of his reign, attempted the reduction of Ethiopia and Arabia Felix. They marched near a thou-

**No. 8.**
1.75M

sand miles to the south of the tropic; but the heat of the climate soon repelled the invaders, and protected the unwarlike natives of those sequestered regions

**No. 9.**
2.00M

The northern countries of Europe scarcely deserved the expense and labor of conquest. The forests and morasses of Germany were

**No. 10.**
2.25M

filled with a hardy race of barbarians who despised life when it was separated from freedom; and though, on the first

**No. 11.**
2.50M

attack, they seemed to yield to the weight of the Roman power, they soon, by a signal

**GRAHAM-FIELD**

Figure 3 Jaeger chart

- **Astigmatism Activity**

The cornea is responsible for the majority of the refracting or bending of light rays so they fall on one spot on the retina. The lens is responsible for fine-tuning the focusing of the light on the retina. When the cornea does not have an even thickness, it does not uniformly bend the light, and the image is not pinpointed on the retina. A person who has an uneven cornea has **astigmatism**. Although the lens correctly focuses the light and visual acuity may be 1, their vision is not crisp. This is often most evident at night when looking at lights. A person with an astigmatism will see streaks of light radiating from the light source rather than a more uniform halo of light. The following clock dial chart (see Figure 4) is one way to test for astigmatism.

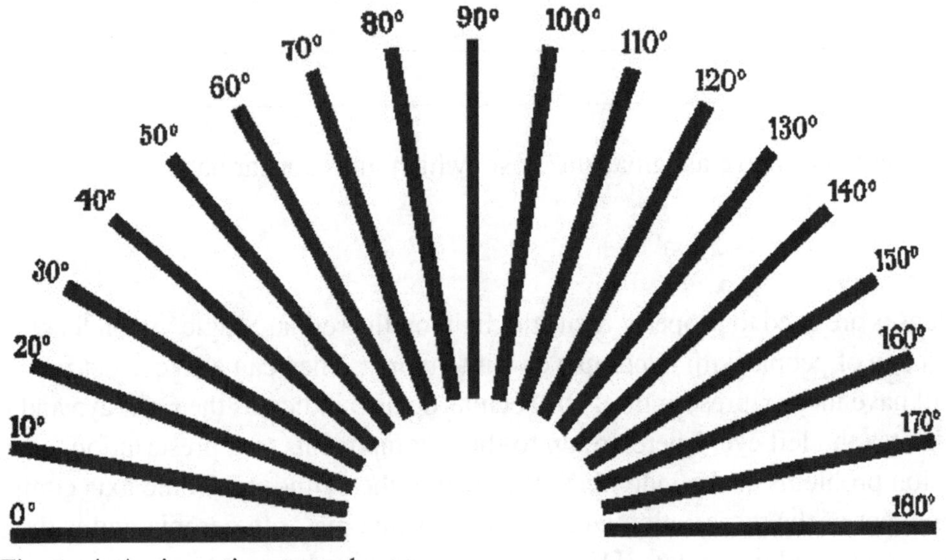


Figure 4 Astigmatism test chart

We will use the similar **Green's chart** (see Figure 5) to test for astigmatism.

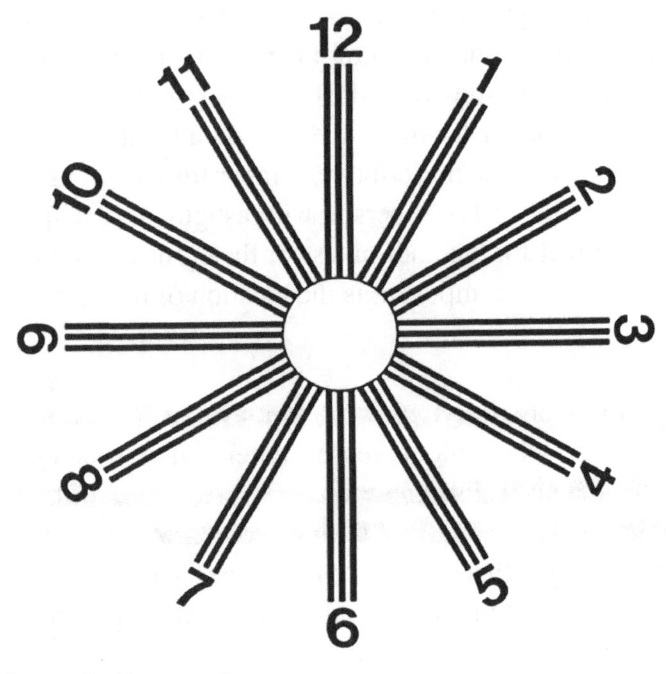


Figure 5 Greens chart

- Stand 20 feet from the Green's astigmatism chart.
- Look at the chart. Close or cover one eye but do not press on the eye.
- If vision is normal, all the lines of the chart should look crisp and have the same darkness. If the subject has astigmatism, some of the lines will appear dark and may appear to stand out from the chart.
- Record your results in the table below.
- Repeat the process for the other eye.

| Eye | Indicate normal vision or identify darker lines |
|---|---|
| Left | |
| Right | |

- Does your subject have astigmatism? If so, which lines appear darker?

Corrective lenses are used to properly focus the light on the retina. Single vision lenses are used to correct the vision of people with hyperopia or with myopia. They can also correct for astigmatism. Each eye will have its own prescription: OD (oculus dexter) indicates the right eye and OS (oculus sinister) indicates the left eye. There are up to three components to a prescription lens depending upon the vision problems of the individual: the sphere, the cylinder, and the axis component. The sphere component of the prescription lens corrects for myopia or hyperopia and is the strength of the correction, measured in diopters (D). Divergent corrective lenses are used to correct the vision of a person with myopia. Divergent corrective lenses spread light out before it enters the eye so the focal point of light is moved further from the cornea and lens. This corrects for the greater length of the eye. As a result light is focused on the retina, and the myopia is corrected for. The prescriptive number for the strength of a divergent lens is negative as in –2.25 diopters (–2.25 D). Convergent corrective lenses are used to correct the vision of a person with hyperopia. Convergent corrective lenses condense light before it enters the eye so the focal point of light is moved closer to the cornea and lens. This corrects for the shorter length of the eye. As a result light is focused on the retina, and the hyperopia is corrected for. The prescriptive number for the strength of a convergent lens is positive as in +2.25 diopters (+2.25 D). A person with astigmatism will also have a cylinder and axis correction. This is to correct for the unevenness of the cornea. The cylinder component is the strength of the correction. The axis component is the location of the correction.

### Blind Spot Activity

In the retina, all the axons from the ganglion cells converge at the **optic disc** to exit the eye as the optic nerve. This is also where blood vessels enter and exit the eye. As a result, there are no sensory receptors in the optic disc and therefore that portion of the field of view is empty for that eye. This is the so called the **blind spot** of your eye. You may have noticed, however, that even if you close one eye, there is no "hole" in the vision for that eye. Your brain is capable of filling in the blind spot in your visual field. In this experiment, you will locate your blind spot.

- Obtain a blind spot test card (see Figure 6)

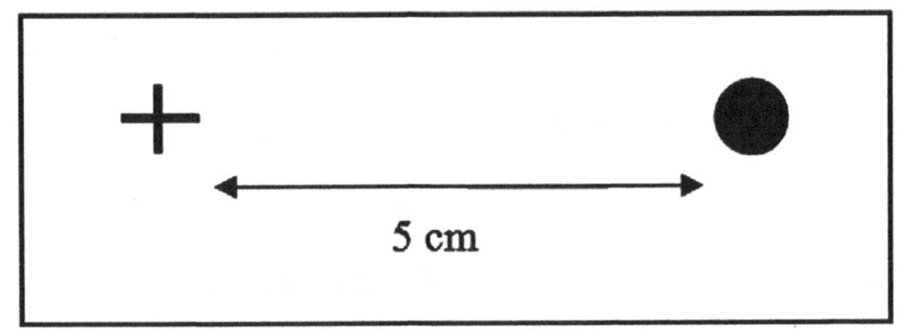


Figure 6 Blind spot testing card

- The card should be positioned so the plus sign is directly in front of your right eye and the dot is further off to your right. Hold the card at arm's length.
- Close your left eye.
- Move the card slowly straight toward you while you look at the plus sign. Although you are looking at the plus sign, the dot should also be visible with your right eye. You must continue to look at the plus sign but pay attention to the dot.
- Stop moving the card forward when the dot disappears. You have just located the blind spot in your right eye. Keep looking straight forward at the plus sign but move the card away. Is the dot still in your blind spot or has it become visible again?

lost vision of dot after it came in contact w/ nose. after I moved it away from nose I saw the dot

- Repeat the procedure for the left eye. You must rotate the card and reposition the card for the left eye and close your right eye.

Same as right

- Try locating the blind spot in your left eye with both eyes open.

- Did you locate the blind spot in the right eye with the left eye closed?

Yes, up against my eye

- Did you locate the blind spot in the left eye with the right eye closed?

^

- Did you locate your blind spot when you left both eyes open? Why or why not?

No, I'm not sure why

- Repeat the procedure with the three additional colored blind spot test cards. You may use either eye, but hold the card so the green side is directly in front of the eye that remains open and look at the + sign. What did you see with additional tests?

Card: Green side + & yellow side ●: same blind spot

Card: Green side + & yellow side ● with the line: same blind spot

Card: Green side + & other side with red and yellow ●: same blind spot

## Macula Screening Test

An **Amsler grid** (see Figure 7) is used to detect changes in the retina as well as in the optic nerve and the pathway of visual information in the brain. The original Amsler grid is black and white. A newer form is in color and consists of the same grid printed in blue and yellow. The newer color form is more sensitive to damage in the retina and the neural pathway. In either case, the grid serves as a tool that can be used by individuals at home to monitor changes in their vision.

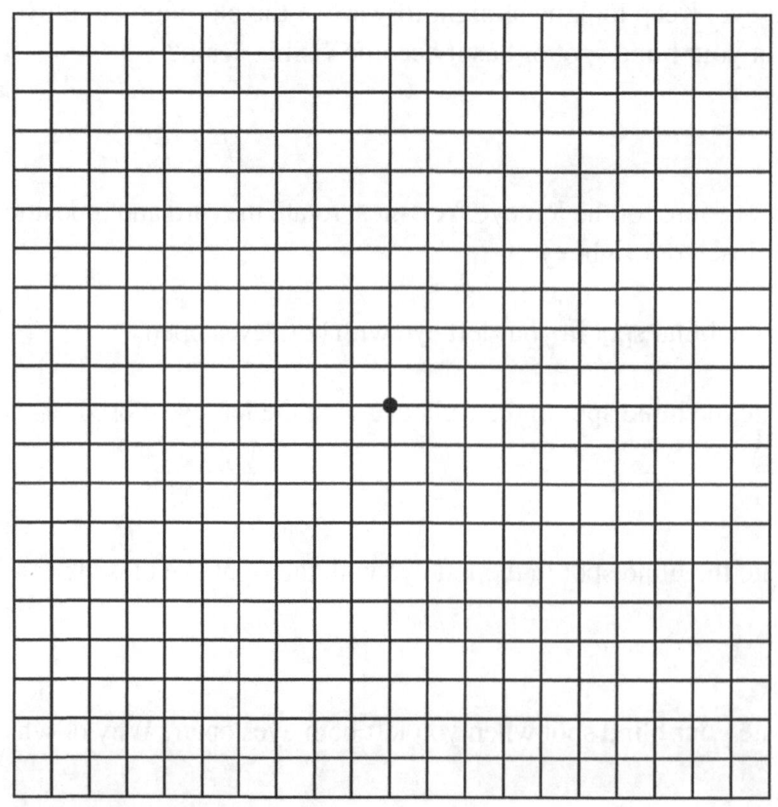

Figure 7 Amsler grid

- If you wear glasses, you should wear them during this test.
- Hold the grid at about the same distance you would hold reading material.
- Close or cover one eye but do not press on the eye. Look at the small dot in the center of the Amsler grid.

- If vision is normal, you should be able to see the center dot and the sides and four corners of the grid. All the lines of the grid should be visible, should look straight, and should be continuous from top to bottom and side to side.
- Record your results in the table below.
- Repeat the process for the other eye.

| Eye | Indicate normal vision or identify the variation |
|---|---|
| Left | |
| Right | |

**Visual Fields Activity**

A person's visual field is approximately 180° and is divided laterally into regions (see Figure 8 and figures in your text or on the lab posters as references). Each eye has a **temporal field** of view that allows you to see toward the lateral side of your body and provides your **peripheral vision**. Each eye also has a **nasal field** of view that allows you to see toward and somewhat past the midline of the body. Because the fields of view of each eye overlap in the nasal fields, you have **binocular vision**. Binocular vision covers approximately 140°. Your binocular vision allows you to integrate information from the two eyes and gives you much better depth perception than **monocular** (single eye) **vision**.

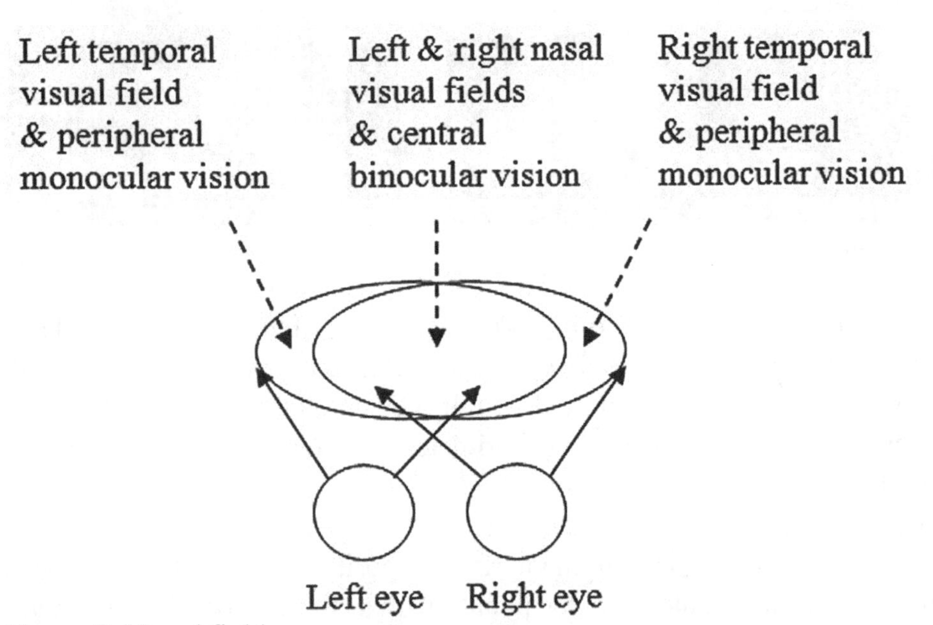


Figure 8 Visual fields

You will use a **perimeter** (see Figures 9 and 10) to identify the field of view for each eye. You will also delineate your color field of view within your black-and-white field of view. Your instructor will demonstrate the use of the perimeter.

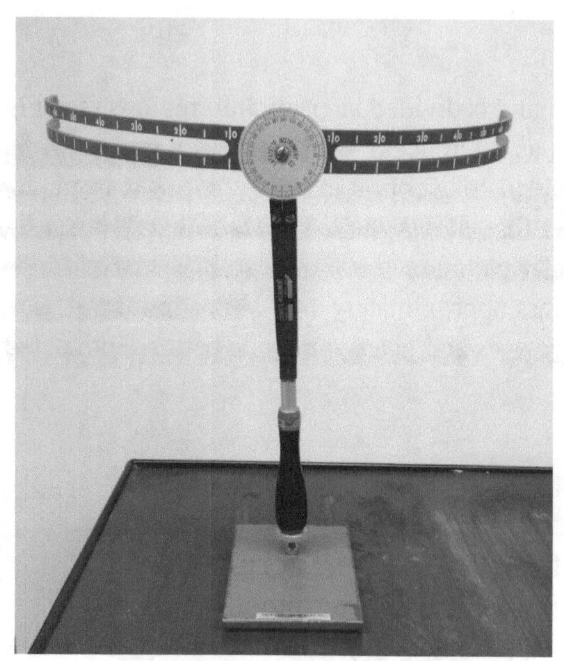

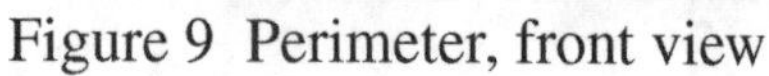

Figure 9 Perimeter, front view

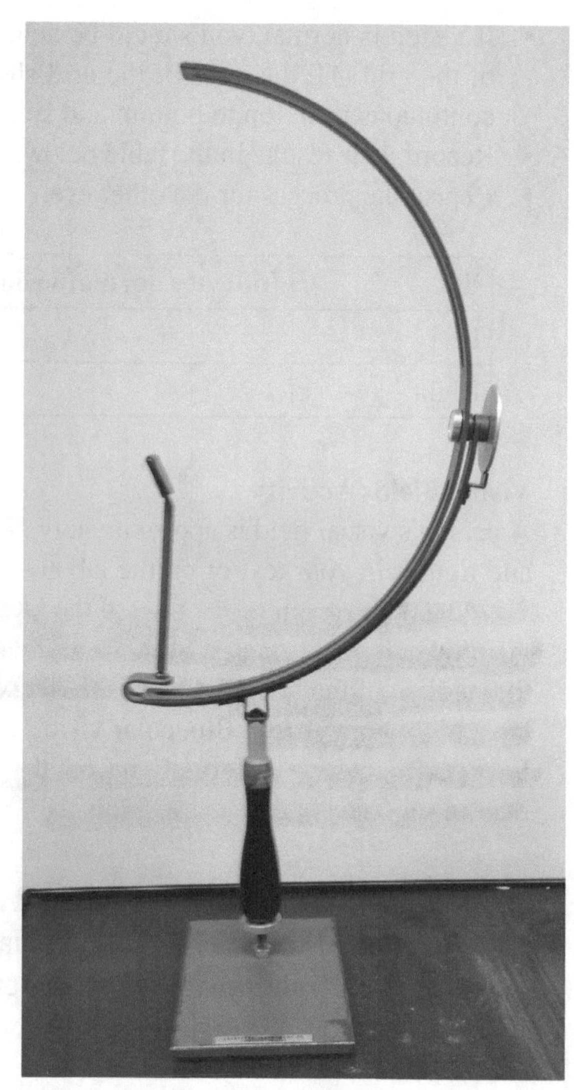

Figure 10 Perimeter, side view

- Obtain a perimeter, pointer, colored disk, and left and right eye perimeter charts.
- The subject should rest the lower edge of the orbit of the eye to be tested on the bar of the perimeter. The other eye should be closed. It is essential that the subject look straight ahead and focus on the central pivot of the perimeter at all times and not turn or shift their gaze toward the stimulus.
- The tester should stand behind the subject.
- With the arc of the perimeter in the horizontal position (0° to 180°), and beginning at the 90° mark, the tester should move the pointer with a colored disk along the outer edge of the perimeter arc.
- The subject should indicate when he/she can first see the disk and then when the color becomes visible. Please note it is important to distinguish between simply seeing the disk (black-and-white or rod vision) and seeing the color of the disk (color or cone vision).
- Mark the degree from the perimeter arc at which the subject first saw the disk and when the subject saw the color of the disk.
- Record these numbers on the appropriate perimeter chart (Figure 11). Be sure to use different colors to mark the points for rod vision and cone vision.

- Reposition the perimeter arc to the vertical position (90° to 270°) and repeat the process.
- Reposition the perimeter arc to an angle (135° to 315°) and repeat the process.
- Reposition the perimeter arc to an angle (45° to 225°) and repeat the process.
- Repeat the entire process for the other eye.

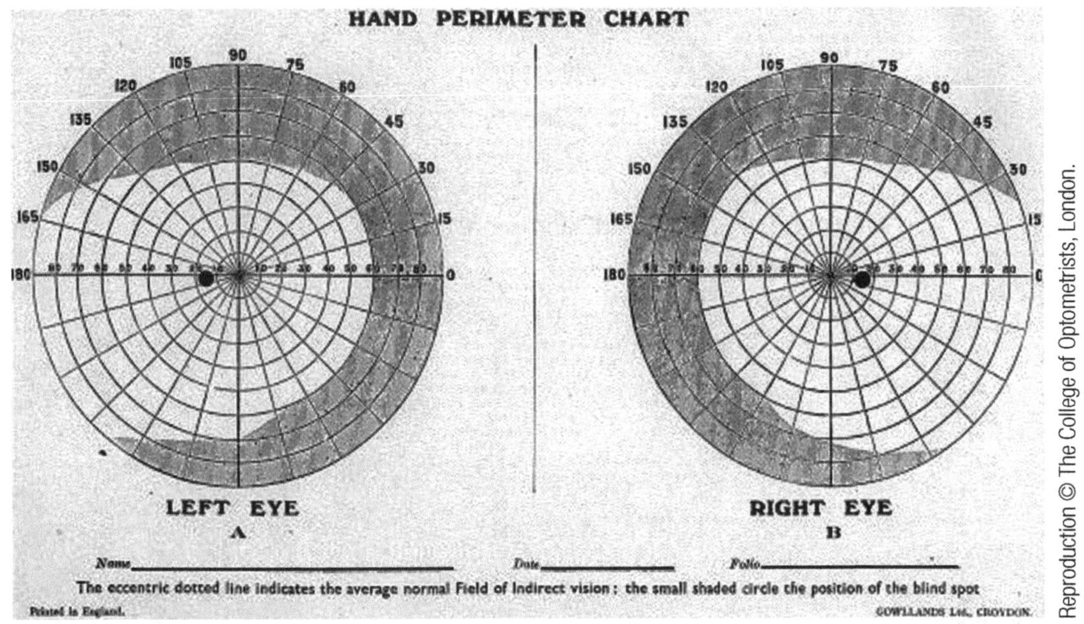


Figure 11 Perimeter chart



- How did the fields of view compare for black-and-white vision versus color vision?

- Knowing how far your nasal field of view extends, explain why your blind spots cannot be found if you keep both eyes open.

## Accommodation reflex: Near Point of Accommodation Activity

The ability to accommodate is the ability to change the focal point of the eye from focusing on a distant object to focusing on a close object. The accommodation of the lens is due to the contraction of the ciliary muscle and a resulting change in the shape of the lens. The **near point of accommodation** is the shortest distance from the eye an object can be brought and still be in focus. Lens accommodation involves the changing of the thickness of the lens by contracting the ciliary muscle. The lens is attached to the ciliary muscle (a circular muscle) by the suspensory ligaments. To focus on objects close to the eye, the ciliary muscle is contracted resulting in the thickening of the lens. When the ciliary muscle contracts, the orifice in the middle of the muscle is decreased in size. This reduces the tension on the suspensory ligaments and allows the lens to bulge and thicken (see Figure 12, A). To focus on objects in the distance, the ciliary muscle is relaxed resulting in the thinning of the lens. When the ciliary muscle is relaxed, the orifice in the middle of the muscle is increased in size. This increases the tension on the suspensory ligaments and causes the lens to thin (see Figure 12, B).

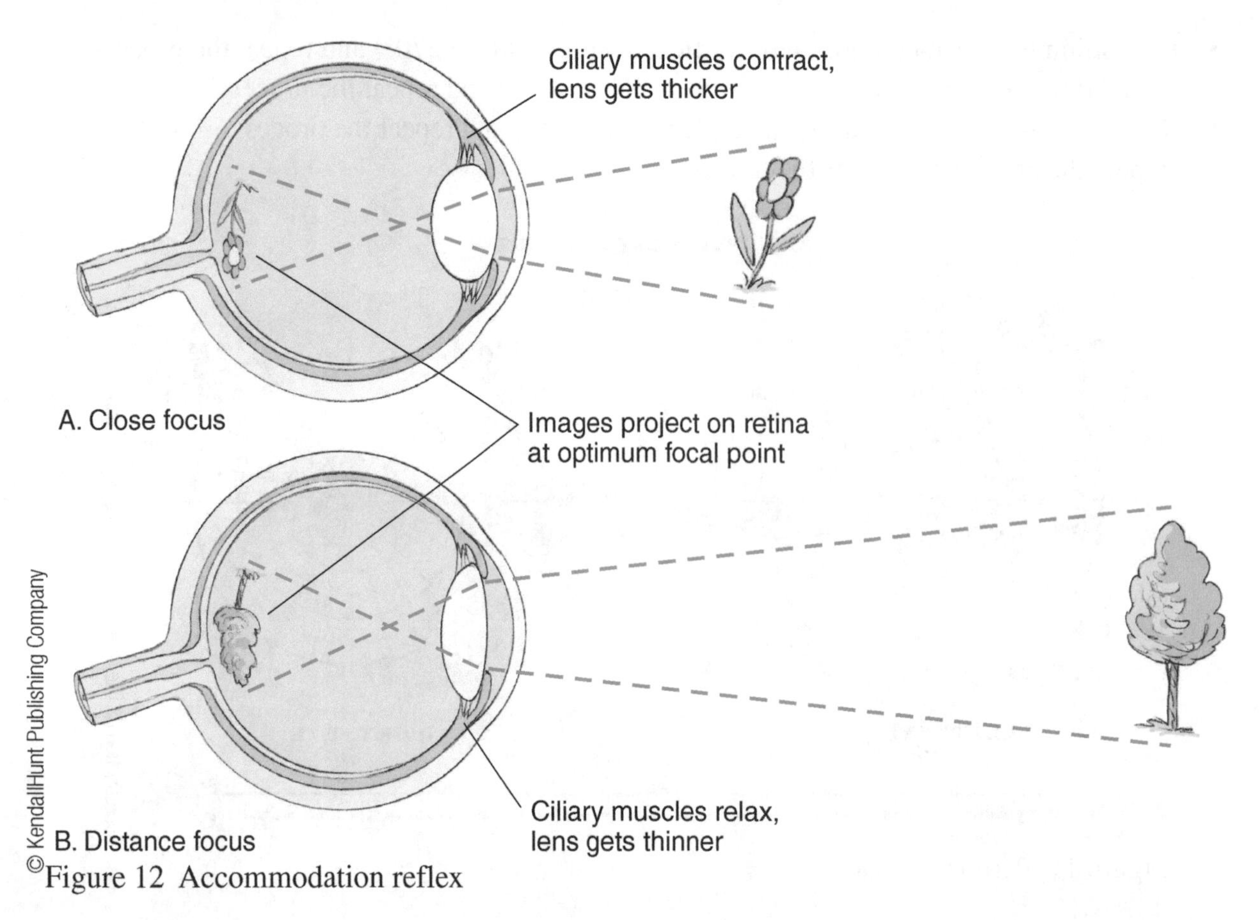



Figure 12 Accommodation reflex

As an individual ages, the lenses lose elasticity and are less able to thicken for close vision. As a result, the distance for near point of accommodation increases. This condition is called **presbyopia,** or "old-age vision." The following chart gives the normal near point of accommodation for various ages.

| Age (years) | Near point (cm) |
|---|---|
| 10 | 7 |
| 20 | 10 |
| 30 | 14 |
| 40 | 22 |
| 50 | 40 |
| 60 | 100 |

- Obtain a meter stick, a card holder, and a card with a letter or word typed on it.
- Hold the end of the meter stick in front of the chin at chin level and perpendicular to the body.
- Place the card holder with the card on the meter stick at arm's length.
- Close your left eye.
- Move the card slowly toward you while keeping your right eye focused on the letter.
- Stop moving the card forward when you can no longer keep the letter or word in focus. Record the distance in the following table.

- Repeat the procedure for the left eye. Record the distance in the table.

| | Right eye | Left eye |
|---|---|---|
| Near point (cm) | | |

- Was the near point of accommodation the same for both eyes?

- Was your near point of accommodation normal?

## Accommodation Reflex: Convergence and Pupil Size Activity

The complete accommodation reflex includes not only changing the shape of the le[illegible] also the movement or vergence of the eye and changing the diameter of the pu[illegible] fo[illegible] activity, you will watch for changes in the position of the eye[illegible] of the pupils as the subject shifts from a more distant foc[illegible] and vice versa.

- Obtain a pencil and hold it at som[illegible] dir[illegible]tly [illegible] subject. Ask the subject to focus on the pen[illegible]
- Slowly [illegible] pe[illegible]il [illegible] wa[illegible]ject. Watch the convergence of the eyes as the pencil is brought close[illegible] subject. At the same time, observe the contraction of the inner pupillary constrictor muscle (circular muscle) of the pupils.
- Repeat the process, but this time start with the pencil close to the subject and move the pencil further away. This time the eyes should diverge and the outer pupillary dilator muscle (radial muscle) contracts.

- Did your subject respond as expected?

- Which cranial nerves are involved in the accommodation reflex?

One of the ways to remember the actions that occur for changing the focal point from a distant one to a close one is to remember the four C's: close, contract, converge, and constrict. To view close images the ciliary muscle contracts, the eyes converge, and the pupils constrict. You can also remember the four D's: distant, don't contract, diverge, and dilate. To view distant objects the ciliary muscle doesn't contract (relaxes), the eyes diverge, and the pupils dilate.

## Ophthalmoscope Activity

The **ophthalmoscope** is a tool used to view the inside of the eye. A trained optometrist would look for abnormalities in the macula and fovea centralis, optic disc, and blood vessels of the eye (see Figure 13). The presence of the red reflex seen when light is shined into the eye is normal. This is often noticed as "red eye" in photos. It is the light reflecting off the surface of the retina. In individuals with a detached retina, the red reflex is absent in the eye with retinal detachment. The use of the ophthalmoscope is voluntary. Your lab instructor will assist you with the use of the ophthalmoscope.

- Obtain an ophthalmoscope and find a darkened area of the room or the hallway.
- Ask the subject to gaze at an object in the distance to facilitate dilation of the pupil.
- Turn on the ophthalmoscope and alter the light source as necessary. The small light is best for constricted pupils and well lit rooms. The medium light is for nondilated eyes in darker rooms. The large light is used if the eyes have been dilated. The half circle is generally used if a portion of the lens is obstructed as with cataracts in order to reduce the amount of light reflected back from the opaque tissue. The slit light is used to observe the contour or shape of the retina, cornea, and lens. The grid is used to measure/estimate sizes or distances within the eye. The green light is also referred to as the red free light. Using this light helps improve contrast so blood vessels and hemorrhages are more visible. The blue light requires fluorescein staining and is used to observe corneal scratches, tears, and ulcers as well as being used to detect foreign bodies in the eye.
- For observers with myopia, while looking through the ophthalmoscope, adjust the focusing wheel from zero through the negative or red numbers until objects in the distance come into focus. For observers with hyperopia, while looking through the ophthalmoscope, adjust the focusing wheel from zero through the positive or green numbers until close objects come into focus. Recall the notation for corrective lenses for myopia (−) and hyperopia (+).
- The observer should try to keep both eyes open during observation, but you will ignore the vision from the eye not being used to look through the ophthalmoscope. Starting about 25 cm away from the subject's eye and at an angle of about 15° from the center of the subject's eye, slowly move in toward the eye, following the red reflex until the blood vessels of the retina become visible, about 10 cm away from the subject's eye.
- While looking in the eye, you should shift the direction of the light so that it pivots up and down and left and right. If you shine the light directly onto the fovea centralis, you will "wash out" the subject's receptor cells.

- What structures of the internal eye did you observe with the ophthalmoscope (see Figure 13)?

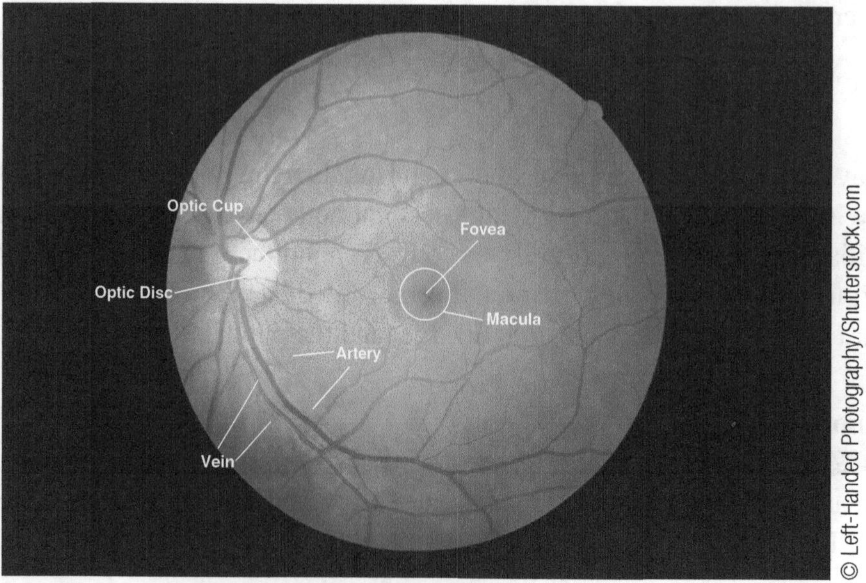




Figure 13 Photograph of the posterior wall of the eye

# Exercise 11

# Hearing and Equilibrium

## Objectives

- Be able to identify the structures of the ear associated with hearing.
- Be able to explain the function of each of the structures of hearing.
- Be able to explain the transmission of sound.
- Be able to define the terms and distinguish between *conductive*, *sensorineural*, and *central hearing loss*.
- Be able to define and explain the cause and effect of each of the following: *Meniere's disease*, *vertigo*, *presbycusis*, *otitis media*, *tympanitis*, and *otosclerosis*.
- Be able to define the terms *frequency*, *pitch*, and *hertz (Hz), amplitude*, *intensity* or *loudness, and decibels (dB)*.
- Be able to graph the results of an *audiometer* hearing test and interpret the data graphed on an *audiogram*.
- Identify the normal range of sound frequencies detected by an adult human.
- Define the terms *air conduction* and *bone conduction*.
- Be able to explain the process and the results from a Weber's hearing test.
- Be able to explain the process and the results from a Rinne's hearing test.
- Be able to define the term *interaural time* and *difference in intensity* and explain their functions with regard to sound localization.
- Be able to identify the structures of the vestibular apparatus associated with equilibrium.
- Be able to explain the function of each of the structures of equilibrium.
- Be able to explain the process and results from a Barany's test.
- Be able to define the term *nystagmus* and explain what occurs during rotational nystagmus and postrotational nystagmus.
- Be able to define and explain *past pointing*.
- Be able to identify the cranial nerve and the individual branches associated with hearing and balance.
- Identify the areas of the brain responsible for processing the sensory information from hearing and equilibrium.

## Introduction

The sense of hearing or audition involves the detection of sound waves. Two characteristics of the sound waves, frequency and intensity, are detected by the sensory receptors in the ear and interpreted in the auditory cortex of the brain. The **cochlear branch** of the **vestibulocochlear nerve** (cranial nerve VIII) is responsible for conducting the impulse from the cochlea to the auditory cortex.

The **frequency** of sound waves is measured as the number of cycles or oscillations per second (**hertz, Hz**). The frequency of a sound determines the pitch of the sound. The higher the frequency of the sound waves, the higher the pitch. Higher frequency sounds deflect the basilar membrane closer to the oval window and stimulate the hair cells in that area of the cochlea. Lower pitch sounds deflect the basilar membrane closer to the apex of the cochlea and stimulate the hair cells in that area of the cochlea. The normal range of audible frequencies is approximately 16 to 20,000 Hz. The **intensity** or loudness of a sound (**decibels, dB**) is determined by the amount of deformation of the hair cells. The higher the decibels, the louder the sound, and the greater the deformation of the hair cells resulting in the generation of more action potentials.

* itis. inflamation of

For the sensory receptors to detect sound, the sound waves must be transported to the cochlea. **Air conduction** transports sound waves through the external auditory canal, tympanic membrane, ossicles, and oval window to the cochlea. **Bone conduction** transports sound waves through the bones of the skull to the cochlea. Disruptions to the pathway of air conduction result in **conductive hearing loss**. Some of the more common causes of conductive hearing loss are otitis media, tympanitis, and otosclerosis. **Otitis media** and **tympanitis** are infections of the middle ear and the tympanic membrane, respectively. **Otosclerosis** is the result of ossification of the ossicle bones. Once the hair cells have been depolarized, neurons of the cochlear branch of cranial nerve VIII carry the impulses to the brain. **Sensorineural hearing loss** is a result of a disruption of the cochlea, the hair cells specifically, or cranial nerve VIII. Some of the more common causes of sensorineural hearing loss are presbycusis and inner ear infections. **Presbycusis** is often associated with old age. As hair cells are exposed to prolonged or loud noises, they may be damaged and no longer able to respond to stimulation. Presbycusis often results in a decrease in the ability to detect sounds in the upper frequencies. Be aware, however, that older individuals are not the only individuals that suffer hearing loss as a result of exposure to noise. **Central hearing loss** is due to defects in the tracts of the CNS associated with the auditory pathway or defects in the auditory cortex. Complications associated with CNS infections, tumors, and oxygen availability are some of the causes of central hearing loss.

The sense of **equilibrium** involves the **vestibular system or apparatus** and is responsible for the detection of body position in space, position relative to gravity, and also acceleration and deceleration. There are multiple sets of sensory receptors involved in equilibrium, one in the ampulla of each of the semicircular canals and one in the macula of each of the utricles and saccules. The receptors in the semicircular canals primarily detect information regarding changes in rotational velocity or acceleration and deceleration. The receptors in the macula detect information regarding the position of the body relative to gravity (utricles: horizontal and saccules: vertical). The **vestibular branch** of the **vestibulocochlear nerve** (cranial nerve VIII) is responsible for conducting the impulses from the vestibular apparatus to the brain, where the information is integrated with additional information from the eyes, proprioceptors, and pressure receptors. The vestibular apparatus is also responsible for collecting the sensory information for the **vestibule-ocular reflex**. This reflex allows the eyes to fix on a reference point in your visual field as your body moves. When the information from the vestibular apparatus does not match the information the brain is receiving from the other senses also involved in equilibrium, the person may experience dizziness and motion sickness. When the sensory receptors of the vestibular apparatus are stimulated by excess rather than by movement, the person may experience **vertigo** or a feeling of spinning. A person who has been drinking alcohol or is experiencing a hangover may experience the sensation of vertigo due to changes in the density of the endolymph. Inner ear infections and **Meniere's disease** are the more common pathologic causes of vertigo. A person with Meniere's disease will also experience **tinnitus** (ringing in the ears) due to overproduction of endolymph throughout the inner ear.

## Auditory Tests

Auditory tests range from very simple function tests that identify general locations of hearing loss to complex audiometer tests of auditory acuity. The goal of a hearing test is early detection of any hearing loss. This is especially important for children, as hearing loss can have an effect on speech development and on learning. The **audiometer** is used to test a person's ability to detect specific frequencies at specific decibel levels. **Weber's test** uses the localization of hearing to detect either conductive or sensorineural hearing loss. **Rinne's test** compares air conduction with bone conduction to detect the location of conductive hearing loss.

- **Audiometer Activity**

  An **audiometer** is used to detect hearing loss by identifying the minimum decibel level at which a particular frequency can be heard. The audiometer can produce a pure tone composed of wavelengths of one frequency. On the scale of the audiometer, the **reference threshold** intensity is set at 0 dB. This is the decibel level at which the sound is expected to be heard by a person with normal hearing. In a hearing test, the **threshold intensity** for hearing a specific frequency is determined for a person by increasing the intensity (decibels, dB) of the sound produced by the audiometer until the person hears the sound. If a person does not hear the sound until the decibel level has been increased to 10 dB, then the person has a 10 dB hearing loss (10 dB – 0 dB = 10 dB) at that particular frequency (Hz). Note that a person may have normal hearing at some frequencies but may have a hearing loss at other frequencies. The person's threshold intensity for each frequency tested is graphed on an **audiogram** (see Figure 1).

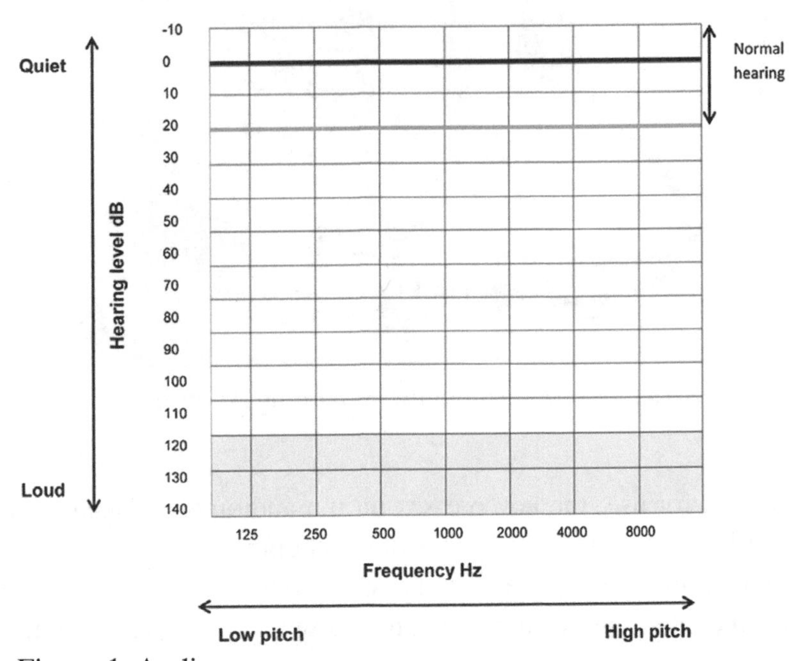


Figure 1 Audiogram

Hearing loss is categorized as follows:

| Hearing category | Child: Intensity sound first detected dB | Adult: Intensity sound first detected dB |
|---|---|---|
| Normal | 0–15 | 0–25 |
| Mild hearing loss | 15–40 | 25–40 |
| Moderate hearing loss | 40–55 | 40–55 |
| Moderately severe hearing loss | 55–70 | 55–70 |
| Severe hearing loss | 70–90 | 70–90 |
| Profound hearing loss | 90 + | 90 + |

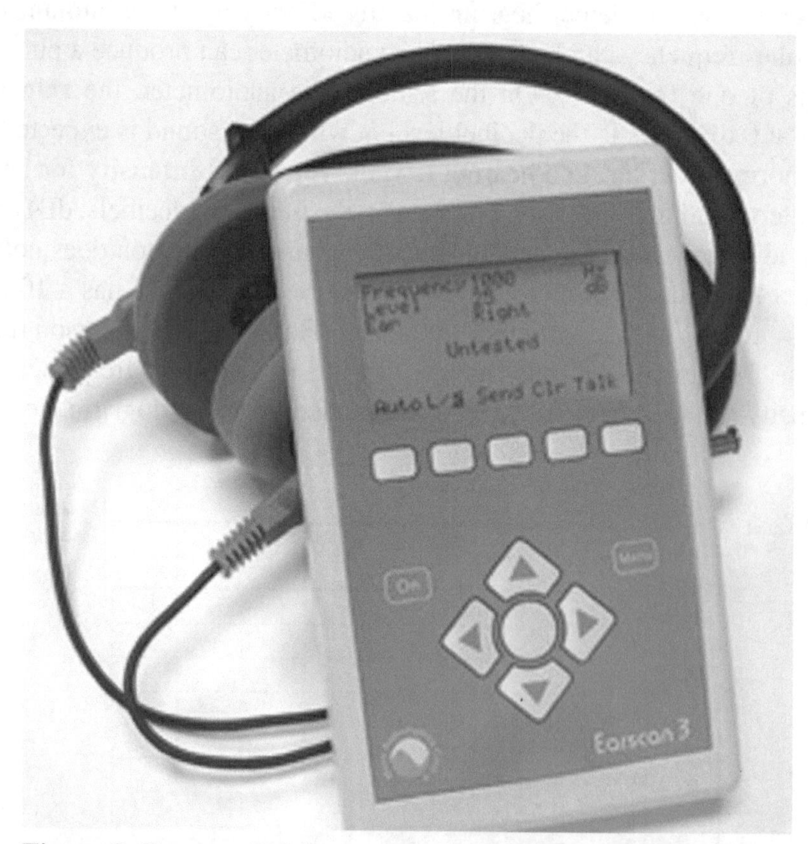


Figure 2 Earscan ® 3 Automatic Screening Audiometer

- In the instructions for use, the key presses on the audiometer appear in brackets: {}. For example, {On} indicates you should press the On button.
- The menus can be obtained by pressing the Menu key, {Menu}. {Menu} will cause you to exit test mode and enter the menu system and allow you to select different menu levels.

- Menu navigation is done by using the arrow keys {◀} / {▶} for left / right respectively, and {▲} / {▼} for up / down respectively, and by using the center circle key {●} for enter.
- The silver key pads below the display screen are used to select options or toggle between options displayed directly above them on the bottom line of the display screen. They are also numbered (not visually) 1 to 5 from left to right.

Manual Audiometry Key Functions

| | | |
|---|---|---|
| {◀}/{▶} | | Select next lower / higher enabled frequency (Hz) |
| {▲}/{▼} | | Increase / decrease sound level by current step size (dB) |
| {1} | Auto | Exit manual test mode and enter automatic test mode |
| {2} | L/R | Toggle between left and right ear |
| {3} | Send | Send the test data to computer or printer |
| {4} | Clr | Clear the current threshold |
| {5} | Talk | Exit test mode and enter talk over mode |
| {Menu} | | Exits manual test mode and enters the menu system |

- {On} The Earscan 3 logo will appear followed by:

  Select an Option:
  New Test
  Load Last Results
- {●} or {1}
- The display should be:

  Frequency 125 Hz
  Level 10 dB
  Ear Right

  Untested

  Auto L/R Send Clr Talk

- Have the subject place the headset on with the blue earpiece on the left ear and the red earpiece on the right ear. The subject should be facing away from the audiometer and should be holding the response switch. Instruct the subject to push the response button as soon as he/she hears the tone.

- Change the Level to 0 dB by {▼}, {▼}
- The display should be:

  Frequency 125 Hz
  Level 0 dB
  Ear Right

  Untested

  Auto L/R Send Clr Talk

- {●} to send tone
- The display should be:

  Frequency 125 Hz
  Level 0 dB
  Ear Right

  On (flashing display)

  Auto L/R Send Clr Talk

- Once a tone has been presented, an 'm' will be displayed after the level, and the display should be:

  Frequency 125 Hz
  Level 0m dB
  Ear Right

  Auto L/R Send Clr Talk

**If the subject does not hear the tone and does not depress the response button, the display will remain the same.**

- Increase the Level to 5 dB by {▲}
- {●} to send tone
- Continue this process until you reach a Level the subject can hear.
- **If the subject hears the tone and pushes the response button**, the display will look like this **while the response button is being pushed**:

  Frequency 125 Hz
  Level 5 dB
  Ear Right
  *

  Auto L/R Send Clr Talk

Note: **The * is only present while the response button is depressed. It is important that you watch for the positive response.**

If the subject indicates the tone was heard, record the intensity level in the following table.

- Increase the Frequency to 250 Hz by {▶}
- Change the Level to 0 dB by {▼} as many times as necessary to return to 0 dB.
- The display should be:

Frequency 250 Hz
Level 0 dB
Ear Right

Untested

Auto L/R Send Clr Talk

- {●} to send tone
- Increase the Level to 5 dB by {▲}
- {●} to send tone
- Continue this process until you reach a Level the subject can hear.

- Repeat this process until you have tested the subject at 500 Hz, 750 Hz, 1000 Hz, 1500 Hz, 2000 Hz, 3000 Hz, 4000 Hz, 6000 Hz, and 8000 Hz.

- {2} to toggle to the left ear.
- Repeat the process for the left ear.
- After completing the test for both ears, {Menu}
- {▼}{▼}{▼}{▼}{▼}{▼} to 'Turn Off', {●}

| | **Threshold Intensity (dB)** | |
|---|---|---|
| **Frequency (Hz)** | **Right ear** | **Left ear** |
| 125 | 30 | 60 |
| 250 | 40 | 40 |
| 500 | 50 | 40 |
| 750 | 50 | 50 |
| 1000 | 30 | 30 |
| 1500 | 30 | 30 |
| 2000 | 20 | 10 |
| 3000 | 20 | 10 |
| 4000 | 10 | 10 |
| 6000 | 20 | 20 |
| 8000 | 0 | 20 |

- Results: any deviation from 0 dB represents a potential hearing loss.

- Does the subject demonstrate any hearing loss that is outside the range considered normal hearing (see Figure 1)? If so, at what frequencies?

  No.

- Use the data to complete the audiogram provided in your lab report.

■ **Weber's Test Activity**

**Weber's test** can be used to detect both conductive hearing loss and sensorineural hearing loss. Obtain a midrange tuning fork (256 Hz or one similar in frequency). Holding only the handle or stem of the tuning fork, strike the prongs of the tuning fork against the palm of your hand or the heel of your shoe. **Do not** strike the tuning fork against the lab bench or any other hard surface. Turn the prongs of the fork toward your ear. You should be able to hear the sound waves being produced by the tuning fork. These sound waves are being transmitted through air (**air conduction**) to the tympanic membrane and the middle ear, where the waves are magnified by the ossicles and transmitted to the oval window and on to the cochlea. Strike the tuning fork again, and place the handle of the tuning fork against your skull. You must not touch the prongs of the tuning fork or you will dampen the vibrations of the prongs and reduce the sound waves being generated. You should be able to hear the sound waves being produced by the tuning fork but this time the sound waves are being conducted through your skull (**bone conduction**) to the cochlea. You are now ready to begin the test.

- Test 1: Strike the tuning fork and place the handle of the tuning fork on the midline of the forehead. Have the subject indicate whether the sound is heard equally well in both ears or heard better in one ear. If it is not heard equally well in both ears, indicate in which ear the sound is heard more distinctly. Record your results by placing a check in the appropriate box in the table.

- Test 2: Have the subject block the pathway of air conduction by placing a finger in the external auditory canal of their right ear. Strike the tuning fork and place the handle of the tuning fork on the midline of the forehead. Have the subject indicate whether the sound is heard equally well in both ears or heard better in one ear. If it is not heard equally well in both ears, indicate in which ear the sound is heard more distinctly. Record your results by placing a check in the appropriate box in the table.

- Test 3: Have the subject block the pathway of air conduction by placing a finger in the external auditory canal of their left ear. Strike the tuning fork and place the handle of the tuning fork on the midline of the forehead. Have the subject indicate whether the sound is heard equally well in both ears or heard better in one ear. If it is not heard equally well in both ears, indicate in which ear the sound is heard more distinctly. Record your results by placing a check in the appropriate box in the table.

| | **Equal in both ears** | **Lateralizes to right ear** | **Lateralizes to left ear** |
|---|---|---|---|
| Test 1: Midline of forehead, both ears unblocked | ✓ | | |
| Test 2: Midline of forehead, right ear blocked | | ✓ | |
| Test 3: Midline of forehead, left ear blocked | | | ✓ |

- Expected Results
  Test 1: The sound should be heard equally well in both ears if there is no conductive or sensorineural damage in either of the ears. If the subject has sensorineural hearing loss, the sound will lateralize to the ear without hearing loss because the cochlea is working more effectively in that ear.

Test 2: The sound should lateralize to the blocked ear. Blocking the ear simulates a hearing loss due to a reduction in air conduction. Because this ear does not carry as much sound from air conduction, the cochlea is more sensitive to the sounds carried by bone conduction. Therefore, this result would indicate a conductive loss.

Test 3: The sound should lateralize to the blocked ear. The reasons are the same as for Test 2.

- Did your subject exhibit sensorineural or conductive hearing loss?

## Rinne Test Activity

A **Rinne test** can be used to detect conductive hearing loss. Obtain a midrange tuning fork (256 Hz or one similar in frequency). Holding only the handle or stem of the tuning fork, strike the prongs of the tuning fork against the palm of your hand or the heel of your shoe. **Do not** strike the tuning fork against the lab bench or any other hard surface. Turn the prongs of the fork toward your ear. You should be able to hear the sound waves being produced by the tuning fork. These sound waves are transmitted through the air to your cochlea (air conduction). Strike the tuning fork again and place the handle of the tuning fork against your mastoid process. You must not touch the prongs of the tuning fork or you will dampen the vibrations of the prongs and reduce the sound waves being generated. You should be able to hear the sound waves being produced by the tuning fork, but this time the sound waves are being conducted through your skull (bone conduction) to the cochlea. You are now ready to begin the test.

- Test 1: Strike the tuning fork and place the handle on the right mastoid process. Have the subject indicate when the sound heard via bone conduction (BC) can no longer be heard. Reposition the tuning fork so the prongs are approximately one inch away from the right external auditory canal. Have the subject indicate if the sound can still be heard via air conduction (AC). Record your results by placing a check in the appropriate box in the table.

- Test 2: Repeat the test with the left mastoid process and ear. Record your results by placing a check in the appropriate box in the table.

| | AC = BC | AC > BC |
|---|---|---|
| Test 1: Right side | | ✓ |
| Test 2: Left side | | ✓ |

- Expected Results
  Test 1 and 2: The sound should be heard via air conduction longer than via bone conduction. If air conduction does not last longer than bone conduction, then the person has conductive hearing loss.

- Did your subject exhibit conductive hearing loss?

## Sound Localization Activity

Unless a sound is directed toward the midsagittal line, there will be a slight time difference between a sound reaching the left ear and the right ear. This difference is the **interaural time difference (ITD)**. Although the time difference is so small (μs) that we can't consciously recognize it, our brain does detect the difference. Another factor that helps locate sound is the **interaural loudness difference (ILD)** detected by each ear. As sound coming from one side of a person reaches the ear on the same side of the head, the sound that reaches the ear on the opposite side of the head must do so by bouncing off of objects and back to that ear. The reflected sound will have a lower intensity (dB). The difference in intensities is then used by the brain to locate the source of the sound. The interaural time difference is more important in locating low-frequency (low-pitch) sounds (0–1600 Hz, especially below 800 Hz) because low-frequency sound waves are large enough to bend around the head. The interaural loudness difference is more important in locating high-frequency (high-pitch) sounds (800–20,000 Hz, especially over 1600 Hz) because the sound waves are absorbed by the head and cannot bend around the head to the opposite ear and must bounce off of objects and be reflected back to that ear.

- Obtain a midrange tuning fork (256 Hz or one similar in frequency).
- Have the subject close his/her eyes.
- Holding only the handle or stem of the tuning fork, strike the prongs of the tuning fork against the palm of your hand or the heel of your shoe. **Do not** strike the tuning fork against the lab bench or any other hard surface.
- Hold the tuning fork to the side of the subject's head and ask the subject to locate the position of the tuning fork.
- Repeat the procedure while holding the tuning fork at different positions around the subject's head.

- Did the subject accurately identify the location of the tuning fork?

Sometimes

- Were one or more positions more difficult to locate than others? What were those positions?

In front

- Based upon the frequency of the tone you used, was the subject using ITD or ILD to locate the sound? ILD

## Vestibular Function

The vestibular system and its interactions with many other sensory mechanisms in the body allow us to respond to changes in position so quickly that we often don't realize that is it working. It's usually only when we have to struggle to maintain or regain balance that we focus on the system. **Nystagmus** is the involuntary but rhythmic movement of the eyes. The movement can be horizontal, vertical, rotational, or a combination, depending on the movement of the body. The vestibulo-ocular reflex can be observed as horizontal nystagmus in a person who is rotating. If a person is rotated to his or her right, as the person is rotated, the eyes will exhibit a slow movement to the person's left followed by a fast movement to the person's right (the direction of rotation). If the person is rotated to the left, the slow and fast movements will be reversed. When a person stops rotating, the person will then exhibit postrotational nystagmus. The eye movements in postrotational nystagmus are the opposite of those in rotational nystagmus. **Past pointing** is the inability of a person to accurately touch the outstretched fingertips of another person. This can occur when the person's vestibular apparatus is not providing accurate information. To understand nystagmus and past pointing, consider the lateral semicircular canals. As the subject begins to rotate (acceleration) clockwise (to his or her right), the semicircular canals move with the person, but the endolymph in the canals doesn't begin to move as quickly, so the endolymph appears to move counterclockwise (to the left) and "pushes" on the ampulla and bends the hair cells. This will result in depolarization of the hair cells on the right side of the head and hyperpolarization of the hair cells on the left side of the head. As a result of the information sent to the brain, the lateral rectus extraocular muscle of the left eye and medial rectus extraocular muscle of the right eye are stimulated. The eyes reflexively attempt to fix on an object and as a result move slowly in the direction opposite rotation (in this case, counterclockwise or to the left). As the body continues to rotate, the eyes must "catch up" with the body and reflexively move quickly in the same direction as rotation (in this case, clockwise or to the right). The eyes then fix on another object, and the process is repeated. This rotational nystagmus would continue until the speed of endolymph movement catches up and matches the speed of rotation. Postrotational nystagmus occurs when the subject stops rotating (deceleration), the semicircular canals stop moving, but the endolymph in the canals continues to move, so it "pushes" on the ampulla in the opposite direction. This bends the hair cells in the opposite direction, resulting in hyperpolarization of the hair cells on the right side of the head and depolarization of the hair cells on the left side of the head. As a result of the information sent to the brain, the opposite lateral and medial rectus extraocular eye muscles are stimulated, and reflexive eye movements will be opposite of those in rotation. Slow eye movement will be in the same direction as the person was rotated (in this case, clockwise and to the right), and fast eye movement will be in the opposite direction (in this case, counterclockwise and to the left). Shortly after the person stops rotating, the endolymph stops moving, and the postrotational nystagmus will stop. Past pointing or the ability to touch a specific point after rotation gives results similar to the results seen with the eyes and nystagmus.

- **Barany's Test and Nystagmus Activity**

  To test for proper functioning of the vestibular apparatus, **Barany's test** and past pointing are often used. As described earlier, the body responds to information the brain receives from the sensory receptors of the vestibular apparatus. Be sure that the subject in these tests does not suffer from any disease or condition that interferes with the vestibular system. Barany's test and past pointing will be performed together. **Please note that the subject should remain seated after rotating until all feelings of spinning or dizziness are gone. Individuals with a balance problem or who experience motion sickness or vertigo should not be the subject.**

- Ask a person to sit in a chair that can be rotated. The subject's head should be bent down so that the chin nearly touches the chest (30° angle).
- Ask the subject to close his/her eyes and quickly rotate the subject to his/her right for approximately 12 rotations. Abruptly stop the rotation.
- Ask the subject to open his/her eyes and watch the eyes for postrotational nystagmus. Pay attention to the direction of the slow and fast movement. Record your results.
- Again ask the subject to close his/her eyes and quickly rotate the subject to his/her right for approximately 12 rotations. Abruptly stop the rotation.
- Ask the subject to open his/her eyes and immediately extend your index fingers and ask the subject to touch them with his/her index fingers.
- After all nystagmus has stopped, extend your index fingers and ask the subject to touch them with his/her index fingers.
- Ask the subject to close his/her eyes and touch your index fingers again. Record your results.

Postrotational nystagmus:

Direction of rotation left

Direction of slow eye movement left

Direction of fast eye movement right

Past pointing (circle the correct response):

Eyes open **during** postrotational nystagmus: touched / missed fingers

Eyes open **after** postrotational nystagmus: touched / missed fingers

Eyes closed **after** postrotational nystagmus: touched / missed fingers

- Expected results for postrotational nystagmus:
  For rotation to the subject's right, postrotational nystagmus eye movements would be slow to the right and fast to the left.

- Expected results for past pointing:
  The expected results for past pointing during the postrotational nystagmus would be for the subject to past point to the right (if rotation of the subject was to the right).

  The expected results for past pointing after postrotational nystagmus with either the eyes open or closed would be to accurately touch the fingers.

# Exercise 12

# iWorx EMG Motor Unit Recruitment and Muscle Fatigue

## Objectives

- Be able to define the term and process of *electromyography* and an *electromyogram* (EMG).
- Be able to define the term *somatic motor neuron.*
- Be able to define the term *muscle fiber* or *myofiber*.
- Be able to define the term *motor unit.*
- Be able to define the term *motor unit recruitment*.
- Be able to define the term *frequency of motor unit stimulation.*
- Be able to define the term *tonus*.
- Be able to define the term *muscle fatigue* and identify factors that can lead to fatigue.
- Be able to explain the relationship between skeletal muscle control and number of motor units and frequency of motor neuron stimulation.
- Explain the relationship between motor unit size and size of area of the cerebral motor cortex used for regulation of a movement.
- Explain the relationship between motor unit size and strength of contraction and precision of movement.
- Be able to calculate the percent difference (increase or decrease) in muscle electrical activity between different levels of recruitment and between the dominant and the nondominant arm.

## Lab Safety Reminders

- Use caution with any electrical equipment.

## Hints for Success

All members of your lab group should feel comfortable operating the iWorx system.

**Never unplug any sensors from the iWorx box!** If you unplug a sensor during an experiment, you will lose all data. You will need to turn the iWorx box off, plug the sensor in and start the lab from the beginning.

On the iWorx recording screens, it is helpful to remember that the blue bars are for data that are being recorded. The green bars are for data that is being calculated based upon the data you have collected.

The use of the electrode gel increases the ability to detect the electrical signals generated by the subject's skeletal muscles. Only a small amount of the gel is needed in the center of the electrode. A large amount will interfere with the ability of the electrode to stick to the subject's skin.

You will have better results if once your subject has picked up the hand dynamometer, the subject continues to hold the hand dynamometer throughout the exercise. Changing the position of the hand and grip on the dynamometer will potentially change your results. Once you have the dynamometer in your hand, you may loosen your grip, but keep the dynamometer in the same position in your hand.

## Introduction

The function of skeletal muscle is to contract and develop force. In doing so, skeletal muscle converts chemical energy, ATP, into mechanical energy, cross-bridge cycling and potential shortening of sarcomeres. When a single muscle fiber contracts as a result of a single stimulus, the result is called a twitch. In order to stimulate a muscle fiber twitch, the stimulus must be above threshold. There are three components to a muscle fiber twitch, the latent period or lag phase, the contraction phase, and the relaxation phase. The latent period is the time between stimulation of the muscle fiber and the development of force. The contraction phase is the time during which the muscle fiber is developing force. The relaxation phase is the time during which muscle fiber returns to its relaxed state. The action of skeletal muscles is directed by the somatic motor system with continuous feedback between the muscle being used and the central nervous system. This feedback is necessary to maintain an appropriate force of contraction. Skeletal muscles are innervated by **somatic motor neurons** and the only response possible at a skeletal muscle neuromuscular junction is excitation. A single somatic motor neuron can innervate several **muscle fibers** or **myofibers**, but each muscle fiber is innervated by only one somatic motor neuron. The somatic motor neuron and all the muscle fibers it innervates make up a **somatic motor unit** (generally shortened to motor unit). You can consider a motor unit to be a functional unit of a skeletal muscle. When the somatic motor neuron propagates an action potential and the neurotransmitter acetylcholine (ACh) is released from the axon terminals, all the muscle fibers innervated by that somatic motor neuron will be stimulated. As a result, one action potential will cause all the muscle fibers in that motor unit to contract. The number of muscle fibers innervated by a single somatic motor neuron determines the size of the motor unit. If a single somatic motor neuron innervates 10 muscle fibers, the motor unit has a size of 10. Motor unit size can also be written as a ratio, 1:10, with 1 referring to the somatic motor neuron and 10 referring to the number of muscle fibers. A size 30 (1:30) motor unit contains 1 somatic motor neuron and 30 muscle fibers. Motor unit sizes vary across the body, depending upon the type of muscle and its function. The closer the neuron-to-muscle fiber ratio is to 1:1, the more precision your brain has over the control of the contraction of that motor unit. You can expect muscles that require precise control, such as those in your fingers, to have a larger number of small motor units. A consequence of this is that the more motor units you have devoted to muscles in an area of the body under precise control, the larger the area of the brain required for regulation of those muscles. Muscles that utilize large motor units require less area of the brain for control of the muscle. However, with larger motor units, a greater force of contraction can be developed from one action potential. For all muscles, the strength of a contraction is proportional or directly related to the number of motor units activated and simultaneously active.

**Motor unit recruitment** occurs when the brain increases the number of motor units activated for a muscle contraction. This allows your body to produce the appropriate amount of force for a particular task. As you begin a contraction, your brain will stimulate a number of motor units. Sensory information regarding the contraction and the amount of force necessary for the task will allow your brain to assess whether additional motor units are necessary for the development of additional force. In addition to the number of motor units stimulated, the brain can also alter the **frequency** of stimulation to each motor unit. When your brain instructs motor units to contract, it will alternate which specific motor units of a muscle are being contracted. This allows the motor units that had been active in the contraction to relax while other motor units are stimulated to contract to maintain the force or tension. Even at rest, your skeletal muscles exhibit a state called **tonus**. This is a slight state of contraction or tension that maintains the muscle in a state of readiness.

One problem associated with skeletal muscle is that it cannot maintain a contraction indefinitely. You are familiar with the inability to maintain maximal force (acute maximal work) in muscles over time. This is referred to as muscle **fatigue**. Although you may be stimulating the motor units at maximal frequency, they cannot maintain a continual, consistent force of contraction. Even muscles asked to develop a weak amount of tension or force over a long period of time (chronic submaximal work) will eventually fatigue. Try standing with your arms stretched out to the sides and held at shoulder height to see how long it takes for muscle fatigue to occur. There are several factors that can contribute to muscle fatigue. These include the temporary depletion of muscle energy (ATP) if energy consumption exceeds energy production and the accumulation of muscle fiber metabolic wastes if blood flow is not adequate to carry the waste products away from the muscle tissue. In this lab, we will define muscle fatigue as a decrease in muscle force to 50% of the maximal force initially developed by the muscle.

The exercises in this lab will illustrate the principles of motor unit recruitment and muscle fatigue. You will be using the technique, **electromyography**. This technique measures the changes in skin voltage as a result of the electrical activity of the skeletal muscles. The amount of force developed by the muscle can be calculated based upon the electrical activity of the skeletal muscle. The actual tracing or recording is called an **electromyogram** or **EMG**.

## Starting the iWorx Software

1. Be sure that the iWorx box is turned on. A small green light will be lit when the power is on. If you need to turn the power on, the on/off switch is on the back of the iWorx box.
2. Click on the **LabScribe3 shortcut** on the computer's desktop to open the program.
   You should see an information box that says "Hardware Found." Click **OK**; this will put you on the Main Window page. If the hardware is not found, check to be sure the iWorx box is turned on.
3. On the Main Window, pull down the **Settings** menu. Scroll to **Human Muscle** then select the **EMG-GripStrength** settings file from the "Human Muscle" list.
   Note: If "Human Muscle" is empty, then complete the following instructions. On the Main Window, pull down the "Settings" menu and select "Load Group." Scroll down to locate the folder that contains the settings group, "IPLMv6Complete.iwxgrp." Select this group and click "Open." At this point, you should be able to then complete step 3.

4. Instructions for the EMG-GripStrength Setup will automatically open. These instructions are not necessary. Close this page by clicking on the close button in the upper right corner of this document.
5. LabScribe will appear on the computer screen as configured by the EMG-GripStrength settings file.
6. For this lab, you do not need any information from the following files. But as a reminder, once the settings file has been loaded, clicking the **Experiment** button on the toolbar will allow you to access any of the following documents:
   - Introduction—introductory material for iWorx Human Muscle lab exercises
   - Appendix—may have additional information and resources
   - Background—may have some background information regarding the lab topic
   - Labs—the lab exercise as written by iWorx
   - Setup (opens automatically)—instructions for setting up this particular iWorx lab. You already closed this out.
7. If you choose to save your data, it should be saved in the **iWorx Saved Files** folder on the **Desktop**. You may create your own folder within the iWorx Saved Files folder.

## EMG Cable and Hand Dynamometer Setup

The EMG cable and hand dynamometer have already been connected to the iWorx box (see Figure 1). Do not unplug any of these cables. Doing so will cause you to lose your data.

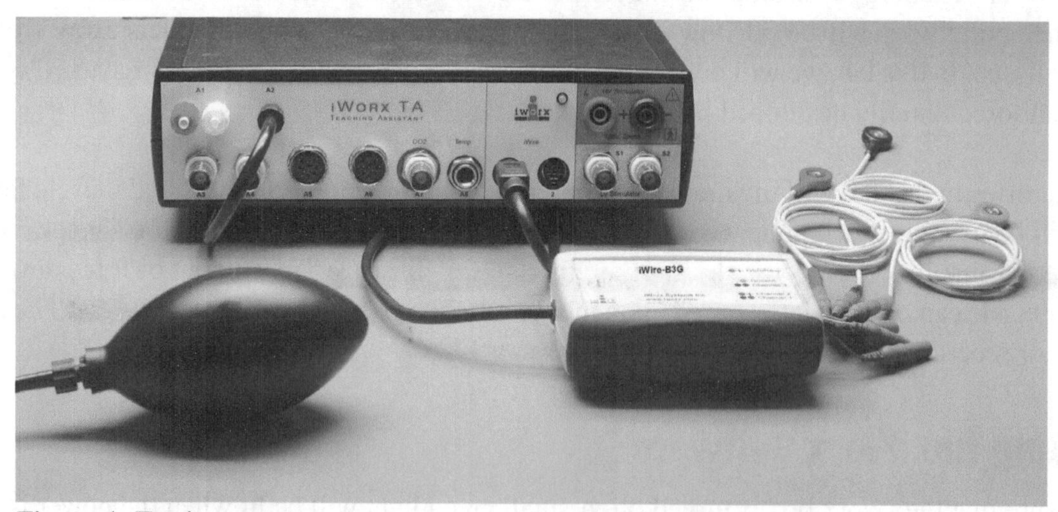

Figure 1 Equipment

For adequate electrical signals, the subject should remove all jewelry from both of their wrists. The subject will first be tested on the nondominant arm then on the dominant arm. Use one or two alcohol swabs to clean the three areas on the anterior surface (recall anatomical position) of each of the subject's forearms where the electrodes will be placed. These areas are the wrist, the middle of the forearm, and approximately 2 inches distal to the elbow (see Figure 2). Let these areas dry completely. Obtain six disposable electrodes and a squeeze bottle of electrode gel. After removing the plastic cover from an electrode, apply one very small drop of electrode gel to the center of the sticky side of the electrode. Apply one prepared disposable electrode to each of the areas previously cleaned with the alcohol swab (see Figure 2). Snap the lead wires onto the electrodes in the following manner:

- Red lead (+1) is attached to the electrode just distal to the elbow on the anterior surface
- Black lead (−1) is attached to the electrode in the mid forearm region on the anterior surface
- Green lead (G) is attached to the electrode at the wrist on the anterior surface
- See Figure 2

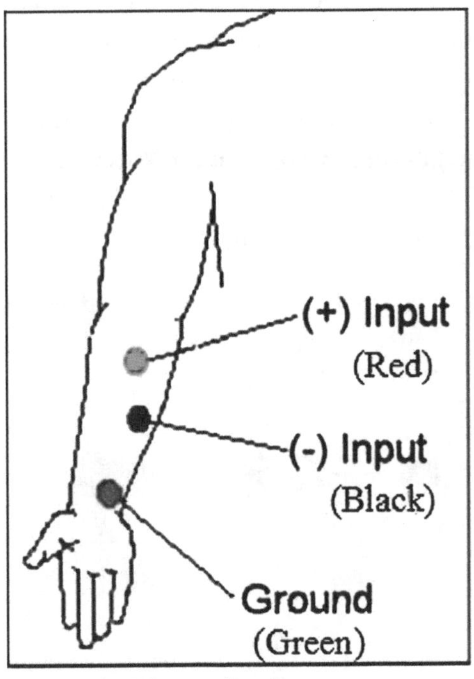


Figure 2 Electrode placement

## Calibrating the Hand Dynamometer

The hand dynamometer will measure in PSI, pounds per square inch. The dynamometer does not need to be calibrated if measuring in PSI.

## Lab Exercises

- **Exercise 1: The EMG and Motor Unit Recruitment in the Nondominant Arm**
  - Read through all instructions before beginning the testing.
  - **Subject Instructions**
    - The subject should sit quietly in a chair with the nondominant forearm resting on the table and the dynamometer resting in his/her hand. The subject will squeeze the dynamometer five times. Each squeeze should last 2 seconds, and each squeeze should be followed by 2 seconds of relaxation with no force exerted on the dynamometer. The first squeeze should be a very light squeeze. The subject should increase the force of the squeeze in each of the subsequent squeezes so that the last squeeze is a maximal squeeze. In each squeeze, the subject should try to maintain a constant level of force (see Figure 3).
  - **Tester Instructions**
    - To label the data for exercise 1, move the cursor to the "Mark" text box to the right of the "Mark" button and click to activate the text box. Type "increasing grip force—nondominant arm" in the "Mark" text box.

- Click the "Record" button to begin recording and press the "Enter" key on the keyboard to enter the mark. Begin the testing and give the subject verbal cues for when to squeeze and when to relax. Use the following testing procedure:
  - 2-second very light squeeze followed by a 2-second relaxation period
  - 2-second light squeeze followed by a 2-second relaxation period
  - 2-second harder squeeze followed by a 2-second relaxation period
  - 2-second harder squeeze followed by a 2-second relaxation period
  - 2-second hardest squeeze followed by a 2-second relaxation period
- After the final 2-second relaxation period, click the "Stop" button to stop recording.
- Click the "Autoscale" buttons for both the "EMG" and the "Muscle Force" channels. The data recording should look similar to Figure 3.

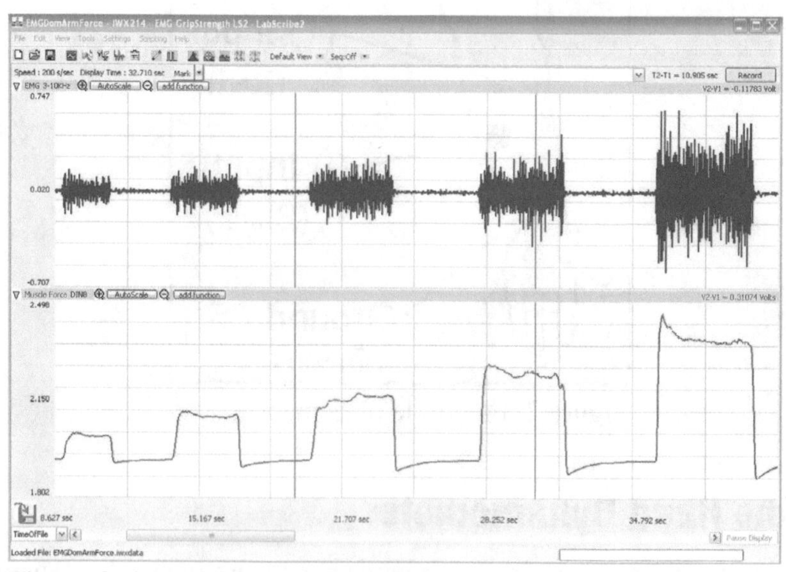

Figure 3 Sample data, exercise 1

- You may choose to save your data at this point or you may continue on with the experiment or data analysis. If you have not yet saved your data, follow the instructions for saving data under exercise 1. If you have previously saved your data, click on the **Save File** icon in the iWorx toolbar.

## Exercise 2: The EMG and Muscle Fatigue in the Nondominant Arm

- Read through all instructions before beginning the testing.
- **Subject Instructions**
  - The subject should sit quietly in a chair with the nondominant forearm resting on the table and the dynamometer resting in his/her hand. The subject will squeeze the dynamometer as hard and as long as possible. The goal is to fatigue the muscles of the forearm. Be sure there are no breaks in the maximal force and that the subject does not shift arm position or hand position on the dynamometer. Continue to squeeze the dynamometer until told to relax.
- **Tester Instructions**
  - To label the data for exercise 2, move the cursor to the "Mark" text box to the right of the "Mark" button and click to activate the text box. Type "muscle fatigue—nondominant arm" in the "Mark" text box.

- Click the "Record" button to begin recording and press the "Enter" key on the keyboard to enter the mark. Allow 5 to 10 seconds of baseline recording, then give the subject the verbal cue to squeeze as hard as possible.
- Record the value of PSI at the minimum or baseline amplitude; this value should be observed just prior to the beginning of the contraction.
- Record the approximate value of PSI at the maximum amplitude of contraction; this value should be observed at or near the beginning of the contraction.
- Calculate 50% of the maximum amplitude of contraction (see Figure 4).

  Minimum amplitude __________ PSI

  Maximum amplitude __________ PSI

  Calculate the following:

  Max. amplitude − [(max. amplitude − min. amplitude)/2] =

  50% of maximum amplitude __________

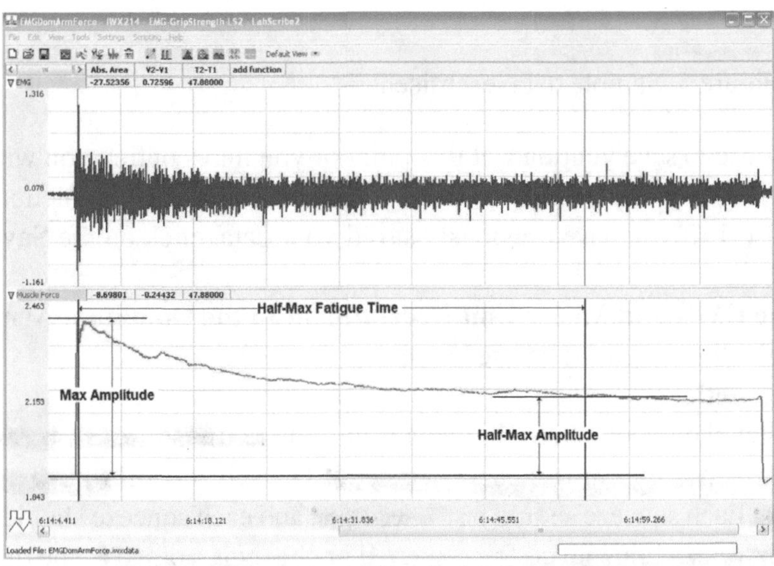


Figure 4 50% maximum amplitude determination

- Over time, the force of contraction will decrease. Watch for the force of contraction to decrease to less than 50% of the maximum amplitude initially generated in this contraction. The amount of time required for this to occur is dependent upon the fitness level of the subject and may vary from about 30 seconds to minutes. When the force of contraction **drops below 50% of maximum amplitude**, give the cue for the subject to relax. Relaxing before this point will invalidate the exercise and you will have to perform this portion of the exercise again.
- Continue to record for approximately 5 seconds after the subject has relaxed. Click the "Stop" button to stop recording.

- Click the "Autoscale" buttons for both the "EMG" and the "Muscle Force" channels. The data recording should look similar to Figure 5.

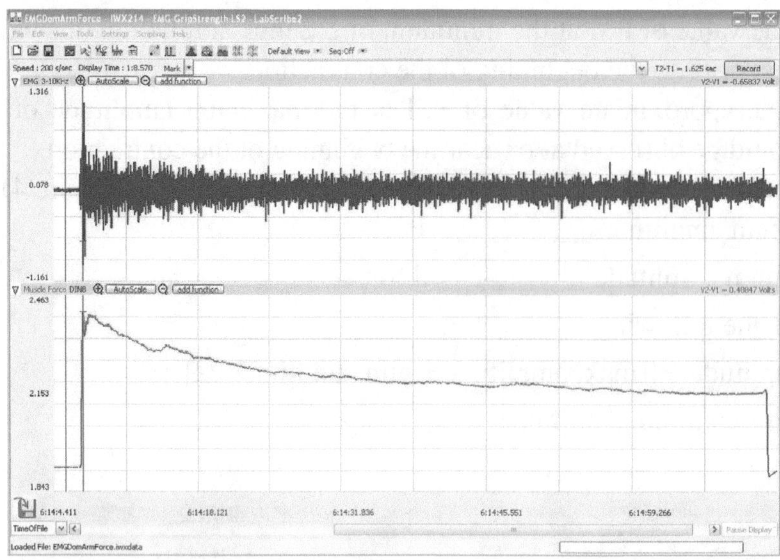

Figure 5 Sample data, exercise 2

- You may choose to save your data at this point or you may continue on with the experiment or data analysis. If you have not yet saved your data, follow the instructions for saving data under exercise 1. If you have previously saved your data, click on the **Save File** icon in the iWorx toolbar.

## Exercise 3: The EMG and Motor Unit Recruitment in the Dominant Arm

- Read through all instructions before beginning the testing.
- **Subject Instructions**
  - The subject should sit quietly in a chair with the dominant forearm resting on the table and the dynamometer resting in his/her hand. The subject will squeeze the dynamometer five times. Each squeeze should last 2 seconds, and each squeeze should be followed by 2 seconds of relaxation with no force exerted on the dynamometer. The first squeeze should be a very light squeeze. The subject should increase the force of the squeeze in each of the subsequent squeezes so that the fifth squeeze is a maximal squeeze. In each squeeze, the subject should try to maintain a constant level of force (see Figure 6).
- **Tester Instructions**
  - To label the data for exercise 3, move the cursor to the "Mark" text box to the right of the "Mark" button and click to activate the text box. Type "increasing grip force—dominant arm" in the "Mark" text box.
  - Click the "Record" button to begin recording and press the "Enter" key on the keyboard to enter the mark. Begin the testing and give the subject verbal cues for when to squeeze and when to relax. Use the following testing procedure:
    - 2-second very light squeeze followed by a 2-second relaxation period
    - 2-second light squeeze followed by a 2-second relaxation period
    - 2-second harder squeeze followed by a 2-second relaxation period
    - 2-second harder squeeze followed by a 2-second relaxation period
    - 2-second hardest squeeze followed by a 2-second relaxation period

- After the final 2-second relaxation period, click the "Stop" button to stop recording.
- Click the "Autoscale" buttons for both the "EMG" and the "Muscle Force" channels. The data recording should look similar to Figure 6.

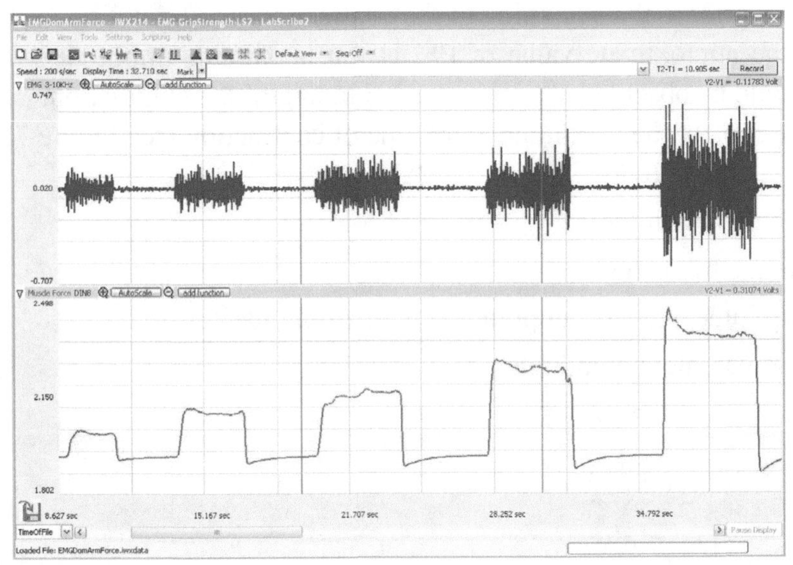

Figure 6 Sample data, exercise 3

- You may choose to save your data at this point or you may continue on with the experiment or data analysis. To save your data, on the Main Window, either
  1. click on the **File** menu. Scroll to **Save As** then save your data, it should be saved in the **iWorx Saved Files** folder on the **Desktop**. You may create your own folder within the iWorx Saved Files folder. Give your file a name and designate the file type as .iwxdata, then click **Save**.

  Or

  2. click on the **Save File** icon in the iWorx toolbar, then save your data, it should be saved in the **iWorx Saved Files** folder on the **Desktop**. You may create your own folder within the iWorx Saved Files folder. Give your file a name and designate the file type as .iwxdata, then click **Save**.

## Exercise 4: The EMG and Muscle Fatigue in the Dominant Arm

- Read through all instructions before beginning the testing.
- **Subject Instructions**
  - The subject should sit quietly in a chair with the dominant forearm resting on the table and the dynamometer resting in his/her hand. The subject will squeeze the dynamometer as hard and as long as possible. The goal is to fatigue the muscles of the forearm. Be sure there are no breaks in the maximal force and that the subject does not shift arm position or hand position on the dynamometer. Continue to squeeze the dynamometer until told to relax.
- **Tester Instructions**
  - To label the data for exercise 4, move the cursor to the "Mark" text box to the right of the "Mark" button and click to activate the text box. Type "muscle fatigue—dominant arm" in the "Mark" text box.

- Click the "Record" button to begin recording and press the "Enter" key on the keyboard to enter the mark. Allow 5 to 10 seconds of baseline recording, then give the subject the verbal cue to squeeze as hard as possible.
- Record the value of PSI at the minimum or baseline amplitude; this value should be observed just prior to the beginning of the contraction.
- Record the approximate value of PSI at the maximum amplitude of contraction; this value should be observed at or near the beginning of the contraction.
- Calculate 50% of the maximum amplitude of contraction (see Figure 7).

  Minimum amplitude __________ PSI

  Maximum amplitude __________ PSI

Calculate the following: 17.468 − 16.981 /2

Max. amplitude − [(max. amplitude − min. amplitude)/2] =

  50% of maximum amplitude 0.2435

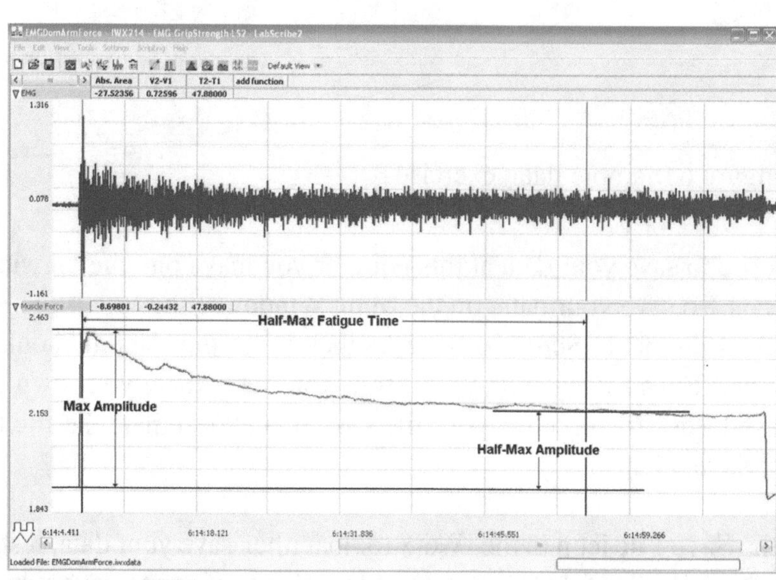


Figure 7 50% maximum amplitude determination

- Over time, the force of contraction will decrease. Watch for the force of contraction to decrease to less than 50% of the maximum amplitude initially generated in this contraction. The amount of time required for this to occur is dependent upon the fitness level of the subject and may vary from about 30 seconds to minutes. When the force of contraction **drops below 50% of maximum amplitude**, give the cue for the subject to relax. Relaxing before this point will invalidate the exercise and you will have to perform this portion of the exercise again.
- Continue to record for approximately 5 seconds after the subject has relaxed. Click the "Stop" button to stop recording.

- Click the "Autoscale" buttons for both the "EMG" and the "Muscle Force" channels. The data recording should look similar to the Figure 8.

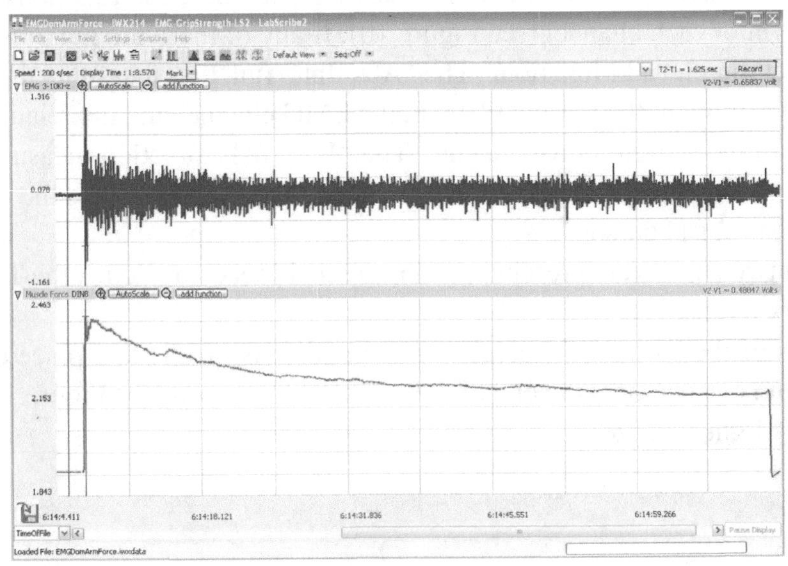

Figure 8 Sample data, exercise 4

- You may choose to save your data at this point or you may continue on with the experiment or data analysis. If you have not yet saved your data, follow the instructions for saving data under exercise 1. If you have previously saved your data, click on the **Save File** icon in the iWorx toolbar.

## Data Analysis Instructions

Figure 9, a diagram of icons in the iWorx toolbar may be helpful.

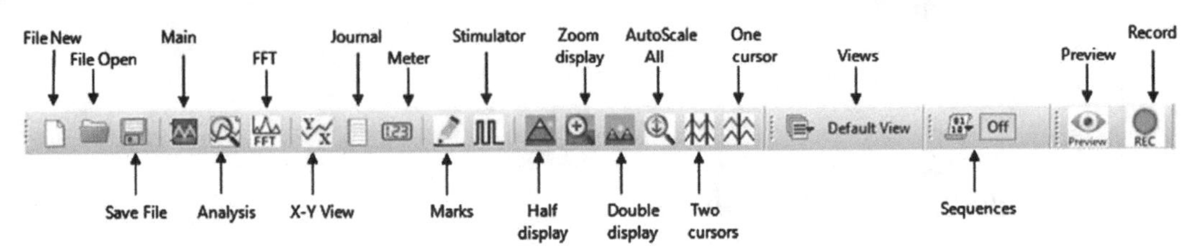


Figure 9 iWorx toolbar icons

- General analysis instructions: Go to the beginning of your recordings. Click on the **Marks** icon to bring up the Mark Dialogue box with a table of the marks created in your lab. Once you have the table of marks, click on the number at the beginning of the row for the mark that identifies the beginning of the data segment you are going to analyze. This will highlight that mark. Next, click on the **Go To Mark** button near the bottom of the Marks Dialogue box. Data at that mark will be displayed on your screen.
- Use the **Half Display** icon, or the **Double Display** icon on the iWorx toolbar to adjust the data visible in the Main Window so that all five of the progressive muscle contractions are visible on the screen. You can also adjust the display by clicking on the **Two Cursors** icon and moving one cursor to the beginning of the first contraction and moving the second cursor to the end of the final contraction. Then click the **Zoom Display** icon.
- Click on the **Analysis** icon in the iWorx toolbar.

- In the Analysis window, a "Function Table" will be displayed above the uppermost channel (EMG channel). The mathematical functions "Abs. Int." (absolute integral), "V2 – V1," and "T2 – T1" should be present. The values for each of these functions are displayed in the top margin of each channel (see Figure 10). If any of these are not present, you can add them by clicking on "add function." The "Abs. Int." function measures the relative amount of electrical activity in the muscle fibers in the EMG channel and the relative strength of the contraction in the Muscle Force channel. The "V2 – V1" function measures the difference in values at cursor 2 minus cursor 1. The "T2 – T1" function measures the difference in time between cursor 2 and cursor 1.
- Be sure that two cursors are present on your screen. If necessary, click the **Two Cursors** icon on the iWorx toolbar.
- For analysis of the motor unit recruitment components of this lab, place one cursor at the beginning and one cursor at the end of the muscle contraction (see Figure 10, but begin with the first very light squeeze).

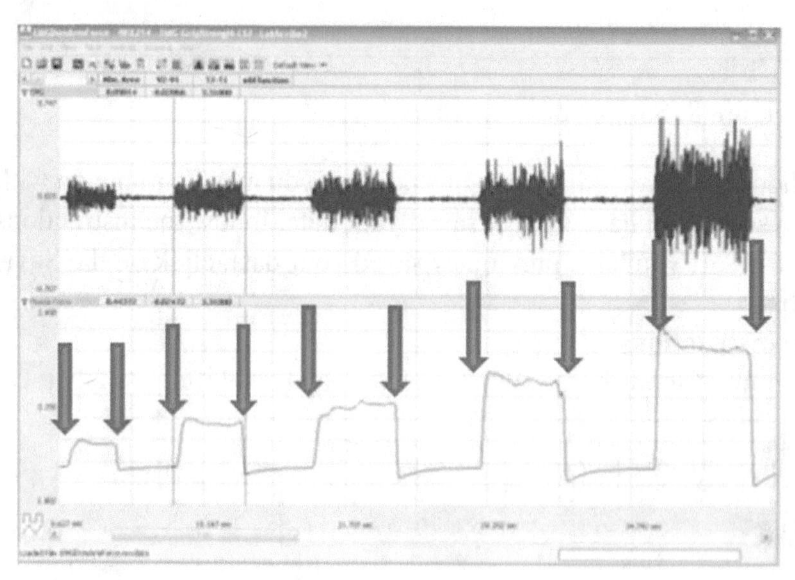

Figure 10 Data analysis

- Record the "Abs. Int." values for both the EMG and the Muscle Force channels in Table 1.
- Repeat the measurement of each of the squeezes using the nondominant arm and record the data in Table 1.
- Repeat the analysis procedure for the motor unit recruitment data collected with the dominant arm, exercise 3. Record the data in Table 1.
- Measure the circumference of both the dominant and the nondominant arm approximately 3 cm below the elbow. Record the data in Table 1.
- For analysis of the muscle fatigue data, exercises 2 and 4, follow the same basic introductory instructions as for analysis of the motor unit recruitment data.
- On the "Muscle Force" channel, place the first cursor on the baseline before the subject began the contraction. The force recorded here will serve as the baseline. Place the second cursor on the maximum force developed by the subject during the contraction. The value for "V2 – V1" on the "Muscle Force" channel is the maximum force of contraction the

**TABLE 1**: Motor Unit Recruitment

Dominant Arm Circumference (cm) 23.5 — (right arm) / left arm

Nondominant Arm Circumference (cm) 22 — right arm / (left arm)

| | Nondominant Arm | | Dominant Arm | |
|---|---|---|---|---|
| **Squeeze** | **EMG Absolute Integral (mV)** | **Muscle Force Absolute Integral (PSI)** | **EMG Absolute Integral (mV)** | **Muscle Force Absolute Integral (PSI)** |
| 1 | 0.120 | 20.778 | 0.091 | 25.946 |
| 2 | 0.144 | 25.303 | 0.145 | 34.528 |
| 3 | 0.248 | 36.478 | 0.190 | 37.204 |
| 4 | 0.292 | 33.426 | 0.277 | 42.242 |
| 5 | 0.398 | 36.744 | 0.509 | 50.162 |

subject developed (see Figure 11 for an illustration of the maximum force). Enter the value in Table 2.

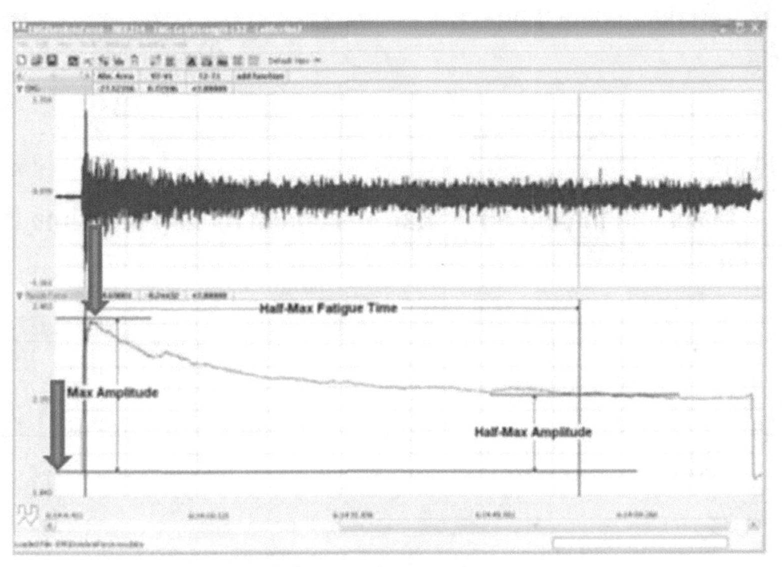


Figure 11 Data analysis maximum force

- To determine half-maximum force, divide the measured maximum force by 2. Enter the value in Table 2.
- To determine the time to fatigue, you will need to determine when the subject reaches their half-maximum force. In the "Muscle Force" channel, keep the first cursor on the baseline before the subject began the contraction. Move the second cursor along the force curve until the absolute value for "V2 – V1" in the "Muscle Force" channel is equal to the half-maximum force you calculated. Leaving the second cursor in place, move the first

the beginning of the contraction (see Figure 12). Enter the value for "T2 – T1," the time to fatigue, in Table 2.

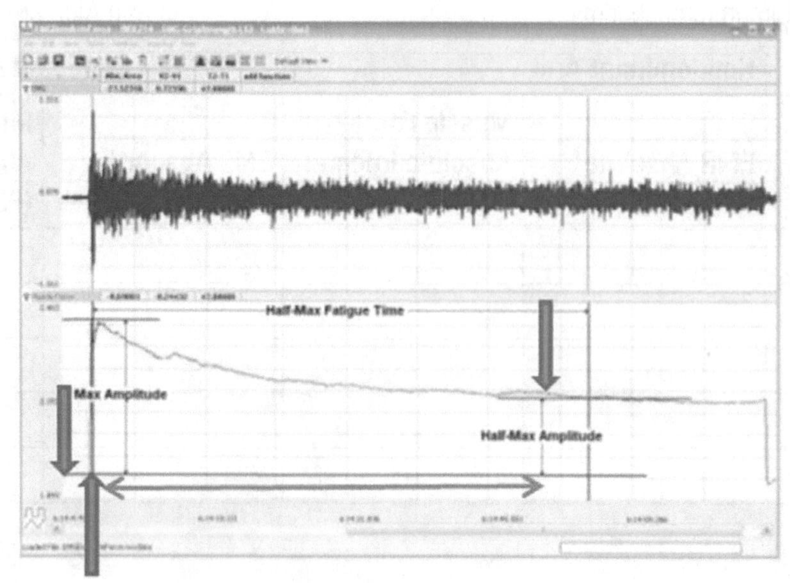


Figure 12 Data analysis time to fatigue

- Repeat the analysis procedure for the muscle fatigue data collected with the dominant arm, exercise 4. Record the data in Table 2.

**TABLE 2**: Muscle Fatigue

| | Maximum force (PSI) | Half-maximum force (PSI) | Time to fatigue (sec) |
|---|---|---|---|
| Nondominant arm right arm / left arm | 17.448 | 0.2435 | 35 |
| Dominant arm right arm / left arm | 20.921 | 1.037 | 50 |

# Exercise 13

# iWorx EMG Activity of Antagonist Muscles: Wrist Flexion & Extension and Arm Wrestling

## Objectives

- Be able to define the term *agonist.*
- Be able to define the term *antagonist.*
- Be able to define the term *flexion.*
- Be able to define the term *extension.*
- Be able to compare level of muscle activity as a result of electrical activity measured in an EMG.
- Be able to calculate the percent difference (increase or decrease) in muscle electrical activity between flexion and extension in the same muscle.
- Be able to calculate the percent difference (increase or decrease) in muscle electrical activity between flexion and extension in the antagonist muscles.
- Be able to identify areas of tonus with in the EMG.

## Lab Safety Reminders

- Use caution with any electrical equipment.

## Hints for Success

All members of your lab group should feel comfortable operating the iWorx system.

**Never unplug any sensors from the iWorx box!** If you unplug a sensor during an experiment, you will lose all data. You will need to turn the iWorx box off, plug the sensor in and start the lab from the beginning.

On the iWorx recording screens, it is helpful to remember that the blue bars are for data that are being recorded. The green bars are for data that is being calculated based upon the data you have collected.

The use of the electrode gel increases the ability to detect the electrical signals generated by the subject's skeletal muscles. Only a small amount of the gel is needed in the center of the electrode. A large amount will interfere with the ability of the electrode to stick to the subject's skin.

## Introduction

Antagonistic muscles are muscles that work in opposition to each other, one working as the agonist and one working as the antagonist. The agonist muscle is the muscle that will give the movement that is desired. The antagonist muscle is the muscle that opposes the desired movement. This lab will be conducted in two parts. The first part of the lab, wrist extension and flexion will illustrate the difference in muscle activity of the agonist and antagonist muscles in a simple muscle movement. In this part of the lab, you will be alternately stimulating the flexor carpi radialis, flexor carpi ulnaris, and palmaris longus for flexion at the wrist and the extensor carpi radialis longus, extensor carpi radialis brevis, and extensor carpi ulnaris for extension at the wrist. In each case, as one group of muscles acts as the agonists, the other group of muscles will act as the antagonists. The second part of the lab, simulated arm wrestling, will use a more complicated but more realistic movement of primarily the forearm involving agonist and antagonist muscles. The pronator teres and the biceps brachii will be the muscles monitored during this portion of the lab exercise. The pectoralis major and flexor carpi ulnaris are also key muscles used to move the forearm in arm wrestling, but these muscles will not be monitored. Other muscles including the deltoid, latissimus dorsii, and the triceps brachii are also used in arm wrestling, but they are used primarily to provide strength. These muscles will not be monitored.

You will be using the technique, **electromyography**. This technique measures the changes in skin voltage as a result of the electrical activity of the skeletal muscles. The amount of force developed by the muscle can be inferred based upon the electrical activity of the skeletal muscle. The actual tracing or recording is called an **electromyogram** or **EMG**.

## Starting the iWorx Software: Wrist Flexion & Extension

1. Be sure that the iWorx box is turned on. A small green light will be lit when the power is on. If you need to turn the power on, the on/off switch is on the back of the iWorx box.
2. Click on the **LabScribe3 shortcut** on the computer's desktop to open the program.
   You should see an information box that says "Hardware Found". Click **OK**; this will put you on the Main Window page. If the hardware is not found, check to be sure the iWorx box is turned on.
3. On the Main Window, pull down the **Settings** menu. Scroll to **Human Muscle**, then select the **EMG-Antagonistic Muscles** settings file from the "Human Muscle" list.
   Note: if "Human Muscle" is empty, then complete the following instructions. On the Main Window, pull down the "Settings" menu and select "Load Group." Scroll down to locate the folder that contains the settings group, "IPLMv6Complete.iwxgrp." Select this group and click "Open." At this point, you should be able to then complete step 3.
4. Instructions for the EMG-Antagonistic Muscles Setup will automatically open. These instructions are not necessary. Close this page by clicking on the close button in the upper right corner of this document.
5. LabScribe will appear on the computer screen as configured by the EMG-Antagonistic Muscles settings file.

6. For this lab, you do not need any information from the following files. But as a reminder, once the settings file has been loaded, clicking the **Experiment** button on the toolbar will allow you to access any of the following documents:
   - Introduction—introductory material for iWorx Human Muscle lab exercises
   - Appendix—may have additional information and resources
   - Background—may have some background information regarding the lab topic
   - Labs—the lab exercise as written by iWorx
   - Setup (opens automatically)—instructions for setting up this particular iWorx lab. You already closed this out.
7. If you choose to save your data, it should be saved in the **iWorx Saved Files** folder on the **Desktop**. You may create your own folder within the iWorx Saved Files folder.

## EMG Cable Setup

The EMG cable has already been connected to the iWorx box (see Figure 1). Do not unplug any of these cables. Doing so will cause you to lose your data.

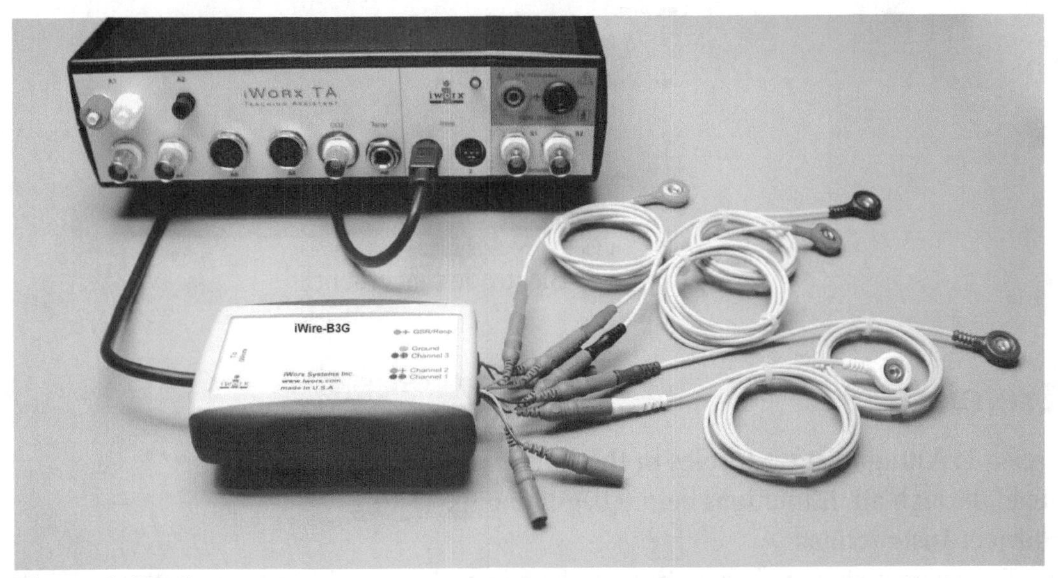


Figure 1 Equipment

For adequate electrical signals, the subject should remove all jewelry from the left wrist. Electrodes will be placed on both the anterior and posterior surfaces of the left forearm (recall anatomical position). Use an alcohol swab to clean the three areas on the subject's anterior left forearm where the electrodes will be placed. Use a second alcohol swab to clean the two areas on the subject's posterior left forearm where the electrodes will be placed (see Figure 2). Let these areas dry completely. Obtain 5 disposable electrodes and a squeeze bottle of electrode gel. After removing the plastic cover from an electrode, apply one very small drop of electrode gel to the center of the sticky side of the electrode. Apply one prepared disposable electrode to each of the areas previously cleaned with the alcohol swab (see Figure 2). Snap the lead wires onto the electrodes in the following manner:

Anterior surface

- Red lead (+1) is attached to the electrode closest to the elbow on the anterior surface over the flexors
- Black lead (−1) is attached to the electrode most distal to the elbow on the anterior surface over the flexors
- Green lead (G) (ground lead) is attached to the electrode intermediate and lateral to the red and black leads on the anterior surface

Posterior surface

- White lead (+2) is attached the electrode closest to the elbow on the posterior surface over the extensors
- Brown lead (−2) is attached the electrode more distal to the elbow on the posterior surface over the extensors
- See Figure 2

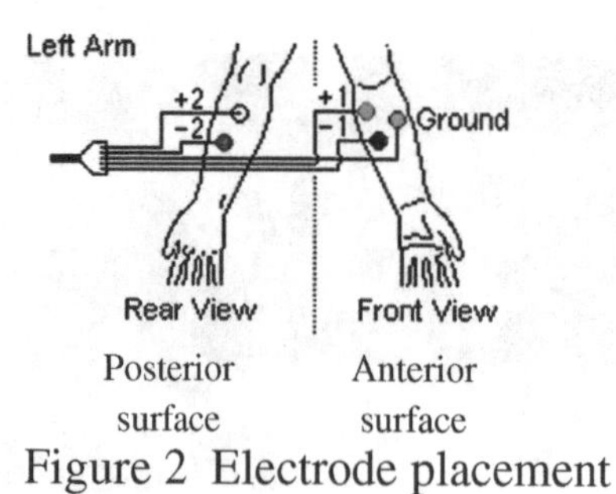


Posterior surface    Anterior surface

Figure 2 Electrode placement

## Lab Exercises

- **Exercise 1: Antagonistic Muscles in the Forearm, Fingers Extended**
  - Read through all instructions before beginning the testing.
  - **Subject Instructions**
    - The subject should sit quietly in a chair in the "neutral position" with the left forearm (electrodes attached) extended in front of his/her body with the palm facing upward and the fingers extended during the entire testing period.
    - The subject will flex at the wrist, moving the hand upward from the neutral position as far as possible (see Figure 3). Hold this position for 4 seconds.
    - Return the arm to the neutral position for 4 seconds.
    - The subject will extend at the wrist, moving the hand downward from the neutral position as far as possible (see Figure 3). Hold this position for 4 seconds.
    - Return the arm to the neutral position for 4 seconds.

- Repeat the entire process two more times.

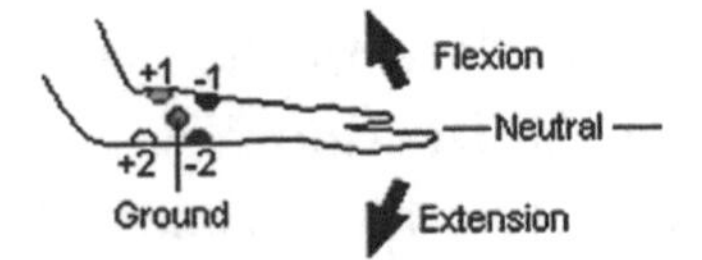


Figure 3 Wrist movements

- **Tester Instructions**
  - To label the data for exercise 1, move the cursor to the "Mark" text box to the right of the "Mark" button and click to activate the text box. Type "neutral position—fingers extended" in the "Mark" text box.
  - Click the "Record" button to begin recording and press the "Enter" key on the keyboard to enter the mark. Begin the testing with the subject in the neutral position and give the subject verbal cues for when to flex at the wrist, when to return to the neutral position, and when to extend at the wrist. During each neutral position time period, type either "flexion" or "extension," whichever is appropriate, into the "Mark" text box and click "Enter" when you give the subject the instruction to either flex or extend at the wrist. Use the following testing procedure:
    - 4-second neutral position
    - 4-second flexion position followed by a 4-second neutral position
    - 4-second extension position followed by a 4-second neutral position
    - 4-second flexion position followed by a 4-second neutral position
    - 4-second extension position followed by a 4-second neutral position
    - 4-second flexion position followed by a 4-second neutral position
    - 4-second extension position followed by a 4-second neutral position
  - After the final 4-second neutral period, click the "Stop" button to stop recording.
  - Click the "Autoscale" buttons for both the "EMG Anterior" and the "EMG Posterior" channels. The data recording should look similar to Figure 4.

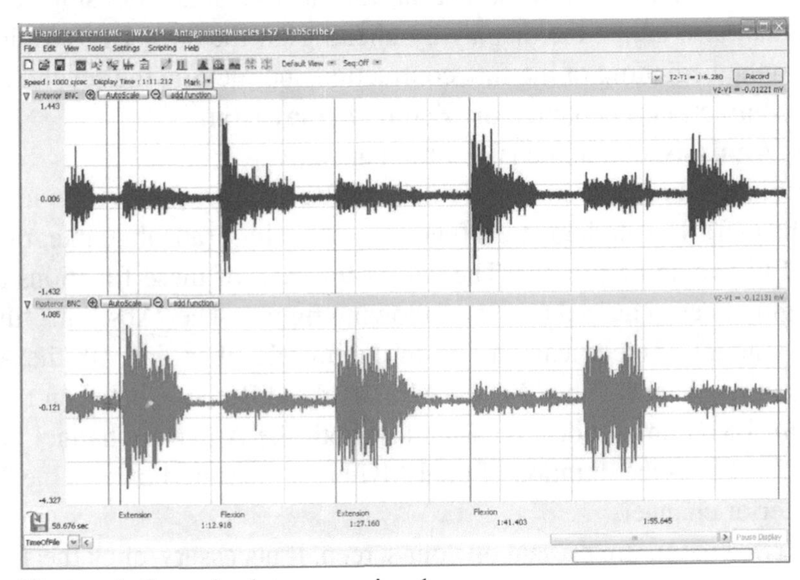

Figure 4 Sample data, exercise 1

- You may choose to save your data at this point or you may continue on with the experiment or data analysis. To save your data, on the Main Window, either
  1. click on the **File** menu. Scroll to **Save As,** then save your data, it should be saved in the **iWorx Saved Files** folder on the **Desktop**. You may create your own folder within the iWorx Saved Files folder. Give your file a name and designate the file type as .iwxdata, then click **Save**.

  Or

  2. 2. click on the **Save File** icon in the iWorx toolbar, then save your data; it should be saved in the **iWorx Saved Files** folder on the **Desktop**. You may create your own folder within the iWorx Saved Files folder. Give your file a name and designate the file type as .iwxdata, then click **Save**.

## Data Analysis Instructions

Figure 5, a diagram of icons in the iWorx toolbar may be helpful.

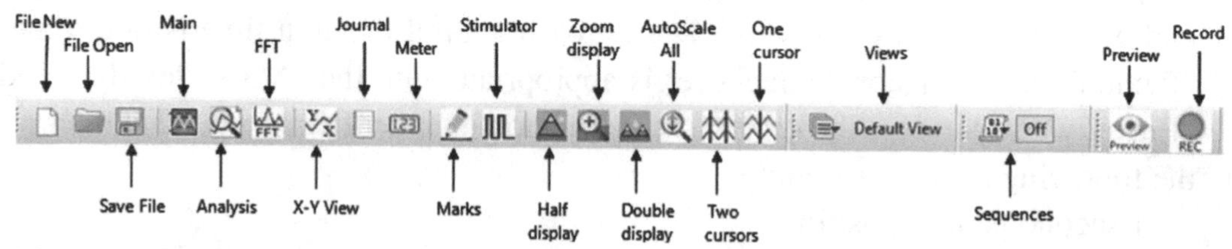


Figure 5 iWorx toolbar icons

- Go to the beginning of your recordings. Click on the **Marks** icon to bring up the Marks Dialogue box with a table of the marks created in your lab. Once you have the table of marks, click on the number at the beginning of the row for the mark that identifies the beginning of the data segment you are going to analyze. This will highlight that mark. Next, click on the **Go To Mark** button near the bottom of the Marks Dialogue box. Data at that mark will be displayed on your screen.
- Use the **Half Display** icon or the **Double Display** icon on the iWorx toolbar to adjust the data visible in the Main Window so that all the flexions and extensions are visible on the screen. You can also adjust the display by clicking on the **Two Cursors** icon and moving one cursor to the beginning of the first contraction and moving the second cursor to the end of the final contraction. Then click the **Zoom Display** icon.
- Click on the **Analysis** icon in the iWorx toolbar.
- In the Analysis window, a "Function Table" will be displayed above the uppermost channel (Anterior channel). The mathematical functions "Abs. Int." (absolute integral), "Max – Min," and "T2 – T1" should be present. The values for each of these functions are displayed in the top margin of each channel (see the following figure). The "Abs. Int." function measures the relative amount of electrical activity in the muscle fibers in both the Anterior channel and the Posterior channel. The "Max – Min" is the difference between the maximum and the minimum EMG amplitudes, measured in both the Anterior channel and the Posterior channel. "T2 – T1" is the duration of the EMG burst measured in both the Anterior channel and the Posterior channel.
- Be sure that two cursors are present on your screen. If necessary, click the **Two Cursors** icon on the iWorx toolbar.

- For analysis of the flexion and extension, place one cursor at the beginning of and one cursor at the end of the EMG burst of the first Anterior flexion (see Figure 6). Record the values for "Abs. Int." and "Max – Min" for both the anterior muscles and the posterior muscles in Table 1. Repeat the procedure for the first Anterior extension. Repeat for the remaining 2 cycles. Calculate the mean for each of the variables measured.

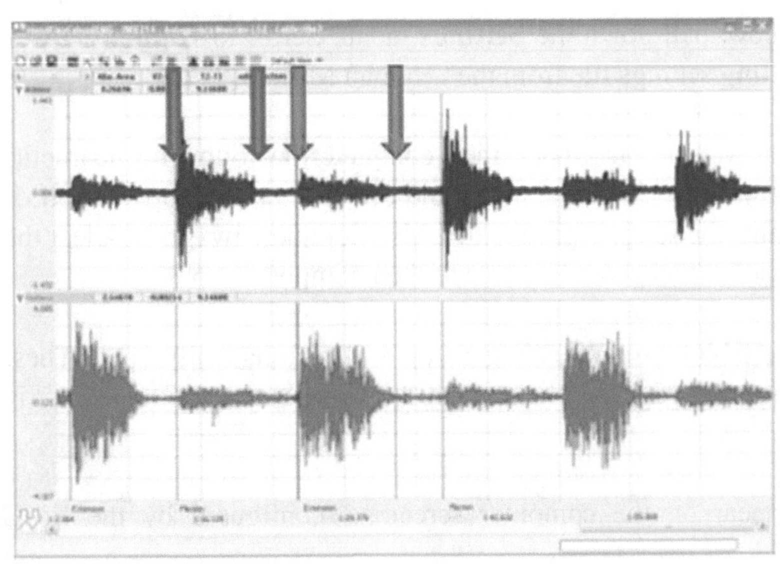

Figure 6 Data analysis wrist extension and flexion

**TABLE 1**: Flexion and Extension at the Wrist

| Wrist Action | | Anterior Muscles | | Posterior Muscles | |
|---|---|---|---|---|---|
| | | Absolute Integral (mV) | Max–Min (mV) | Absolute Integral (mV) | Max – Min (mV) |
| Flexion | Cycle 1 | 0.124 | 1.425 | 0.083 | 0.709 |
| | Cycle 2 | 0.112 | 0.948 | 0.066 | 0.708 |
| | Cycle 3 | 0.127 | 0.410 | 0.062 | 0.424 |
| | Mean | 0.122 | 0.928 | 0.703 | 0.614 |
| Extension | Cycle 1 | 0.122 | 0.713 | 0.622 | 3.872 |
| | Cycle 2 | 0.084 | 0.447 | 0.583 | 4.115 |
| | Cycle 3 | 0.089 | 0.445 | 0.508 | 3.422 |
| | Mean | 0.098 | 0.601 | 0.571 | 3.803 |

## Starting the iWorx Software: Antagonistic Muscle Action During Arm Wrestling

Follow the same general instructions as for the Wrist Flexion & Extension Exercise. The Arm Wrestling Lab is a separate lab in iWorx.

On the Main Window, pull down the **Settings** menu. Scroll to **Human Muscle**, then select the **EMG-Arm Wrestling** settings file from the "Human Muscle" list.

Note: if "Human Muscle" is empty, then complete the following instructions. On the Main Window, pull down the "Settings" menu and select "Load Group." Scroll down to locate the folder that contains the settings group, "IPLMv6Complete.iwxgrp." Select this group and click "Open." At this point, you should be able to then complete step 3.

Instructions for the EMG-Arm Wrestling Setup will automatically open. These instructions are not necessary. Close this page by clicking on the close button in the upper right corner of this document.

LabScribe will appear on the computer screen as configured by the EMG-Arm Wrestling settings file.

If you choose to save your data, it should be saved in the **iWorx Saved Files** folder on the **Desktop**. You may create your own folder within the iWorx Saved Files folder.

## EMG Cable Setup

The EMG cable has already been connected to the iWorx box and has the same setup as for the Wrist Flexion & Extension Exercise (see Figure 7). Do not unplug any of these cables. Doing so will cause you to lose your data.

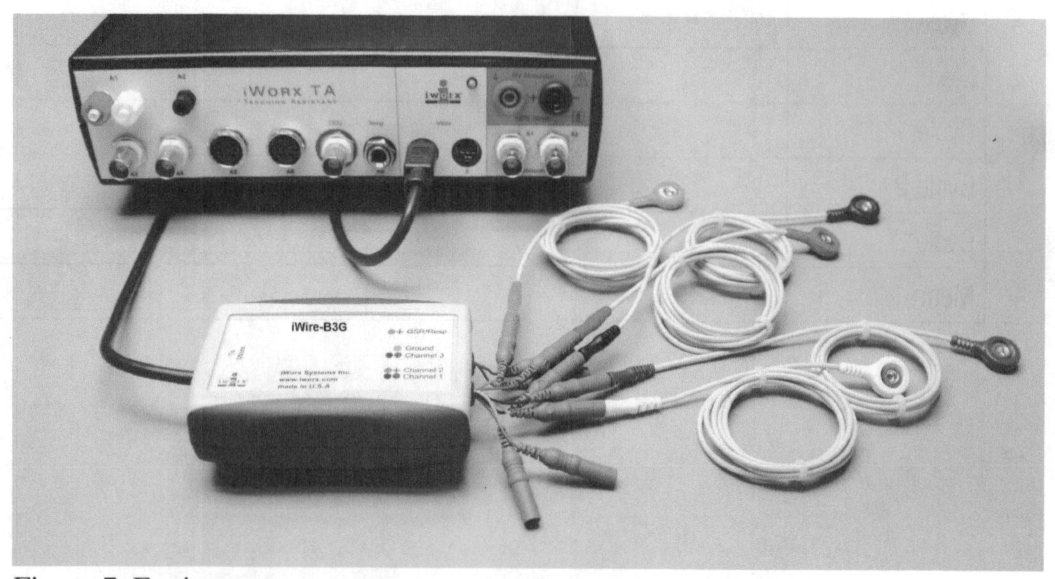


Figure 7 Equipment

For adequate electrical signals, the subject should remove all jewelry from the left wrist. Electrode placement is different for the Arm Wrestling Exercise compared to the Wrist Flexion and Extension Exercise. Electrodes will be placed on the anterior surface (recall anatomical position) of the right forearm and upper arm; one electrode will also be placed on the lower right abdomen. Use an alcohol swab to clean the two areas on the subject's anterior right forearm where the electrodes will be placed. Use a second alcohol swab to clean the two areas on the subject's anterior upper arm where the electrodes will be placed (see Figure 8). Use a third alcohol swab to clean an area on the subject's lower right abdomen. Let these areas dry completely. Obtain 5 disposable electrodes and a squeeze bottle of electrode gel. After removing the plastic cover from an electrode, apply one very small drop of electrode gel to the center of the sticky side of the electrode. Apply one prepared disposable electrode to each of the areas previously cleaned with the alcohol swab (see Figure 8). Snap the lead wires onto the electrodes in the following manner:

- Red lead (+1) is attached to the electrode on the lateral edge of the anterior surface of the forearm over the pronator teres, closer to the wrist
- Black lead (–1) is attached to the electrode in the middle of the anterior surface of the forearm over the pronator teres, closer to the elbow
- White lead (+2) is attached to the electrode on the anterior surface over the biceps brachii, near the elbow
- Brown lead (–2) is attached to the electrode on the anterior surface over the biceps brachii, proximal to the white +2 lead
- Green lead (G) is attached to the electrode on the lower right abdomen.
- See Figure 8.

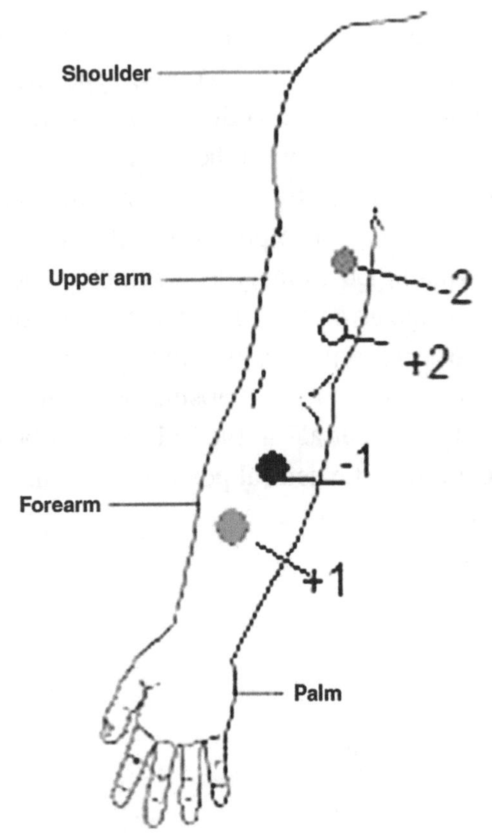


Figure 8 Electrode placement

## Lab Exercises

- **Exercise 2: Simulated Arm Wrestling**
  - Read through all instructions before beginning the testing. Note that this is NOT an actual arm wrestling match. The subject and opponent should use some muscle force, but they should not use maximal effort during the activity. The goal is to observe the difference in muscle activity when a particular muscle is acting as the agonist and when it is acting as the antagonist.
  - **Subject Instructions**
    - The subject and opponent should get into the neutral position for arm wrestling: right hands clasped, elbows bent roughly at a 90° angle with elbows on the bench top, the right arms of both individuals should be roughly perpendicular with the bench top.
    - The subject will slowly move from the neutral position to the winning position by pushing the opponent's hand toward the bench top. The opponent should provide some resistance, but not maximal resistance. Do not go past half way to the bench top surface. Hold this position for 4 seconds.
    - Return the arms to the neutral position for 4 seconds.
    - The subject will slowly move from the neutral position to the losing position as the opponent pushes the subject's hand toward the bench top. The subject should provide some resistance, but not maximal resistance. Do not go past half way to the bench top surface. Hold this position for 4 seconds.
    - Return the arms to the neutral position for 4 seconds.
    - Repeat the entire process two more times.
  - **Tester Instructions**
    - To label the data for exercise 2, move the cursor to the "Mark" text box to the right of the "Marks" button and click to activate the text box. Type "neutral" in the "Mark" text box.
    - Click the "Record" button to begin recording and press the "Enter" key on the keyboard to enter the mark. Begin the testing with the subject in the neutral position. While in the recording period of the neutral position, type "winning" in the "Mark" text box and give the subject verbal cue to change arm position to the winning position. While in the recording period of the winning position, type "neutral" in the "Mark" text box and give the subject verbal cue to change arm position to the neutral position. While in the recording period of the neutral position, type "losing" in the "Mark" text box and give the subject verbal cue to change arm position to the losing position. While in the recording period of the losing position, type "neutral" in the "Mark" text box and give the subject verbal cue to change arm position to the neutral position. Use the following testing procedure:
      - 4 second neutral position
      - 4 second winning position
      - 4 second neutral position

  - 4 second losing position
  - 4 second neutral position
  - 4 second winning position
  - 4 second neutral position
  - 4 second losing position
  - 4 second neutral position
  - 4 second winning position
  - 4 second neutral position
  - 4 second losing position
  - 4 second neutral position
- After the final 4 second neutral period, click the "Stop" button to stop recording.
- Click the "Autoscale" buttons for both the "Pronator teres" and the "Biceps brachii" channels. The data recording should look similar to Figure 9.

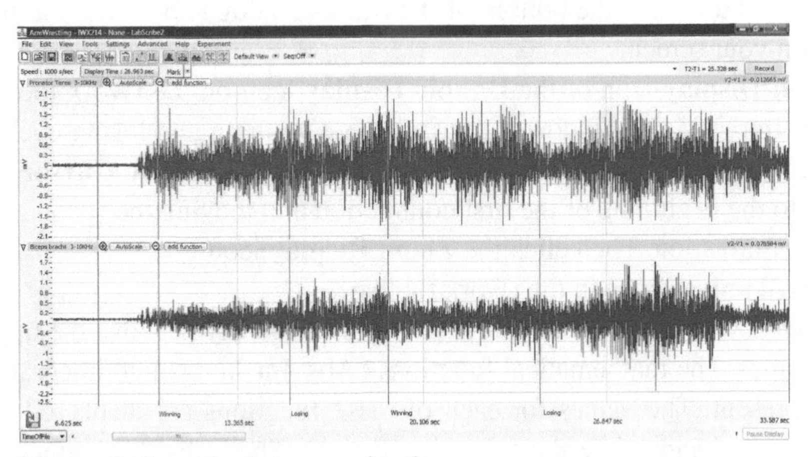

Figure 9 Sample data exercise 2

- You may choose to save your data at this point or you may continue on with the experiment or data analysis. To save your data, on the Main Window, either
  1. click on the **File** menu. Scroll to **Save As,** then save your data; it should be saved in the **iWorx Saved Files** folder on the **Desktop**. You may create your own folder within the iWorx Saved Files folder. Give your file a name and designate the file type as .iwxdata, then click **Save**.

  Or

  2. click on the **Save File** icon in the iWorx toolbar, then save your data; it should be saved in the **iWorx Saved Files** folder on the **Desktop**. You may create your own folder within the iWorx Saved Files folder. Give your file a name and designate the file type as .iwxdata, then click **Save**.

## Data Analysis Instructions

Figure 10, a diagram of icons in the iWorx toolbar may be helpful.

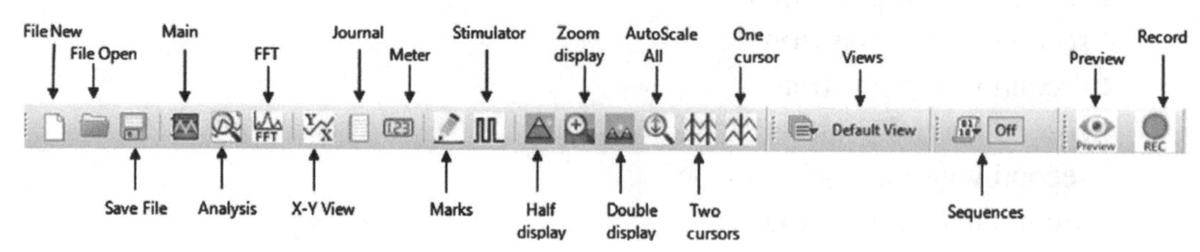


Figure 10 iWorx toolbar icons

- Go to the beginning of your recordings. Click on the **Marks** icon to bring up the Marks Dialogue box with a table of the marks created in your lab. Once you have the table of marks, click on the number at the beginning of the row for the mark that identifies the beginning of the data segment you are going to analyze. This will highlight that mark. Next, click on the **Go To Mark** button near the bottom of the Marks Dialogue box. Data at that mark will be displayed on your screen.
- Use the **Half Display** icon or the **Double Display** icon on the iWorx toolbar to adjust the data visible in the Main Window so that all three winning and losing cycles are visible on the screen. You can also adjust the display by clicking on the **Two Cursor** icon and moving one cursor to the beginning of the first contraction and moving the second cursor to the end of the final contraction. Then click the **Zoom Display** icon.
- Click on the **Analysis** icon in the iWorx toolbar.
- In the Analysis window, a "Function Table" will be displayed above the uppermost channel (Pronator teres). The mathematical functions "Abs. Int." (absolute integral) and "T2 – T1" should be present. The values for each of these functions are displayed in the in the top margin of each channel (see the following figure). The "Abs. Int." function measures the relative amount of electrical activity in the muscle fibers in both the Pronator teres channel and the Biceps brachii channel. "T2 – T1" is the duration of the EMG burst measured in both the Pronator teres channel and the Biceps brachii channel.
- Be sure that two cursors are present on your screen. If necessary, click the **Two Cursors** icon on the iWorx toolbar.
- For analysis of the winning event, place one cursor at the beginning of and one cursor at the end of the EMG burst of the first winning recording (see Figure 11). Record the values for "Abs. Int." for both the pronator teres and the biceps brachii in Table 2. Repeat the procedure for the first losing event. Repeat for the remaining 2 cycles. Calculate the mean for each of the variables measured.
- IMPORTANT: when analyzing your data, be sure you are making the correct comparisons. Do not compare the pronator teres to the bicep brachii. You should compare the pronator teres in the winning position versus the pronator teres in the losing position. You should compare the biceps brachii in the winning position versus the biceps brachii in the losing position. This allows you to compare the activity of the specific muscle between each condition (winning or losing).

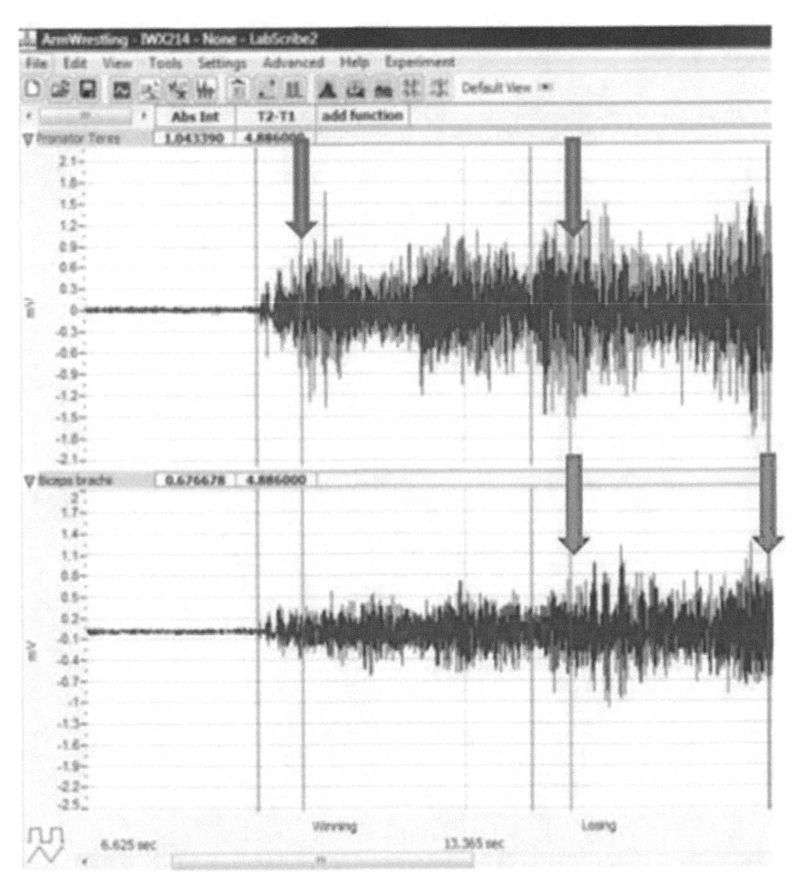


Figure 11 Data analysis arm wrestling

**TABLE 2**: Arm Wrestling

| Arm Action | | Pronator teres | Biceps brachii |
|---|---|---|---|
| | | Absolute Integral (mV) | Absolute Integral (mV) |
| Winning | Cycle 1 | 0.240 | 0.177 |
| | Cycle 2 | 0.206 | 0.198 |
| | Cycle 3 | 0.191 | 0.185 |
| | Mean | 0.212 | 0.187 |
| Losing | Cycle 1 | 0.138 | 0.049 |
| | Cycle 2 | 0.112 | 0.074 |
| | Cycle 3 | 0.094 | 0.060 |
| | Mean | 0.115 | 0.061 |

# Exercise 14

# iWorx Electrocardiogram—ECG and Pulse

## Objectives

- Be able to define the process of *electrocardiography* and an *electrocardiogram* (ECG).
- Be able to define the term *frequency* as it applies to a waveform.
- Understand the basic components of the cardiac cycle: *diastole* and *systole* of both the atria and the ventricles.
- Understand the terms *myogenic* and *autorhythmic* (or *autorhythmicity*) and the implications for the initiation of a cardiac cycle.
- Define the terms *normal sinus rhythm, tachycardia, bradycardia,* and *sinus arrhythmia.*
- Understand the conduction system (or pacemaker system) of the heart. Be able to identify the location of each of the components as well as the function of each of the components.
- Be able to identify each of the *standard bipolar leads* and the proper placement of electrodes for each: *lead I, lead II, and lead III.*
- Know the standard scale for recording an ECG: 1 mm = 0.1 mV (amplitude or vertical scale) and 1 mm = 0.04 seconds (time or horizontal scale). Knowing these scales, you should be able to determine heart rate when given the distance or time for one full cardiac cycle.
- Know what each of the following components of an ECG represents with regard to electrical and mechanical activity of the cardiac cycle: isoelectric line, P-wave, P-R interval, P-R segment, QRS complex, S-T segment, T-wave, and Q-T interval.
- Be able to associate a peripheral pressure wave with an ECG cycle.
- Be able to explain how a pressure wave relates to a peripheral pulse.
- Define the terms *apical pulse, peripheral pulse*, and *pulse deficit.*

## Lab Safety Reminders

- Use caution with any electrical equipment.

## Hints for Success

All members of your lab group should feel comfortable operating the iWorx system. **Never unplug any sensors from the iWorx box!** If you unplug a sensor during an experiment, you will lose all data. You will need to turn the iWorx box off, plug the sensor in, and start the lab from the beginning.

On the iWorx recording screens, it is helpful to remember that the blue bars are for data that are being recorded. The green bars are for data that is being calculated based upon the data you have collected.

Recall that an EMG measures voltage changes associated with muscle stimulation. If a subject moves skeletal muscles in the area of the electrodes or even in the body in general, the electrical changes from the muscle activity will be picked up in the ECG. It is important that the subject remains as motionless as possible during the periods of data recording. The use of the electrode gel increases the ability to detect the electrical signals generated by the subject's heart. Only a small amount of the gel is needed in the center of the electrode. A large amount will interfere with the ability of the electrode to stick to the subject's skin. After clipping the appropriate leads to the appropriate electrodes, be sure the lead cables are not pulling on the electrodes.

While using the iWorx, it is important that you do not get so caught up with obtaining the correct numbers from your data that you forget the function or process you are trying to observe. As you are using the iWorx, keep in mind the principles from the lecture being illustrated in the lab exercise.

## Introduction

An electrocardiogram (ECG or EKG) is a sum of all the electrical activity that occurs in one cardiac cycle (see Figure 1). In this lab, you will be using the iWorx system to learn to identify the components of an ECG and relate them to the electrical and resultant mechanical activity of the heart. Although a true diagnostic ECG would most likely use multiple precordial unipolar chest leads, the basic bipolar leads that we will be using are quite adequate for our experimentation. A diagnostic ECG can provide important information regarding the electrical activity of the heart and the functioning of the conduction system as well as the functioning of the myocardium itself.

The electrical activity of the heart is **myogenic**, meaning that it is initiated by the heart itself. The heartbeat is also **autorhythmic**, meaning that is sets its own rate. This rate can be modified by the autonomic system, but the autonomic system cannot initiate a heartbeat. The electrical activity of the heart is initiated by the SA (sinoatrial) node, also known as the pacemaker. In **normal sinus rhythm** (NSR), the rhythm is regular and originates from the SA node. In adults, the rate set by the SA node is generally 60–100 bpm. A sinus rhythm above 100 bpm is considered **tachycardia**. A sinus rhythm below 60 bpm is considered **bradycardia**. Be aware that some variation in sinus rhythm due to age, sex, activity, and cardiac fitness levels is considered normal. **Sinus arrhythmia** occurs when there is a variation in the P-P interval of 120 milliseconds or more in a person with normal P waves, or a change of at least 10% between the shortest and longest P-P intervals. The action potential generated by the SA node spreads through the atria to the atrial cardiac muscle cells, resulting in atrial depolarization (0.08–0.1 sec), the **P-wave** in an ECG tracing (see Figure 1). This electrical activity leads to the mechanical activity of atrial **systole** or contraction. It also leads to the stimulation of the AV (atrioventricular) node. The action potential generated in the AV node is conducted through the bundle of His, the right and left bundle branches, and the conduction or Purkinje fibers to the ventricular cardiac muscle cells resulting in ventricular depolarization (0.06–0.1 sec), the **QRS complex** in an ECG tracing (see Figure 1). This electrical activity leads to the mechanical activity of ventricular **systole** or contraction. **Diastole** or relaxation of the cardiac muscle occurs as a result of electrical repolarization. The atria repolarize during the QRS complex, resulting in atrial **diastole**. The ventricles repolarize during the **T-wave** (see Figure 1), resulting in ventricular diastole (variable duration).

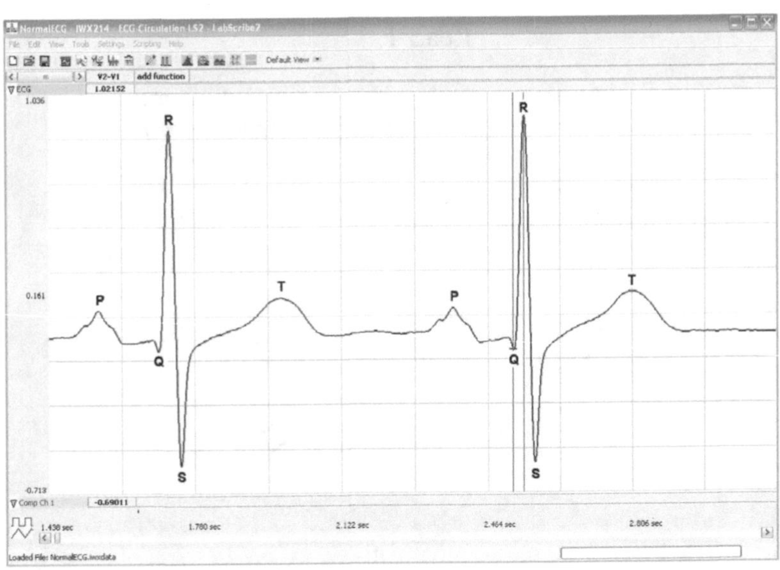

Figure 1 ECG tracing

In an ECG, a lead is used to measure the total electrical activity of the heart relative to the direction of the lead (see Figure 2). The three bipolar leads are:

Lead I: right arm (–) to left arm (+)
Lead II: right arm (–) to left leg (+)
Lead III: left arm (–) to left leg (+)

Because lead II most closely follows the normal electrical axis of the heart, this is the lead that is often used to assess the electrical activity of the heart. The following is a generalized explanation of the direction of the segments of the waveforms seen in an ECG when using lead II. When the electrical activity of the heart has the same polarity as the lead, lead II flows from right arm (–) to left leg (+), the amplitude of the wave is positive. When the electrical activity of the heart has the opposite polarity as the lead, the wave is negative. The more the electrical polarity of the heart parallels or opposes the electrical direction of the lead, the more positive or negative, respectively, the waveform will be. With this in mind, which parts of an ECG have electrical current flowing in parallel with lead II (see Figure 1)? Which parts of an ECG have electrical current flowing opposite lead II (see Figure 1)?

The following is a list of the major components of the ECG:

Isoelectric line: no electrical activity is being generated in the heart
P-wave: depolarization of the atria as a result of the activation of the SA node
QRS complex or interval: depolarization of the ventricles as a result of the activation of the AV node (the atria also repolarize during this period, but this is not visible in an ECG)
T-wave: repolarization of the ventricles
Note that ventricular repolarization is slower than ventricular depolarization.
U wave: small positive wave that sometimes follows the T-wave, final remnants of ventricular repolarization

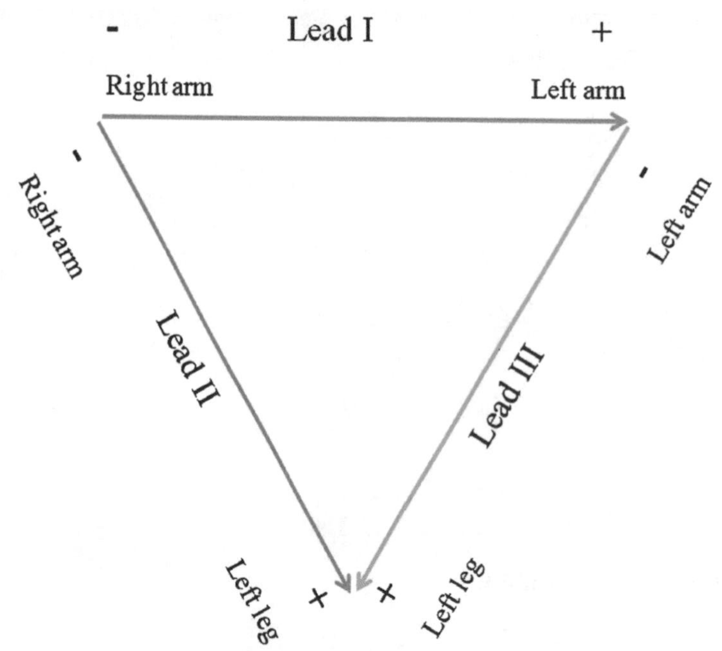


Figure 2 Bipolar limb leads

In addition to the major components of the ECG, there are other components that are important diagnostically. The duration of these components is used to assess the conduction of electricity through the heart.

Additional diagnostic components of the ECG (see Figure 3):

- P-R interval: beginning of P-wave to the beginning of QRS complex (0.12–0.2 sec)
  - It is the interval between activation of the SA node and activation of the AV node.
  - It is used to assess the conduction of electrical activity to the ventricles; a prolonged P-R interval could indicate interference of conduction.
- P-R segment: end of P-wave to the beginning of QRS complex
  - It is the interval between atrial depolarization and ventricular depolarization.
  - Similar to the P-R interval, a prolonged P-R segment could indicate interference in conduction of electrical activity to the ventricles.
- S-T segment: end of S wave to the beginning of the T-wave
  - It should be at the isoelectric line because the ventricles are uniformly excited and there is no change in charge. It corresponds roughly to the plateau phase of the ventricular muscle cell action potential.
  - An elevated or depressed ST segment may indicate ventricular ischemia or hypoxia.
- Q-T interval: beginning of the QRS complex to the end of the T-wave (0.2–0.4 sec)
  - It is the period of time it takes for the ventricles to depolarize and repolarize, and it corresponds roughly to the average ventricular action potential.

You will also be measuring the peripheral pulse in this lab. The signal recorded in the Pulse channel is the rate of change of blood pressure entering the fingertip of the subject. This change in pressure is also referred to as a pressure wave. This change in pressure or pressure wave is what is felt when

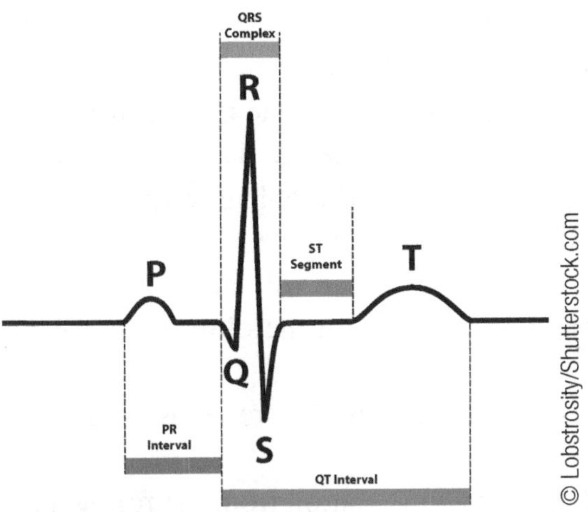


Figure 3 ECG components

you take a person's pulse in the wrist, ankle, neck, or anywhere one can be felt in a relatively superficial artery. It should be noted that the flow of blood is slower than the transmission of the pressure wave through the arteries. As a result, you are not feeling the movement of blood pumped out of the heart in a peripheral pulse wave but rather the reverberation of the artery as a result of the ventricular contraction. The correct term for a pulse that is counted in the peripheral arteries of a person is **peripheral pulse**. The **apical pulse** is the number of ventricular contractions counted. When you are counting the apical and peripheral pulses simultaneously, you would expect the counts for the apical and peripheral pulses to be the same; there should be one pressure wave for each ventricular contraction. However, there are instances when a person exhibits a **pulse deficit**, meaning that not all ventricular contractions produce a pressure wave that can be counted. In a pulse deficit, the peripheral pulse counted will be lower than the apical pulse that was counted. Several factors can influence the strength of a pressure wave and thus peripheral pulse. These include strength of ventricular contraction, body position, body temperature, metabolic activity of the surrounding tissue, and artery health. Abnormalities with the electrical conduction and the cardiac cycle such as atrial fibrillation, fast ventricular tachycardia (VT), ventricular fibrillation or very premature ectopic beats such as premature ventricular complexes (PVC or VPC) can lead to a pulse deficit. Shortened cardiac cycles with little time for ventricular filling can also lead to a pulse deficit and in some cases no pulse at all. When no pulse results from electrical activity in the heart, the abnormality is called pulseless electrical activity (PEA). PEA could result from a number of causes, including hypovolemia due to blood loss, excessive use of diuretics, or severe dehydration, potassium imbalance of either hyper- or hypokalemia, thrombosis, trauma, and others. A small pulse deficit is not considered a medical problem.

## Starting the iWorx Software

1. Be sure that the iWorx box is turned on. A small green light will be lit when the power is on. If you need to turn the power on, the on/off switch is on the back of the iWorx box.
2. Click on the **LabScribe3 shortcut** on the computer's desktop to open the program.
   You should see an information box that says "Hardware Found." Click **OK**; this will put you on the Main Window page. If the hardware is not found, check to be sure the iWorx box is turned on.

3. On the Main Window, pull down the **Settings** menu. Scroll to **Human Heart** then select the **ECG-Circulation** settings file from the "Human Heart" list.

   Note: If "Human Heart" is empty, then complete the following instructions. On the Main Window, pull down the "Settings" menu and select "Load Group." Scroll down to locate the folder that contains the settings group, "IPLMv6Complete.iwxgrp." Select this group and click "Open." At this point, you should be able to then complete step 3.
4. Instructions for the ECG-Circulation Setup will automatically open. These instructions are not necessary. Close this page by clicking on the close button in the upper right corner of this document.
5. LabScribe will appear on the computer screen as configured by the ECG-Circulation settings file.
6. For this lab, you do not need any information from the following files. But as a reminder, once the settings file has been loaded, clicking the **Experiment** button on the toolbar will allow you to access any of the following documents:
   - Introduction—introductory material for iWorx Human Heart lab exercises
   - Appendix—may have additional information and resources
   - Background—may have some background information regarding the lab topic
   - Labs—the lab exercise as written by iWorx
   - Setup (opens automatically)—instructions for setting up this particular iWorx lab. You already closed this out.
7. If you choose to save your data, it should be saved in the **iWorx Saved Files** folder on the **Desktop**. You may create your own folder within the iWorx Saved Files folder.

## ECG Cable and Pulse Transducer Setup

The ECG cables and pulse plethysmograph have already been connected to the iWorx box (see Figure 4). Do not unplug any of these cables. Doing so will cause you to lose your data.

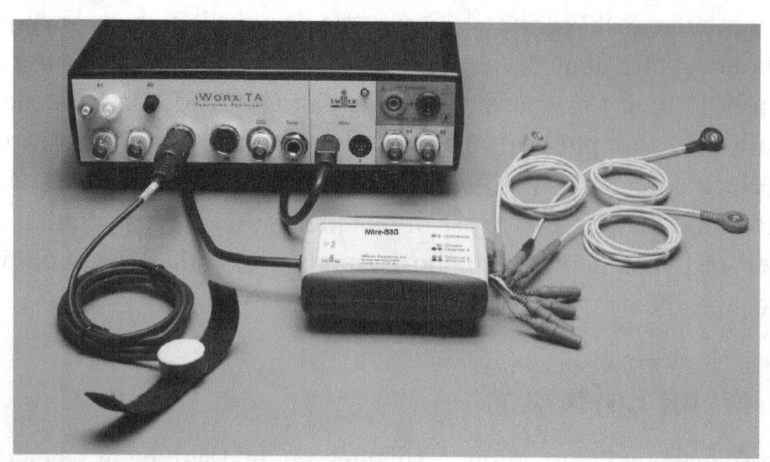


Figure 4 Equipment

For adequate electrical signals, the subject should remove all jewelry from the wrists and ankles. Electrodes will be placed on the anterior surfaces of both forearms, just proximal to the wrists. Use an alcohol swab to clean the anterior surface of the subject's right wrist. Use additional alcohol swabs to clean the medial surfaces of the left and right lower legs, just proximal to the medial malleolus. Let these areas dry completely. Obtain three disposable electrodes and a squeeze bottle of electrode gel. After removing the plastic cover from an electrode, apply one **very small** drop of electrode gel to the center of the sticky side of the electrode. Apply one prepared disposable electrode to each of the areas previously cleaned with the alcohol swab. Snap the lead wires onto the electrodes in the following manner:

- Red lead (+1) is attached to the electrode just proximal to the medial malleolus of the left leg
- Black lead (−1) is attached to the electrode on the anterior surface of the forearm just proximal to the right wrist
- Green lead (G – ground) is attached just proximal to the medial malleolus of the right leg
  Note: The bipolar limb leads use two electrodes for comparison of electrical flow in the heart. The third electrode is the ground. It is not the third component of Einthoven's triangle. For the bipolar limb lead I, the left wrist is the location of the positive (+) electrode. For the bipolar limb leads II and III, the left leg is the location of the positive (+) electrode. The right leg is the location of the ground for all three bipolar limb leads.

Place the sensor surface of the plethysmograph on the volar (fingerprint) surface of the distal phalange of the subject's middle finger or thumb. Wrap the Velcro™ strap around the end of the finger to hold the sensor firmly in place, but not so tight as to restrict blood flow. **Do not squeeze the plethysmograph**.

The subject should sit quietly and relaxed with both hands in his/her lap, palms facing upward. Any extraneous movement by the subject may cause the ECG to shift off the top or bottom of the screen. The subject will also activate muscle activity and the electrical activity of the muscles will be recorded with the ECG. The ECG tracing will contain electrical noise from the muscle activity making the tracing less distinct and difficult to interpret.

## Lab Exercises

- **Exercise 1: ECG and Pulse in the Resting Subject**
  - Read through all instructions before beginning the testing. You will perform a short test run to be sure you are getting a proper recording. Once your recording looks appropriate, you will begin the actual recording
  - To label the actual data for exercise 1, move the cursor to the "Mark" text box to the right of the "Mark" button and click to activate the text box. Type the subject's name followed by "resting ECG" in the "Mark" text box. Insert the mark when you begin the recording period.
  - **Subject Instructions**
    - The subject should sit quietly with both hands in his/her lap, **palms facing upward**. There should be no extraneous muscle activity.
  - **Tester Instructions**
    - You will conduct a short test run to be sure your data is being collected and recorded properly.
    - When you are ready to begin your test run, click the "Record" button in the upper right corner of your screen. Instruct the subject that the testing has begun and to remain still. After 10 to 20 seconds, click "Stop."

- Click the "Autoscale" buttons for the "ECG ," "Pulse," and the "Pulse Integral" channels. The data recording should look similar to Figure 5.

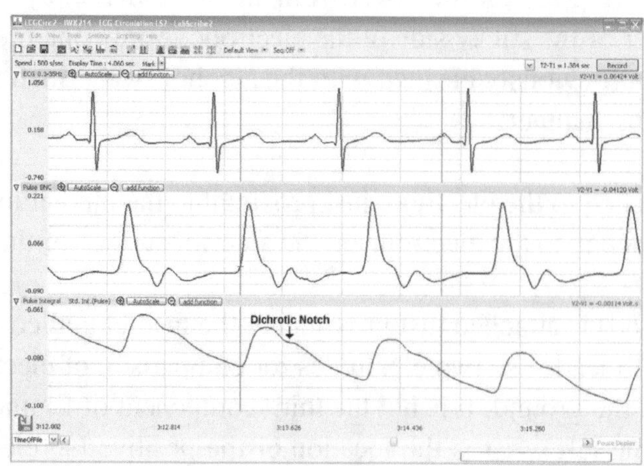


Figure 5 Sample data

  - If the signal on either the "ECG" or the "Pulse" channel is upside down when compared to the figure, click the downward arrow to the left of the channel title and select the "Invert" function.
  - If the pulse signal is small or noisy, adjust the tension on the plethysmograph strap. Be sure the subject has the hands placed **palms up** in the lap.
  - When your tracing looks appropriate, begin your actual data collection. Click the "Record" button in the upper right corner of your screen. Insert your mark when you begin the recording. Instruct the subject that the testing has begun and to remain still. Continue to record for 30 second to 1 minute. Click "Stop." You may record longer if your tracings do not look normal or you do not have 6 to 7 good sequential cardiac cycles.
- You may choose to save your data at this point or you may continue on with the experiment or data analysis. To save your data, on the Main Window, either
  1. click on the **File** menu. Scroll to **Save As**, then save your data, it should be saved in the **iWorx Saved Files** folder on the **Desktop**. You may create your own folder within the iWorx Saved Files folder. Give your file a name and designate the file type as .iwxdata then click **Save**.

     Or
  2. click on the **Save File** icon in the iWorx toolbar, then save your data, it should be saved in the **iWorx Saved Files** folder on the **Desktop**. You may create your own folder within the iWorx Saved Files folder. Give your file a name and designate the file type as .iwxdata, then click **Save**.
- **Data Analysis Instructions**

  Figure 6, a diagram of icons in the iWorx toolbar may be helpful.

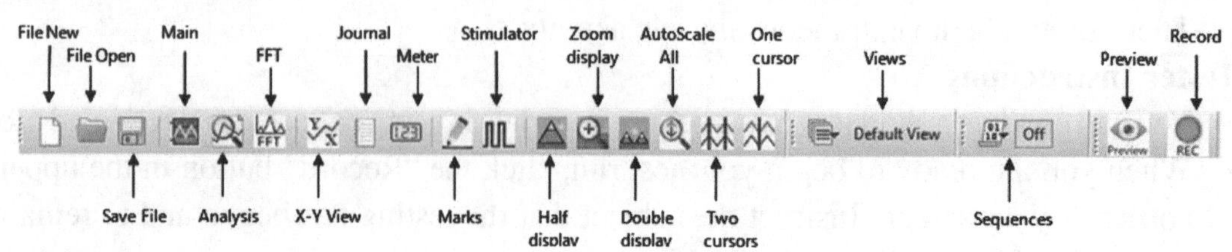


Figure 6 iWorx toolbar icons

- Click on the **Analysis** icon in the iWorx toolbar.
- Go to the beginning of your recordings. Click on the **Marks** icon to bring up the Marks Dialogue box with a table of the marks created in your lab. Once you have the table of marks, click on the number at the beginning of the row for the mark that identifies the beginning of the experimental data segment you are going to analyze. This will highlight that mark. Next, click on the **Go To Mark** button near the bottom of the Marks Dialogue box. Data at that mark will be displayed on your screen.
- Be sure that two cursors are present on your screen. If necessary, click the **Two Cursors** icon on the iWorx toolbar.
- Scroll through your experimental tracing to find a section of data with a succession of 5 to 6 very good ECG/Pulse cycles.
- Place the cursors on either side of the 5 to 6 ECG/Pulse cycles, then click the **Double Display** icon in the iWorx toolbar. This will expand the data section to the width of the Main Window. Adjust your data section as necessary with the display time icons or the zoom icons (see Figure 7).
- Click on the **Analysis** icon in the iWorx toolbar.
- In the Analysis window, a "Function Table" will be displayed above the uppermost channel (ECG channel). The mathematical functions "V2 – V1," and "T2 – T1" should be present. The values for each of these functions are displayed in the top margin of each channel (see the following figure). Only the values from the "T2 – T1" function will be used. The "T2 – T1" function measures the difference in time between cursor 2 and cursor 1.

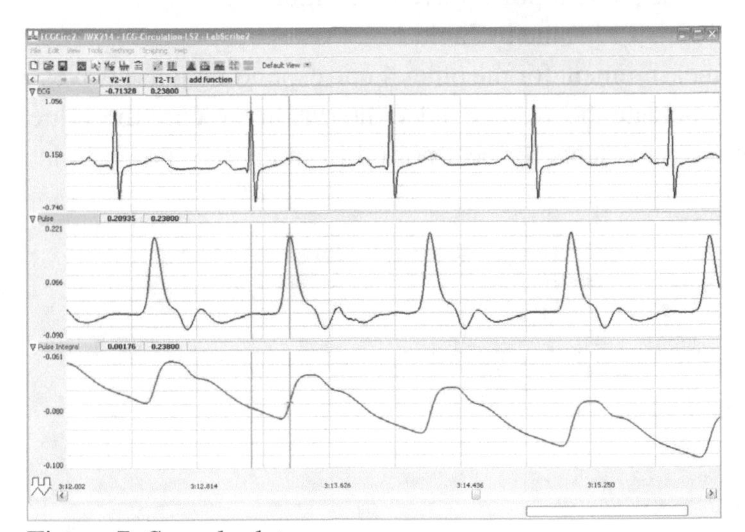


Figure 7 Sample data

- For the following measurements, **you must use the corresponding ECG and pulse cycles.** The trial numbers should correspond among all five analyses.
- Apical pulse or beat period analysis.
  - For analysis of the beat period (time per heart beat), in the **ECG channel**, place one cursor at the peak of an R-wave and the second cursor at the peak of the sequential R-wave (see Figure 8).
  - Record the "T2 – T1" value in Table 1.
  - Repeat this measurement for 4 more successive R-waves and record the values in Table 1. Calculate the average beat period and record the value in Table 1.

- Obtain the measurements for beat period from another group in lab and record these in Table 1.

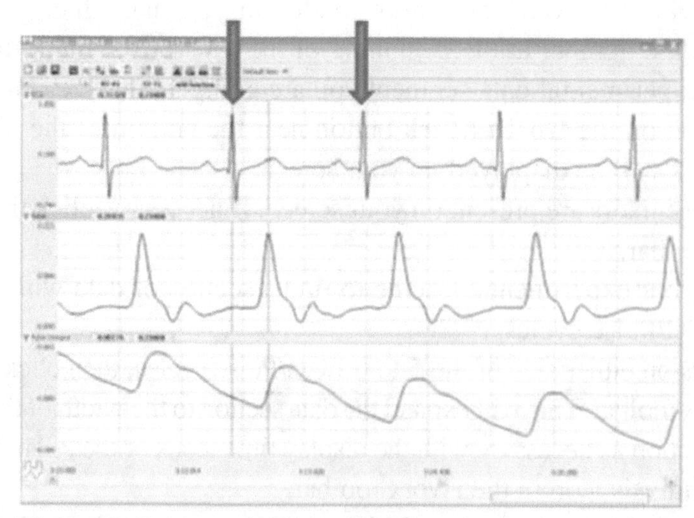

Figure 8 Data analysis beat period

- Peripheral pulse interval analysis.
  - For analysis of the peripheral pulse interval or time between peripheral pulse waves, in the **Pulse channel** place one cursor at the peak of a pulse wave and the second cursor at the peak of the sequential pulse wave (see Figure 9).
  - Record the "T2 – T1" value in Table 1.
  - Repeat this measurement for the other 4 corresponding pulse waves and record the values in Table 1. Calculate the average pulse interval and record the value in Table 1.
  - Obtain the measurements for beat period from another group in lab and record these in Table 1.

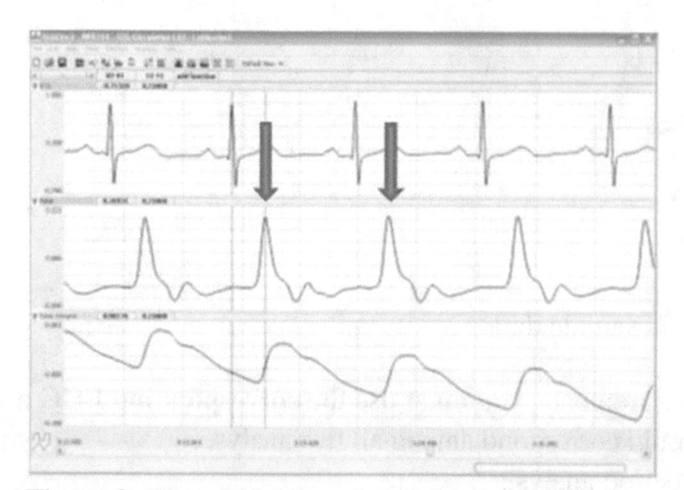

Figure 9 Data analysis peripheral pulse interval

- R-wave—peripheral pulse interval analysis.
  - For analysis of the R-wave—peripheral pulse interval or time between ventricular depolarization and the pulse pressure wave, in the **ECG channel**, place one cursor at the peak of an R-wave and in the **Pulse channel**; place the second cursor at the peak of the associated pulse wave that just follows the R-wave (see Figure 10).

- Record the "T2 – T1" value in Table 1.
- Repeat this measurement for the other 4 corresponding R-wave—peripheral pulse waves and record the values in Table 1. Calculate the average R-wave—pulse interval and record the value in Table 1.
- Obtain the measurements for beat period from another group in lab and record these in Table 1.

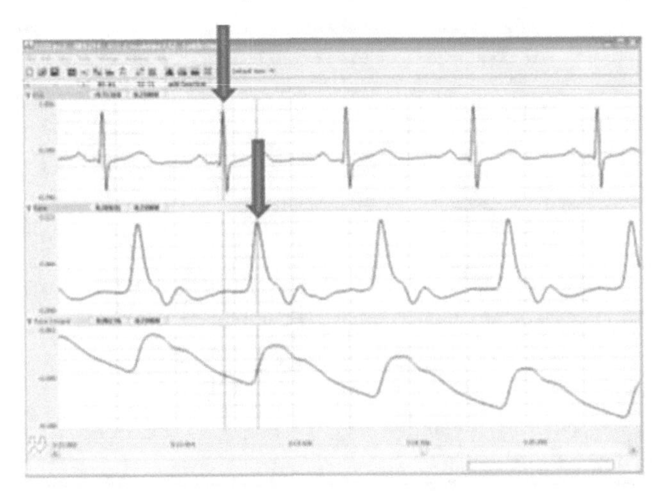

Figure 10 Data analysis R-wave peripheral pulse interval

**TABLE 1:** Beat Periods, Pulse Intervals, and R-Pulse Intervals for Two Subjects

| | | **Beat Period: time between two adjacent R-waves** | **Pulse–Pulse Interval: time between the peak of a pulse wave and the peak of the following pulse wave** | **R-Pulse Interval: time between the peak of an R-wave and the peak of the following pulse wave** |
|---|---|---|---|---|
| | | **T2 – T1 (sec/beat)** | **T2 – T1 (sec/pulse)** | **T2 – T1 (sec)** |
| Subject 1 | Trial 1 | | | |
| | Trial 2 | | | |
| | Trial 3 | | | |
| | Trial 4 | | | |
| | Trial 5 | | | |
| | Average | | | |
| Subject 2 | Trial 1 | | | |
| | Trial 2 | | | |
| | Trial 3 | | | |
| | Trial 4 | | | |
| | Trial 5 | | | |
| | Average | | | |

# Exercise 15

# iWorx Electrocardiogram and Heart Sounds

## Objectives

- Be able to define and describe *unidirectional flow* in the circulatory system.
- Identify the valves associated with the heart, their locations, and their functions.
- Be able to describe the path of blood flow in pulmonary circulation.
- Be able to describe the path of blood flow in systemic circulation.
- Be able to define a *cardiac cycle*.
- Be able to identify each of the four sounds, $S_1$, $S_2$, $S_3$, and $S_4$ associated with the cardiac cycle and explain the action within the heart that causes each of the sounds.
- Be able to define the terms *diastole*, *systole*, *isovolumetric contraction*, *ejection*, and *isovolumetric relaxation*, *passive filling*, and *final filling* in reference to actions in the heart, including the actions of the valves.
- Be able to identify the locations on the chest for listening to the individual valve sounds.
- Be able to explain the difference between a *stenotic valve* and an *insufficient valve*.
- Be able to explain the difference between a *diastolic murmur* and a *systolic murmur* and the valves involved in each.
- Be able to identify the following electrical activities of the heart: atrial depolarization and P-wave, ventricular depolarization and QRS complex, and ventricular repolarization and T-wave.
- Be able to identify the following mechanical activities of the heart: atrial systole and diastole; ventricular systole and diastole; opening and closing of atrioventricular and semilunar valves; first ($S_1$), second ($S_2$), third ($S_3$), and fourth ($S_4$) heart sounds, atrial and ventricular filling and emptying, and increase and decrease in pressure in the atria, ventricles, and aorta.
- Be able to relate the electrical activities of the heart to the mechanical activities of the heart.

## Lab Safety Reminders

- Use caution with any electrical equipment.

## Hints for Success

All members of your lab group should feel comfortable operating the iWorx system.

**Never unplug any sensors from the iWorx box!** If you unplug a sensor during an experiment, you will lose all data. You will need to turn the iWorx box off, plug the sensor in and start the lab from the beginning.

On the iWorx recording screens, it is helpful to remember that the blue bars are for data that are being recorded. The green bars are for data that is being calculated based upon the data you have collected.

Recall that an EMG measures voltage changes associated with muscle stimulation. If a subject moves skeletal muscles in the area of the electrodes or even in the body in general, the electrical changes from the muscle activity will be picked up in the ECG. It is important that the subject remains as motionless as possible during the periods of data recording. The use of the electrode gel increases the ability to detect the electrical signals generated by the subject's heart. Only a small amount of the gel is needed in the center of the electrode. A large amount will interfere with the ability of the electrode to stick to the subject's skin. After clipping the appropriate leads to the appropriate electrodes, be sure the lead cables are not pulling on the electrodes.

While using the iWorx, it is important that you do not get so caught up with obtaining the correct numbers from your data that you forget the function or process you are trying to observe. As you are using the iWorx, keep in mind the principles from the lecture being illustrated in the lab exercise.

## Introduction

In this lab, you will be listening for the two heart sounds, $S_1$ and $S_2$, associated with unidirectional blood flow associated with the closure of the heart valves. There are two more heart sounds, $S_3$ and $S_4$, which are the result of turbulent blood flow into the ventricles. These sounds are usually not audible with a stethoscope. You will be using the iWorx system to record an ECG and listening for the two most audible heart sounds resulting from the closing of the atrioventricular and the semilunar valves.

The sequence of all the electrical and mechanical events associated with a heartbeat constitutes a **cardiac cycle**. The electrical events are the electrical changes you have already observed in an ECG. The mechanical events include the contraction of the heart chambers, the closure and opening of the heart valves as a result of pressure changes in the heart chambers and the great arteries, and the movement of blood into, through, and out of the heart. It is important to remember that in the cardiac cycle, the electrical stimulation and activity of the heart is what drives the mechanical activity of the heart. Without the electrical stimulation of the nodal tissue and the conduction system of the heart, the heart would not contract. Without the contracting action of the heart, blood would not move, pressure in the chambers and great arteries would not change, and valves would not function. The mechanical events that occur during a cardiac cycle are all closely tied together. For example, the atrioventricular valves will not close until the pressure in the ventricles increases and becomes greater than the pressure in the atria; the semilunar valves will not open until the pressure in the ventricles becomes greater than the pressure in the great arteries; the semilunar valves will not close until the pressure in the ventricles is reduced to a pressure less than the pressure in the great arteries; and the atrioventricular valves will not open until the pressure in the ventricles is reduced to a pressure less than the pressure in the atria. Therefore, blood will not move through the heart unless the pressure gradients are appropriate and the valves are operating properly. The pressure within a chamber of the heart is directly related to the volume of blood in the chamber and the force of contraction being exerted on the blood by the chamber.

When listening to the heart sounds resulting from the closing of the valves, the closing of the atrioventricular valves (AV) is the first heart sound, $S_1$ (lub). $S_1$ is heard very shortly after the beginning of ventricular systole. The electrical activity of the QRS complex (see Figure 1) results in the mechanical activity of ventricular contraction. Isovolumetric contraction occurs during the initial portion of ventricular systole. During *isovolumetric contraction*, pressure builds up in the ventricles. This results in the closure of the atrioventricular valves but not enough pressure to eject blood. As isovolumetric contraction continues, pressure continues to build in the ventricles. When the pressure in the ventricles exceeds the pressure in the great arteries, the ventricles enter the *ejection* phase of ventricular systole and blood is ejected from the ventricles into the great arteries. The closing of the semilunar valves (SL) is the second sound, $S_2$ (dub). $S_2$ is heard very shortly after the beginning of ventricular diastole. The electrical activity of the T-wave (see the following figure) results in the mechanical activity of ventricular relaxation. *Isovolumetric relaxation* occurs during the initial portion of ventricular diastole. During isovolumetric relaxation, the pressure in the ventricles begins to drop. In addition to the sounds associated with the closure of valves, $S_1$ and $S_2$, there are also third and fourth heart sounds, both the result of the heart filling with blood. The third sound, $S_3$, is the result of passive filling of the ventricles with blood flow from the vena cavas and atria into the ventricles. Passive filling occurs immediately after the pressure in the ventricles drops below the pressure in the atria. As a result of the decreased ventricular pressure, atrioventricular valves open. This occurs following $S_2$ but prior to the P-wave (see Figure 1). The electrical activity of the P-wave (see Figure 1) results in the mechanical activity of atrial systole and the final filling of the ventricles, the source of the fourth heart sound, $S_4$. Of the four heart sounds, $S_1$, the first sound, is the loudest. The AV valves are larger than the SL valves, and they snap shut with a greater force due to the high pressure developed as a result of ventricular systole. The third and fourth sounds are difficult to detect with a nonelectrical stethoscope.

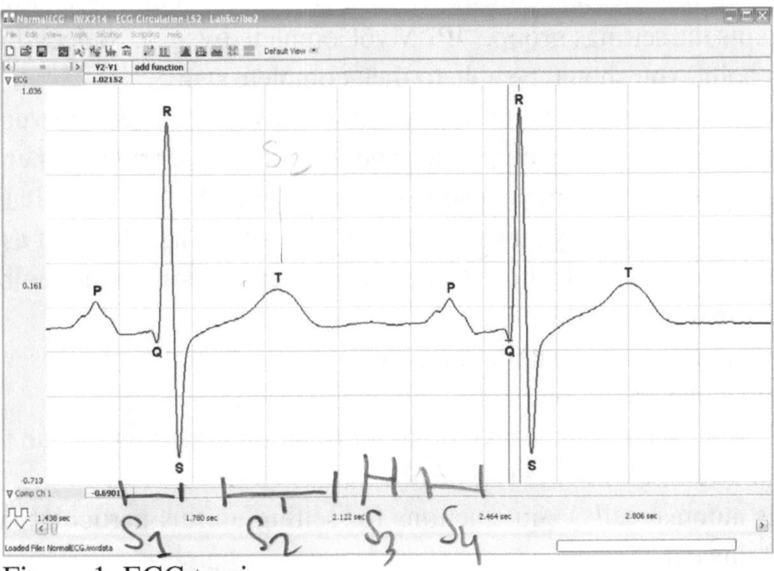


Figure 1 ECG tracing

The function of the heart valves is to cause blood to flow only in one direction through the heart, **unidirectional flow**. The direction of blood flow should be from the great veins into the atria, from atria through the atrioventricular valves to the ventricles, and from the ventricles through the semilunar valves to the great arteries. When valves do not work properly, they interrupt the laminar flow of blood. This can occur when there is a regurgitation or backflow of blood through an **insufficient valve** or a valve that does not close completely. A murmur can also occur when a valve is **stenotic** or has an opening that is reduced in size so that blood flow through the open valve is restricted. A valve that should be closed during systole but allows some backflow of blood or a valve that should be open during systole but has a reduced opening would cause a **systolic murmur**. A valve that should be closed during diastole but allows some backflow of blood or a valve that should be open during diastole but has a reduced opening would cause a **diastolic murmur**. We are more familiar with insufficient valve murmurs that would be due to an insufficient atrioventricular valve in a systolic murmur and an insufficient semilunar valve in a diastolic murmur. Auscultation, or listening to the heart sounds, often provides the first information regarding a potential problem with valve function and blood flow through the heart. Additional and more detailed information can be obtained through the use of echocardiograms and Doppler examinations. These ultrasonic tools provide much more detailed information for accurate diagnosis of the problem.

## Starting the iWorx Software

1. Be sure that the iWorx box is turned on. A small green light will be lit when the power is on. If you need to turn the power on, the on/off switch is on the back of the iWorx box.
2. Click on the **LabScribe3 shortcut** on the computer's desktop to open the program.
   You should see an information box that says "Hardware Found." Click **OK**; this will put you on the Main Window page. If the hardware is not found, check to be sure the iWorx box is turned on.
3. On the Main Window, pull down the **Settings** menu. Scroll to **Human Heart**, then select the **Auscultation** settings file from the "Human Heart" list.
   Note: If "Human Heart" is empty, then complete the following instructions. On the Main Window, pull down the "Settings" menu and select "Load Group." Scroll down to locate the folder that contains the settings group, "IPLMv6Complete.iwxgrp." Select this group and click "Open." At this point, you should be able to then complete step 3.
4. Instructions for the Auscultation Setup will automatically open. These instructions are not necessary. Close this page by clicking on the close button in the upper right corner of this document.
5. LabScribe will appear on the computer screen as configured by the Auscultation settings file.
6. For this lab, you do not need any information from the following files. But as a reminder, once the settings file has been loaded, clicking the **Experiment** button on the toolbar will allow you to access any of the following documents:
   - Introduction—introductory material for iWorx Human Heart lab exercises
   - Appendix—may have additional information and resources
   - Background—may have some background information regarding the lab topic
   - Labs—the lab exercise as written by iWorx
   - Setup (opens automatically)—instructions for setting up this particular iWorx lab. You already closed this out.
7. If you choose to save your data, it should be saved in the **iWorx Saved Files** folder on the **Desktop**. You may create your own folder within the iWorx Saved Files folder.

## ECG Cable and Heart Sounds Microphone Setup

The ECG cables and heart sounds microphone have already been connected to the iWorx box (see Figure 2). Do not unplug any of these cables. Doing so will cause you to lose your data.

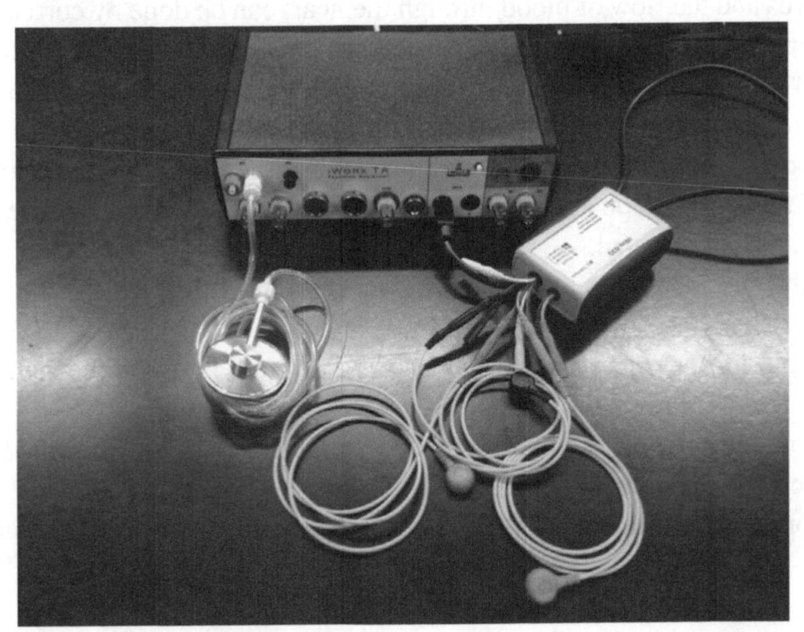

Figure 2 Equipment

For adequate electrical signals, the subject should remove all jewelry from the wrists and ankles. Electrodes will be placed on the anterior surfaces of both forearms, just proximal to the wrists. Use an alcohol swab to clean the anterior surface of the subject's right wrist. Use additional alcohol swabs to clean the medial surfaces of the left and right lower legs, just proximal to the medial malleolus. Let these areas dry completely. Obtain three disposable electrodes and a squeeze bottle of electrode gel. After removing the plastic cover from an electrode, apply one **very small** drop of electrode gel to the center of the sticky side of the electrode. Apply one prepared disposable electrode to each of the areas previously cleaned with the alcohol swab. Snap the lead wires onto the electrodes in the following manner:

- Red lead (+1) is attached to the electrode just proximal to the medial malleolus of the left leg
- Black lead (−1) is attached to the electrode on the anterior surface of the forearm just proximal to the right wrist
- Green lead (G—ground) is attached just proximal to the medial malleolus of the right leg

  Note: The bipolar limb leads use two electrodes for comparison of electrical flow in the heart. The third electrode is the ground. It is not the third component of Einthoven's triangle. For the bipolar limb lead I, the left wrist is the location of the positive (+) electrode. For the bipolar limb leads II and III, the left leg is the location of the positive (+) electrode. The right leg is the location of the ground for all three bipolar limb leads.

The subject should sit quietly with both hands in his or her lap, palms facing upward. Any extraneous movement by the subject may cause the ECG to shift off the top or bottom of the screen. The subject will also activate muscle activity and the electrical activity of the muscles will be recorded with the ECG. The ECG tracing will contain electrical noise from the muscle activity, making the tracing less distinct and difficult to interpret.

# Lab Exercises

## Exercise 1: Listening to the Heart Valves

- Auscultation or listening to the heart sounds associated with the movement of individual heart valves and the flow of blood through the heart can be done by correctly positioning the stethoscope so that the sound from a particular valve is heard. The auscultatory areas are not the actual location of the valves. The auscultation locations are points where the directional flow of blood can best be heard for each particular valve.
- For this portion of the exercise, you will be using a traditional stethoscope. Note that the ear pieces of the stethoscope are directional. Insert the ear pieces so that the sound is directed into the external ear canal.
- Identify the ausculation points on yourself and listen for the sound of each of the four heart valves:
  - Tricuspid (T) or right atrioventricular valve located between the right atrium and right ventricle—left lower sternal border
  - Bicuspid, mitral (M), or left atrioventricular valve located between the left atrium and left ventricle—fifth intercostal space at the apex of the heart
  - Pulmonic (P) semilunar valve located between the right ventricle and the pulmonary trunk—second intercostal space near the left sternal border
  - Aortic (A) semilunar valve located between the left ventricle and the aorta—second intercostal space near the right sternal border
  - Erb's point or the third intercostal space near the left sternal border is the optimal location for hearing $S_2$

## Exercise 2: ECG and Heart Sounds in the Resting Subject

- Read through all instructions before beginning the testing. You will perform a short test run to be sure you are getting a proper recording. Once your recording looks appropriate, you will begin the actual recording
- To label the actual data for exercise 1, move the cursor to the "Mark" text box to the right of the "Mark" button and click to activate the text box. Type the subject's name followed by "resting ECG" in the "Mark" text box. Insert the mark when you begin the recording period.
- **Subject Instructions**
  - The subject should sit quietly with both hands in his or her lap, palms facing upward. There should be no extraneous muscle activity.
- **Tester Instructions**
  - Begin by placing the head of the stethoscope in the location for auscultation of the tricuspid or the bicuspid valve, near the left sternal border near the fourth or fifth intercostal space, and listen for the heart sounds. Move the stethoscope in the general area until the heart sounds are heard clearly. Once you have located a good area for detecting heart sounds, put the heart sounds microphone in place. It is very important that the microphone head does not move across the subject's clothing or skin or you will have a large amount of background noise. The background noise will make hearing the heart sounds difficult. You may either hold the microphone in place or hold it in place by wrapping the subject with an elastic bandage.
  - You will conduct a short test run to be sure your data is being collected and recorded properly.

- When you are ready to begin your test run, click the "Record" button in the upper right corner of your screen. Instruct the subject that the testing has begun and to remain still. After 10 to 20 seconds, click "Stop."
- Click the "Autoscale" buttons for the "ECG" and the "Heart Sounds" channels. The ECG data recording should look similar to Figure 3.
  - If the signal on the "ECG" channel is upside down when compared to the figure, click the downward arrow to the left of the channel title and select the "Invert" function.

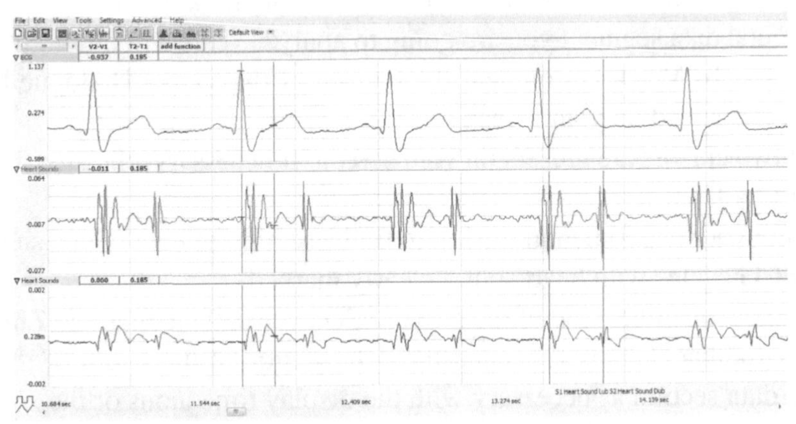

Figure 3 Sample data

  - Be sure the subject has the hands placed palm up in the lap.
  - If there is a large amount of noise in the tracing in the "Heart Sounds" channel, either hold the microphone head more tightly against the skin or wrap the subject more tightly with the elastic bandage.
  - Record another test run if necessary.
  - When the tracing looks appropriate, begin your actual data collection, following the same procedure used in the test run. Click the "Record" button in the upper right corner of your screen. Insert your mark when you begin the recording. Instruct the subject that the testing has begun and to remain still. Continue to record for 30 seconds to 1 minute. Click "Stop."
- You may choose to save your data at this point or you may continue on with the experiment or data analysis. To save your data, on the Main Window, either
  1. click on the **File** menu. Scroll to **Save As**, then save your data; it should be saved in the **iWorx Saved Files** folder on the **Desktop**. You may create your own folder within the iWorx Saved Files folder. Give your file a name and designate the file type as .iwxdata, then click **Save**.

  Or

  2. click on the **Save File** icon in the iWorx toolbar, then save your data; it should be saved in the **iWorx Saved Files** folder on the **Desktop**. You may create your own folder within the iWorx Saved Files folder. Give your file a name and designate the file type as .iwxdata, then click **Save**.
- **Data Analysis Instructions**

  Figure 4, a diagram of icons in the iWorx toolbar may be helpful.
- Click on the **Analysis** icon in the iWorx toolbar.
- Go to the beginning of your recordings. Click on the **Marks** icon to bring up the Marks Dialogue box with a table of the marks created in your lab. Once you have the table of marks,

Figure 4 iWorx toolbar icons

click on the number at the beginning of the row for the mark that identifies the beginning of the experimental data segment you are going to analyze. This will highlight that mark. Next, click on the **Go To Mark** button near the bottom of the Marks Dialogue box. Data at that mark will be displayed on your screen.

- Be sure that two cursors are present on your screen. If necessary, click the **Two Cursors** icon on the iWorx toolbar.
- Scroll through your experimental tracing to find a section of data with a succession of 5 to 6 very good ECG cycles with heart sound clearly marked.
- Place the cursors on either side of the 5 to 6 ECG cycles, then click the **Zoom Display** icon in the iWorx toolbar. This will expand the data section to the width of the Main Window. Adjust your data section as necessary with the display time icons or the zoom icons.
- Click on the **Analysis** icon in the iWorx toolbar.
- In the Analysis window, a "Function Table" will be displayed above the uppermost channel (ECG channel). The mathematical functions "V2 – V1," and "T2 – T1" should be present. The values for each of these functions are displayed in the top margin of each channel. Only the values from the "T2 – T1" function will be used. The "T2 – T1" function measures the difference in time between cursor 2 and cursor 1.
- For the following measurements, **you must use the same ECG cycle for each of the measurements within a cycle**. The trial numbers should correspond between all three analyses.
- Identification of the $S_1$ and $S_2$ sounds.
  - The beginning of $S_1$ closely follows the peak of the R-wave in a cardiac cycle. In the **Heart Sounds channel** (see Figure 5), the cursors are placed at the beginning and end of $S_1$ of the second ECG cycle. The $S_2$ sound of the second ECG is represented by the following deflection of the tracing in the **Heart Sounds channel**.
  - The markers are located in the middles of $S_1$ and $S_2$ of the fourth ECG cycle.
- Apical pulse or beat period analysis.

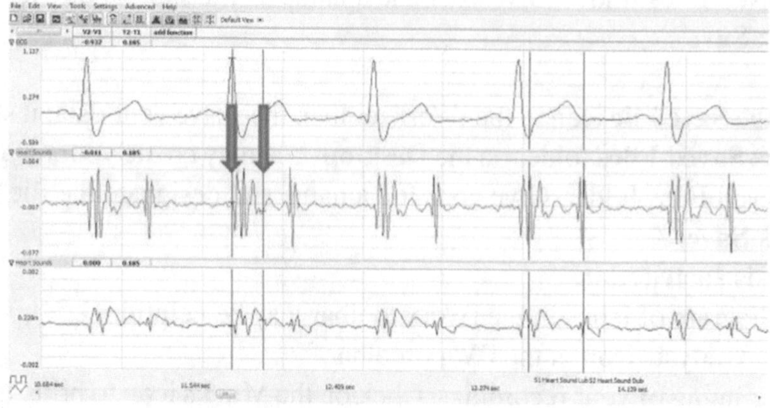

Figure 5 Data analysis S1 and S2

- For analysis of the beat period (time per heart beat), in the **ECG channel** place one cursor at the peak of an R-wave and the second cursor at the peak of the sequential R-wave (see Figure 6).
- Record the "T2 – T1" value in Table 1.
- Repeat this measurement for four more successive R-waves and record the values in Table 1. Calculate the average beat period and record the value in Table 1.

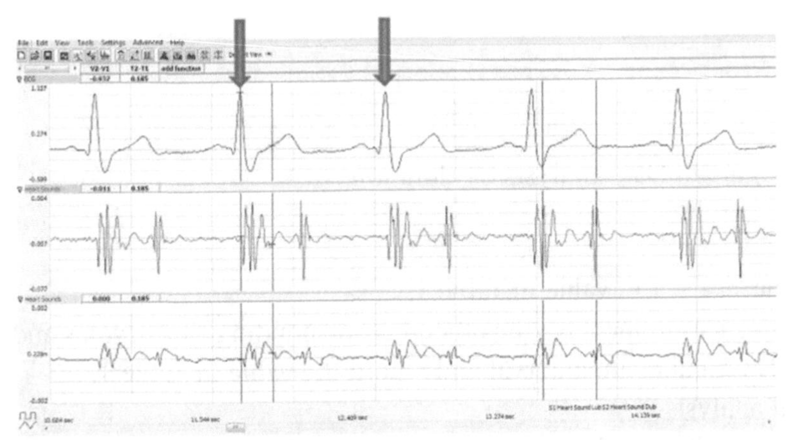

Figure 6 Data analysis beat period

- R-$S_1$ interval analysis.
  - Use the same ECG tracings used for the apical pulse analysis.
  - For analysis of the R-$S_1$ interval, in the **ECG channel** place one cursor at the peak of an R-wave and in the **Heart Sounds channel**, place the second cursor at the beginning of the $S_1$ (see Figure 7).
  - Record the "T2 – T1" value in Table 1.
  - Repeat this measurement for the other 4 corresponding ECG cycles and record the values in Table 1. Calculate the average and record the value in Table 1.

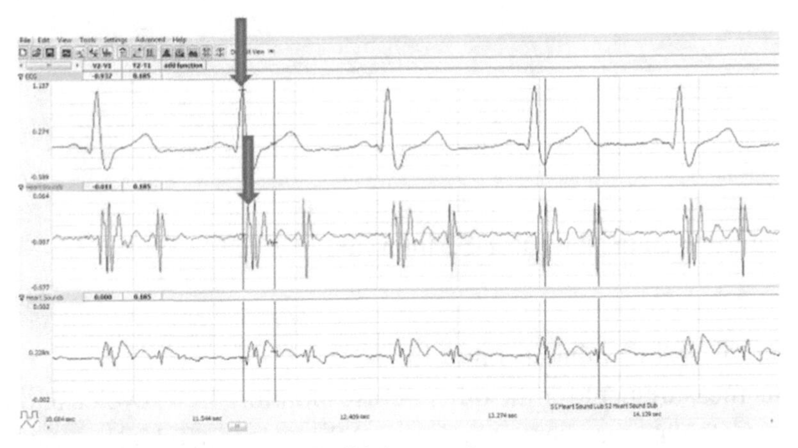

Figure 7 Data analysis R-S1 interval

- $S_1$ duration.
  - Use the same ECG tracings used for the apical pulse analysis.
  - For analysis of the $S_1$ duration, in the **Heart Sounds channel** place one cursor at the beginning of the $S_1$ and place the second cursor at the end of the $S_1$ (see Figure 8).

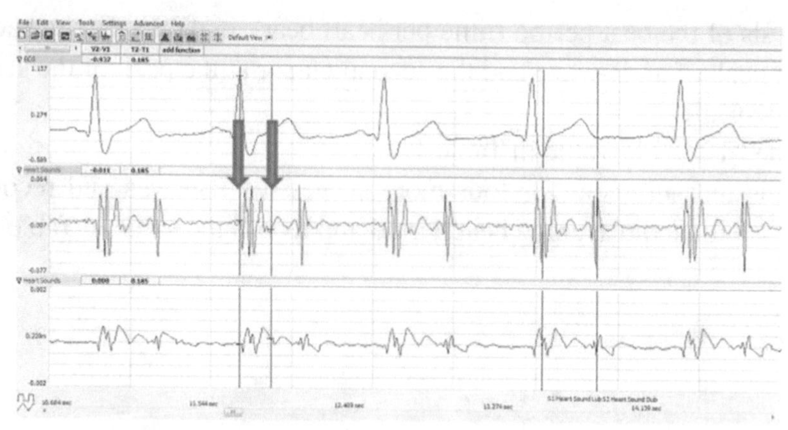

Figure 8 Data analysis S1 duration

- Record the "T2 – T1" value in Table 1.
- Repeat this measurement for the other 4 corresponding ECG cycles and record the values in Table 1. Calculate the interval and record the value in Table 1.

- T-$S_2$ interval analysis.
  - Use the same ECG cycles used for the apical pulse analysis.
  - For analysis of the T-$S_2$ interval, in the **ECG channel** place one cursor at the peak of a T-wave and in the **Heart Sounds channel**, place the second cursor at the beginning of the $S_2$ (see Figure 9).

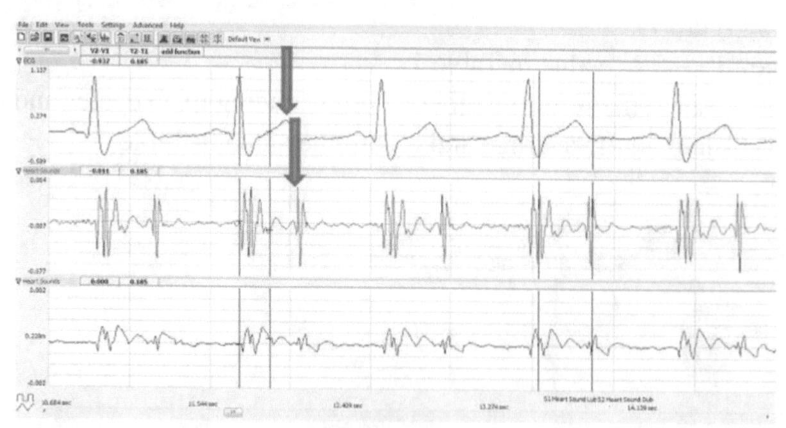

Figure 9 Data analysis T-S2 interval

  - Record the "T2 – T1" value in Table 1.
  - Repeat this measurement for the other 4corresponding ECG cycles and record the values in Table 1. Calculate the average and record the value in Table 1.
- $S_2$ duration.
  - Use the same ECG tracings used for the apical pulse analysis.
  - For analysis of the $S_2$ duration, in the **Heart Sounds channel** place one cursor at the beginning of the $S_2$ and place the second cursor at the end of the $S_2$ (see Figure 10).

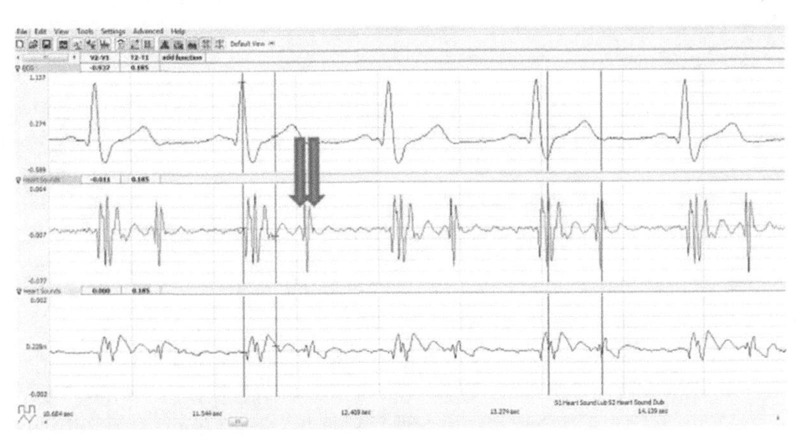

Figure 10 Data analysis S2 duration

- Record the "T2 – T1" value in Table 1.
- Repeat this measurement for the other 4 corresponding ECG cycles and record the values in Table 1. Calculate the interval and record the value in Table 1.

- $S_1$-$S_2$ interval analysis.
  - Use the same ECG tracings used for the apical pulse analysis.
  - For analysis of the $S_1$-$S_2$ interval, in the **Heart Sounds channel** place one cursor at the beginning of the $S_1$ and place the second cursor at the beginning of the $S_2$ (see Figure 11).
  - Record the "T2 – T1" value in Table 1.

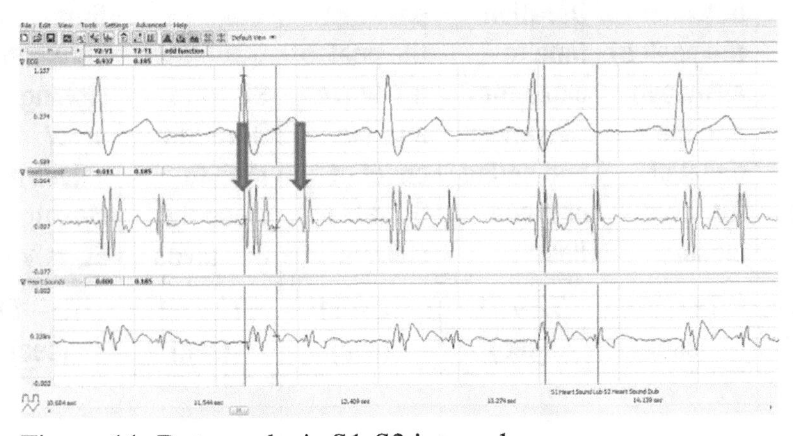

Figure 11 Data analysis S1-S2 interval

- Repeat this measurement for the other 4 corresponding ECG cycles and record the values in Table 1. Calculate the interval and record the value in Table 1.

- $S_2$-$S_1$ interval analysis.
  - Use the same ECG tracings used for the apical pulse analysis.
  - For analysis of the $S_2$-$S_1$ interval, in the **Heart Sounds channel** place one cursor the beginning of the $S_2$ and place the second cursor at the beginning of the following $S_1$ (see Figure 12).
  - Record the "T2 – T1" value in Table 1.
  - Repeat this measurement for the other 4 corresponding ECG cycles and record the values in Table 1. Calculate the average and record the value in Table 1.

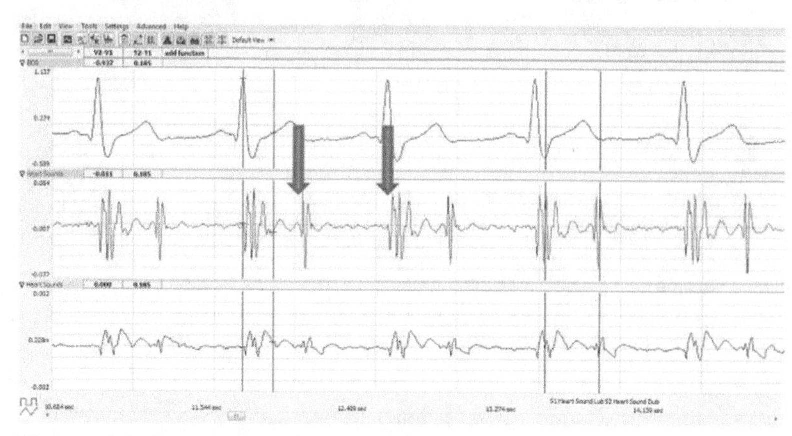

Figure 12 Data analysis S2 -S1 interval

- $S_3$ and $S_4$.
    - Using your ECG cycle tracing as a reference for timing, look at the heart sounds tracing. Can you see any regular deflections in the tracing that might indicate $S_3$ and $S_4$? Do you observe then in each of the ECG cycles? Why are they so much less prevalent than $S_1$ and $S_2$?

**TABLE 1**: Beat Periods and Sound Intervals

| | Beat Period: time between 2 adjacent R-waves | R-S1 Interval: time between the peak of a R-wave and the onset of AV valve closure | S1 Duration: time to close the AV valves and sound of blood flow | T-S2 Interval: time between the peak of a T-wave and the onset of SL valve closure | S2 Duration: time to close the SL valves | S1-S2 Interval: time between closing of AV valves and closing of SL valves | S2-S1 Interval: time between closing of SL valves and closing of AV valves |
|---|---|---|---|---|---|---|---|
| | T2 – T1 (sec/beat) | T2 – T1 (sec) | T2 – T1 (sec) | T2 – T1 (sec) | T2 – T1 (sec) | T2 – T1 (sec) | T2 – T1 (sec) |
| Trial 1 | | | | | | | |
| Trial 2 | | | | | | | |
| Trial 3 | | | | | | | |
| Trial 4 | | | | | | | |
| Trial 5 | | | | | | | |
| Avg. | | | | | | | |

# Exercise 16

# iWorx Electrocardiogram and Exercise

## Objectives

- Be able to define the process of *electrocardiography* and an *electrocardiogram* (ECG).
- Be able to define the term *frequency* as it applies to a waveform.
- Understand the basic components of the cardiac cycle: *diastole* and *systole* of both the atria and the ventricles.
- Be able to identify each of the *standard bipolar leads* and the proper placement of electrodes for each: *lead I, lead II, and lead III.*
- Know the standard scale for recording an ECG: 1 mm = 0.1 mV (amplitude or vertical scale) and 1 mm = 0.04 seconds (time or horizontal scale). Knowing these scales, you should be able to determine heart rate when given the distance or time for one full cardiac cycle.
- Be able to associate a peripheral pressure wave with an ECG cycle.
- Know what each of the following components of an ECG represents with regard to electrical and mechanical activity of the cardiac cycle: isoelectric line, P-wave, P-R interval, P-R segment, QRS complex, S-T segment, T-wave, and Q-T interval.
- Be able to explain the effects of exercise on the ECG and heart rate.
- Be able to explain the changes, as a result of exercise, in each of the following components of a cardiac cycle and the resulting peripheral pressure waves and circulation: R-wave amplitude, P-R interval, Q-T interval, T-R interval, R-wave–pulse interval, and pulse wave amplitude.
- Be able to explain the effects of parasympathetic and sympathetic stimulation on heart rate.

## Lab Safety Reminders

- Use caution with any electrical equipment.

## Hints for Success

All members of your lab group should feel comfortable operating the iWorx system.
**Never unplug any sensors from the iWorx box!** If you unplug a sensor during an experiment, you will lose all data. You will need to turn the iWorx box off, plug the sensor in and start the lab from the beginning.

On the iWorx recording screens, it is helpful to remember that the blue bars are for data that are being recorded. The green bars are for data that is being calculated based upon the data you have collected.

Recall that an EMG measures voltage changes associated with muscle stimulation. If a subject moves skeletal muscles in the area of the electrodes or even in the body in general, the electrical changes from the muscle activity will be picked up in the ECG. It is important that the subject remains as motionless as possible during the periods of data recording. The use of the electrode gel increases the ability to detect the electrical signals generated by the subject's heart. Only a small amount of the gel is needed in the center of the electrode. A large amount will interfere with the ability of the electrode to stick to the subject's skin. After clipping the appropriate leads to the appropriate electrodes, be sure the lead cables are not pulling on the electrodes.

For the experimental component of this lab, the subject is asked to exercise, then reconnect the leads to the electrodes and once again record the ECG. To collect valid data, your subject must **exercise to the point that the heart rate has truly increased and will remain elevated long enough to record the data**. It is important that after the exercise you **quickly** reconnect the leads and begin your recording. Delaying too long between exercise and the recording will allow your subject's heart rate to return to normal, and you will not see the effects of exercise on the cardiac cycle.

While using the iWorx, it is important that you do not get so caught up with obtaining the correct numbers from your data that you forget the function or process you are trying to observe. As you are using the iWorx, keep in mind the principles from the lecture being illustrated in the lab exercise.

## Introduction

In this lab, you will be using the iWorx system to identify and measure the effects of exercise on the heart rate. You will measure the amplitude (mV) and duration (seconds) of various components of the cardiac cycle. The goal is to be able to identify each of the components of the ECG (see Figure 1) and to compare the tracings when recorded under the resting and the experimental conditions. The control for this experiment will be when the subject is sitting and resting. Experimental data will be collected postexercise. The data from the experimental conditions will then be compared to the control data to observe any differences. During rest (bpm = 72), the ventricles are expected to be in systole for 40% of a cardiac cycle and diastole for the remaining 60% of a cardiac cycle. During strenuous exercise

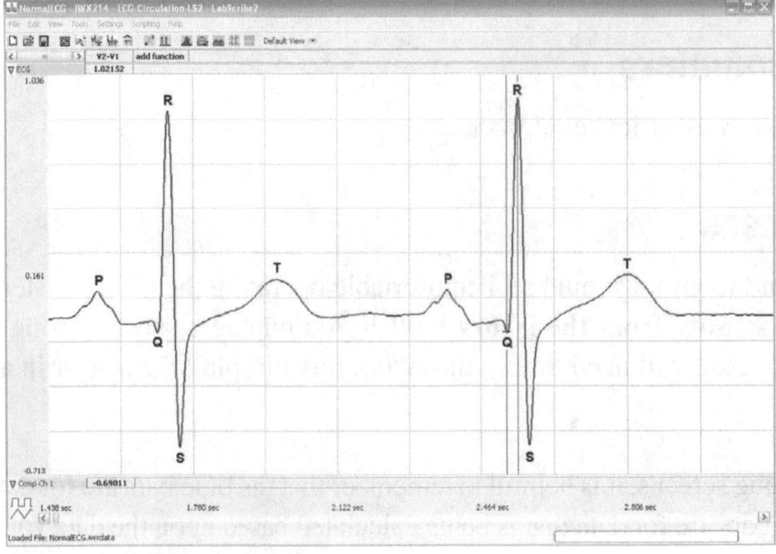


Figure 1 ECG tracing

(bpm = 200), the ventricles are expected to be in systole for 65% of a cardiac cycle and diastole for the remaining 35% of a cardiac cycle. Due to the change in relative time spent in systole versus diastole, less blood enters the ventricle as a result of shorter time for passive filling. The amount of blood contributed to end diastolic volume (EDV) as a result of atrial systole (final filling) is proportionally greater during exercise than when at rest.

Review the major components and the additional diagnostic components of the ECG:
Isoelectric line: no electrical activity is being generated in the heart
P-wave: depolarization of the atria as a result of the activation of the SA node
QRS complex or interval: depolarization of the ventricles as a result of the activation of the AV node
T-wave: repolarization of the ventricles
Note that ventricular repolarization is slower than ventricular depolarization.

In addition to the major components of the ECG, there are other components that are important diagnostically. The duration of these components is used to assess the conduction of electricity through the heart.

Additional diagnostic components of the ECG (see Figure 2):
P-R interval: beginning of P-wave to the beginning of QRS complex (0.12–0.2 sec)
It is the interval between activation of the SA node and activation of the AV node.
It is used to assess the conduction of electrical activity to the ventricles, and a prolonged P-R interval could indicate interference of conduction.
P-R segment: end of P-wave to the beginning of QRS complex
It is the interval between atrial depolarization and ventricular depolarization.
Similar to the P-R interval, a prolonged P-R segment could indicate interference in conduction of electrical activity to the ventricles.
S-T segment: end of S wave to the beginning of the T-wave
It should be at the isoelectric line because the ventricles are uniformly excited and there is no change in charge. It corresponds roughly to the plateau phase of the ventricular muscle cell action potential. An elevated or depressed ST segment may indicate ventricular ischemia or hypoxia.
Q-T interval: beginning of the QRS complex to the end of the T-wave (0.2–0.4 sec)
It is the period of time it takes for the ventricles to depolarized and repolarize, and it corresponds roughly to the mean ventricular action potential.

The normal heart rate of an individual at rest is 60–100 bpm. This rate is at or lower than the normal rate established by the SA node. When the heart rate is lower than the inherent rate of the SA node, the parasympathetic system is regulating the activity of the SA node resulting in a slower heart rate. Recall that acetylcholine (ACh) is the neurotransmitter released from the parasympathetic postganglionic fiber. ACh activates muscarinic type ACh receptors at the neuroeffector junction. When the parasympathetic system is active, release of ACh at the SA node results in a slowing of the pacemaker potential, lowering the heart rate. When the heart rate exceeds the inherent rate of the SA node, the sympathetic system is influencing the activity of the heart. Recall that the norepinephrine (NE) is the neurotransmitter released from the sympathetic postganglionic fiber. NE activates β-1 adrenergic receptors at the neuroeffector junction. Epinephrine (E) released from the adrenal medulla as a hormone also activates the β-1 adrenergic receptors. β-1 adrenergic receptors are located in the atria and the ventricles. An increase in sympathetic stimulation of the

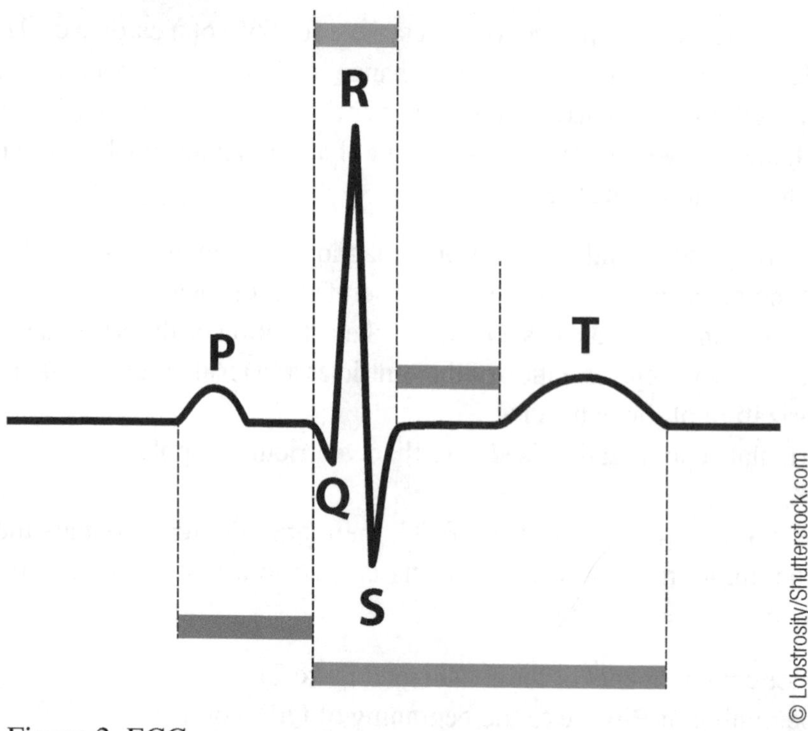


Figure 2 ECG components

heart increases both heart rate and heart contractility. Exercise activates the sympathetic system resulting in the release of both NE and E. You will be using the "Resting Subject" data to simulate parasympathetic regulation of the heart. You will be using the "Post-exercise Subject" to simulate sympathetic regulation of the heart.

## Starting the iWorx Software

1. Be sure that the iWorx box is turned on. A small green light will be lit when the power is on. If you need to turn the power on, the on/off switch is on the back of the iWorx box.
2. Click on the **LabScribe2 shortcut** on the computer's desktop to open the program. You should see an information box that says "Hardware Found." Click **OK**; this will put you on the Main Window page. If the hardware is not found, check to be sure the iWorx box is turned on.
3. On the Main Window, pull down the **Settings** menu. Scroll to **Human Heart**, then select the **Exercise-ECG-Circulation-LS2** settings file from the "Human Heart" list.
   Note: if "Human Heart" is empty, then complete the following instructions. On the Main Window, pull down the "Settings" menu and select "Load Group." Scroll down to locate the folder that contains the settings group, "IPLMv4Complete.iwxgrp." Select this group and click "Open." At this point, you should be able to then complete step 3.
4. Instructions for the Exercise-ECG-Circulation-LS2 Setup will automatically open. These instructions are not necessary. Close this page by clicking on the close button in the upper right corner of this document.
5. LabScribe will appear on the computer screen as configured by the Exercise- ECG-Circulation-LS2settings file.
6. For this lab, you do not need any information from the following files. But as a reminder, once the settings file has been loaded, clicking the **Experiment** button on the toolbar will allow you to access any of the following documents:

- Introduction—introductory material for iWorx Human Heart lab exercises
- Appendix—may have additional information and resources
- Background—may have some background information regarding the lab topic
- Labs—the lab exercise as written by iWorx
- Setup (opens automatically)—instructions for setting up this particular iWorx lab. You already closed this out.

7. If you choose to save your data, it should be saved in the **iWorx Saved Files** folder on the **Desktop**. You may create your own folder within the iWorx Saved Files folder.

## ECG Cable and Pulse Transducer Setup

The ECG cables and pulse plethysmograph have already been connected to the iWorx box (see Figure 3). Do not unplug any of these cables. Doing so will cause you to lose your data.

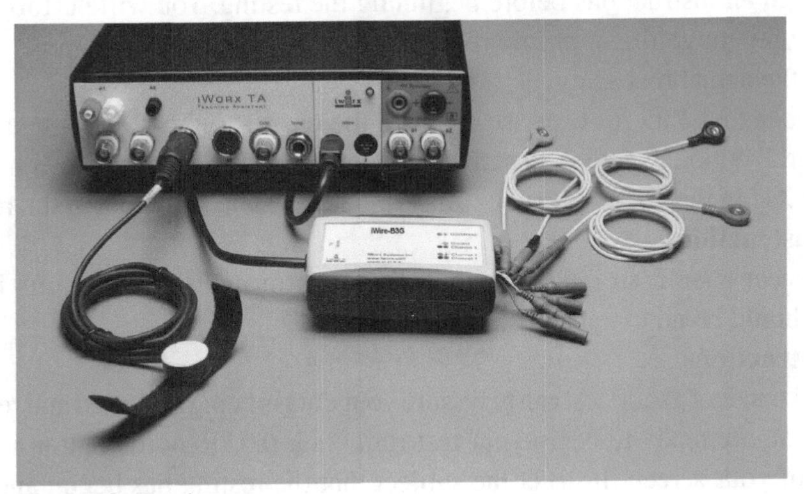

Figure 3 Equipment

For adequate electrical signals, the subject should remove all jewelry from the wrists and ankles. Electrodes will be placed on the anterior surfaces of both forearms, just proximal to the wrists. Use an alcohol swab to clean the anterior surface of the subject's right wrist. Use additional alcohol swabs to clean the medial surfaces of the left and right lower legs, just proximal to the medial malleolus. Let these areas dry completely. Obtain three disposable electrodes and a squeeze bottle of electrode gel. After removing the plastic cover from an electrode, apply one **very small** drop of electrode gel to the center of the sticky side of the electrode. Apply one prepared disposable electrode to each of the areas previously cleaned with the alcohol swab. Snap the lead wires onto the electrodes in the following manner:

- Red lead (+1) is attached to the electrode just proximal to the medial malleolus of the left leg
- Black lead (−1) is attached to the electrode on the anterior surface of the forearm just proximal to the right wrist
- Green lead (G—ground) is attached just proximal to the medial malleolus of the right leg
  Note: The bipolar limb leads use two electrodes for comparison of electrical flow in the heart. The third electrode is the ground. It is not the third component of Einthoven's triangle. For the bipolar limb lead I, the left wrist is the location of the positive (+) electrode. For the bipolar limb leads II and III, the left leg is the location of the positive (+) electrode. The right leg is the location of the ground for all three bipolar limb leads.

Place the sensor surface of the plethysmograph on the volar (fingerprint) surface of the distal phalange of the subject's middle finger or thumb. Wrap the Velcro™ strap around the end of the finger to hold the sensor firmly in place, but not so tight as to restrict blood flow. **Do not squeeze the plethysmograph.**

The subject should sit quietly and relaxed with both hands in his/her lap, palms facing upward. Any extraneous movement by the subject may cause the ECG to shift off the top or bottom of the screen. The subject will also activate muscle activity and the electrical activity of the muscles will be recorded with the ECG. The ECG tracing will contain electrical noise from the muscle activity, making the tracing less distinct and difficult to interpret.

## Lab Exercises

- **Exercise 1: ECG and Pulse in the Resting Subject**
  - Read through all instructions before beginning the testing. You will perform a short test run to be sure you are getting a proper recording. Once your recording looks appropriate, you will begin the actual recording
  - To label the actual data for exercise 1, move the cursor to the "Mark" text box to the right of the "Mark" button and click to activate the text box. Type the subject's name followed by "resting ECG" in the "Mark" text box. Insert the mark when you begin the recording period.
  - **Subject Instructions**
    - The subject should sit quietly with both hands in his/her lap, **palms facing upward**. There should be no extraneous muscle activity.
  - **Tester Instructions**
    - You will conduct a short test run to be sure your data is being collected and recorded properly.
    - When you are ready to begin your test run, click the "Record" button in the upper right corner of your screen. Instruct the subject that the testing has begun and to remain still. After 10 to 20 seconds, click "Stop."
    - Click the "Autoscale" buttons for the "ECG," "Pulse," and the "Pulse Integral" channels. The data recording should look similar to Figure 4.

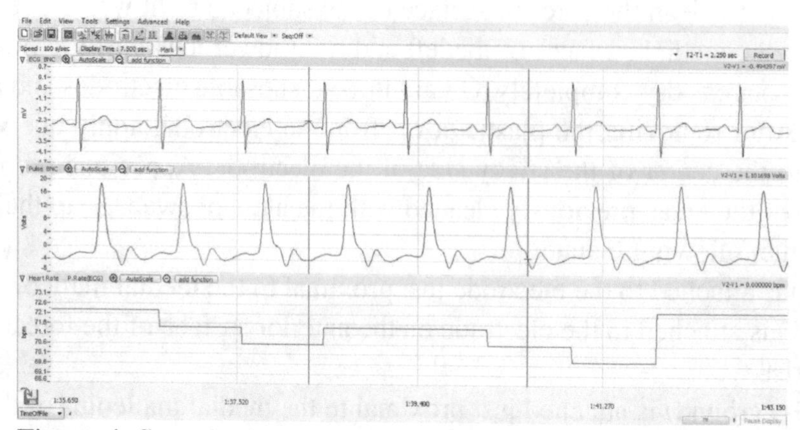

Figure 4 Sample data

- If the signal on either the "ECG" or the "Pulse" channel is upside down when compared to the figure, click the downward arrow to the left of the channel title and select the "Invert" function.

- If the pulse signal is small or noisy, adjust the tension on the plethysmograph strap. Be sure the subject has the hands placed **palms up** in the lap.
- When your tracing looks appropriate, begin your actual data collection. Click the "Record" button in the upper right corner of your screen. Insert your mark when you begin the recording. Instruct the subject that the testing has begun and to remain still. Continue to record for 30 second to 1 minute. Click "Stop." You may record longer if your tracings do not look normal or you do not have 6 to 7 good sequential cardiac cycles.

- You may choose to save your data at this point or you may continue on with the experiment or data analysis. To save your data, on the Main Window, either
  1. click on the **File** menu. Scroll to **Save As**, then save your data; it should be saved in the **iWorx Saved Files** folder on the **Desktop**. You may create your own folder within the iWorx Saved Files folder. Give your file a name and designate the file type as .iwxdata, then click **Save**.

  Or

  2. click on the **Save File** icon in the iWorx toolbar, then save your data; it should be saved in the **iWorx Saved Files** folder on the **Desktop**. You may create your own folder within the iWorx Saved Files folder. Give your file a name and designate the file type as .iwxdata, then click **Save**.

**Exercise 2: ECG and Pulse in the Post-exercise Subject**

- Read through all instructions before beginning the testing.
- To label the actual data for exercise 2, move the cursor to the "Mark" text box to the right of the "Mark" button and click to activate the text box. Type the subject's name followed by "postexercise ECG" in the "Mark" text box. Insert the mark when you begin the recording period.
- **Subject Instructions**
  - Disconnect the snap leads from the subject and remove the pulse plethysmograph from the subject. **Do not unplug the wires from iWorx box**; you will lose your data and must begin the exercise again.
  - The subject must exercise the leg muscles for a minimum of three minutes. Activities such as jogging in place or vigorously walking stairs are appropriate.
  - Immediately after the exercise period, the subject must be reconnected to the leads and the pulse plethysmograph.
  - The subject should sit quietly with both hands in his/her lap, palms facing upward. There should be no extraneous muscle activity.
- **Tester Instructions**
  - As soon as possible after the exercise period and reconnecting the subject to the leads and pulse plethysmograph, click the "Record" button in the upper right corner of your screen. Instruct the subject that the testing has begun and to remain still. Insert your new mark when you begin the recording. Instruct the subject that the testing has begun and to remain still. Continue to record until the subject's breathing rate and heart rate have returned to normal. Your recording period may last from 2 to 10 minutes, depending on the physical condition of your subject and the intensity of exercise.
  - Enter marks at 30-second intervals during your recording.
  - Click "Stop" when the subject has returned to preexercise breathing and heart rates or 4 minutes of data have been recorded, whichever comes first.

- You may choose to save your data at this point or you may continue on with the data analysis. If you have not yet saved your data, follow the instructions for saving data under exercise 1. If you have previously saved your data, click on the **Save File** icon in the iWorx toolbar.
- **Data Analysis Instructions**

  Figure 5, a diagram of icons in the iWorx toolbar may be helpful.

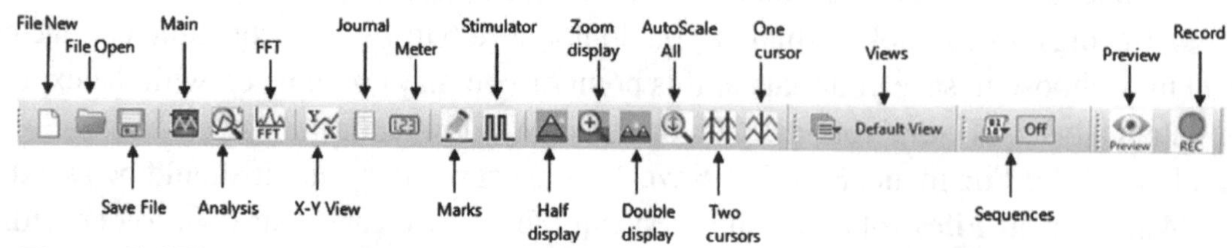


Figure 5 iWorx toolbar icons

- Click on the **Analysis** icon in the iWorx toolbar.
- Go to the beginning of your recordings. Click on the **Marks** icon to bring up the Marks Dialogue box with a table of the marks created in your lab. Once you have the table of marks, click on the number at the beginning of the row for the mark that identifies the beginning of the experimental data segment you are going to analyze. This will highlight that mark. Next, click on the **Go To Mark** button near the bottom of the Marks Dialogue box. Data at that mark will be displayed on your screen.
- Be sure that two cursors are present on your screen. If necessary, click the **Two Cursors** icon on the iWorx toolbar.
- Scroll through your experimental tracing to find a section of data with a succession of 5 to 6 very good ECG/Pulse cycles.
- Place the cursors on either side of the 5 to 6 ECG/Pulse cycles, then click the **Zoom Display** icon in the iWorx toolbar. This will expand the data section to the width of the Main Window. Adjust your data section as necessary with the **Display Time** icon or the **Zoom** icon (see Figure 6).
- Click on the **Analysis** icon in the iWorx toolbar.
- In the Analysis window, a "Function Table" will be displayed above the uppermost channel (ECG channel). The mathematical functions "V2 – V1," and "T2 – T1" should be present. The values for each of these functions are displayed in the top margin of each channel (see Figure 6). The "V2 – V1" function measures the difference in amplitude between cursor 2 and cursor 1.The "T2 – T1" function measures the difference in time between cursor 2 and cursor 1.

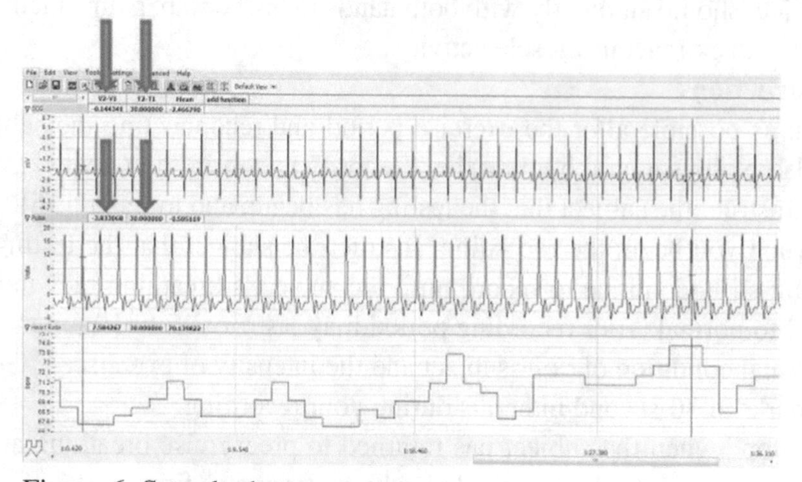

Figure 6 Sample data

- You will be measuring the following values for resting or preexercise data, 0 to 30 seconds postexercise data, 60 to 90 seconds postexercise data, and 120 to 150 seconds postexercise data.
- For the following measurements, **you must use the corresponding ECG and pulse cycles**. The trial numbers should correspond between all five analyses. Begin with the 0 to 30 seconds postexercise data.
- Apical pulse or beat period analysis.
  - For analysis of the beat period (time per heart beat), in the **ECG channel**, place one cursor at the peak of an R-wave and the second cursor at the peak of the sequential R-wave (see Figure 7).
  - Record the "T2 – T1" value in Table 1.
  - Repeat this measurement for four more successive R-waves and record the values in Table 1. Calculate the mean beat period and record the value in Table 1.
- R-wave amplitude analysis.
  - For analysis of the R-wave amplitude, in the **ECG channel** place one cursor at the beginning of the Q wave preceding the R-wave and the second cursor at the peak of the R-wave (see Figure 7).
  - Record the "V2 – V1" value in Table 1.
  - Repeat this measurement for 4 more successive R-waves and record the values in Table 1. Calculate the mean R-wave amplitude and record the value in Table 1.
- P-R interval analysis.
  - For analysis of the P-R interval, in the **ECG channel** place one cursor at the beginning of the P-wave and place the second cursor at the beginning of the Q of the QRS complex (see Figure 7).
  - Record the "T2 – T1" value in Table 1.
  - Repeat this measurement for the other 4 corresponding P-R intervals and record the values in Table 1. Calculate the mean P-R interval and record the value in Table 1.
- Q-T interval analysis.
  - For analysis of the Q-T interval, in the **ECG channel**, place one cursor at the beginning of the QRS complex and place the second cursor at the end of the T-wave (see Figure 7).
  - Record the "T2 – T1" value in Table 1.
  - Repeat this measurement for the other 4 corresponding Q-T intervals and record the values in Table 1. Calculate the mean Q-T interval and record the value in Table 1.
- T-R interval analysis.
  - For analysis of the T-R interval, in the **ECG channel** place one cursor at the end of the T-wave and place the second cursor at the beginning of the subsequent QRS complex (see Figure 7).
  - Record the "T2 – T1" value in Table 1.
  - Repeat this measurement for the other 4 corresponding T-R intervals and record the values in Table 1. Calculate the mean T-R interval and record the value in Table 1.
- R-wave–pulse interval analysis.
  - For analysis of the R-wave–pulse interval, in the **ECG channel**, place one cursor at the peak of an R-wave, and in the **Pulse channel**, place the second cursor at the peak of the associated pulse wave that just follows the R-wave (see Figure 7).
  - Record the "T2 – T1" value in Table 1.

- Repeat this measurement for the other 4 corresponding R-wave–peripheral pulse waves and record the values in Table 1. Calculate the mean R-wave–pulse interval and record the value in Table 1.

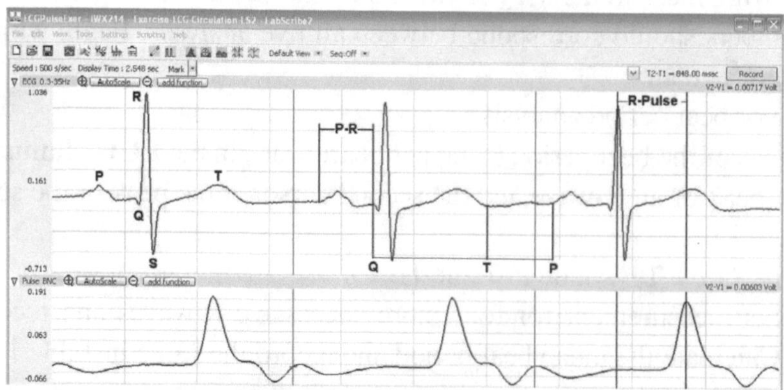


Figure 7 Data analysis ECG

- Pulse wave amplitude analysis.
  - For analysis of the pulse wave amplitude, in the **Pulse channel** place one cursor on the baseline that precedes the pulse wave and the second cursor at the peak of the pulse wave (see Figure 8).
  - Record the "V2 – V1" value in Table 1.
  - Repeat this measurement for 4 more successive pulse waves and record the values in Table 1. Calculate the mean pulse wave amplitude and record the value in Table 1.

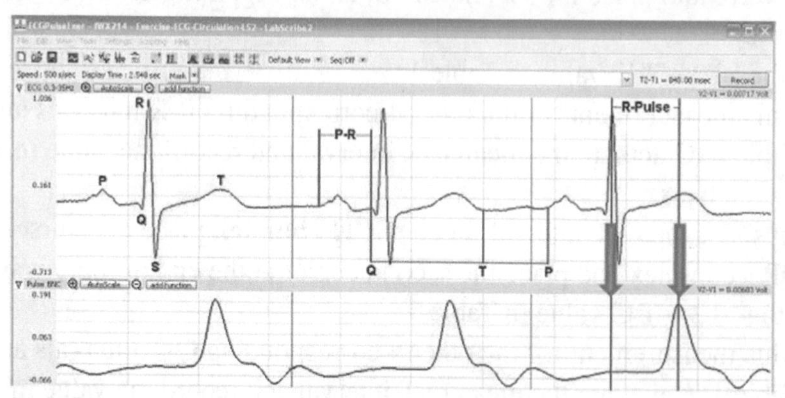


Figure 8 Data analysis pulse wave amplitude

- Repeat the measurements for the 60 to 90 and 120 to 150 seconds postexercise data.

**TABLE 1**: ECG Measurements for Pre- and Postexercise

| | | Trial 1 | Trial 2 | Trial 3 | Trial 4 | Trial 5 | Mean |
|---|---|---|---|---|---|---|---|
| Pre-exercise | Beat period (sec) | | | | | | |
| | R-wave amplitude (mV) | | | | | | |
| | P-R Interval (sec) | | | | | | |
| | Q-T Interval (sec) | | | | | | |
| | T-R Interval (sec) | | | | | | |
| | R-Pulse Interval (sec) | | | | | | |
| | Pulse Wave Amplitude (mV) | | | | | | |
| Post-exercise<br>0-30 sec | Beat period (sec) | | | | | | |
| | R-wave amplitude (mV) | | | | | | |
| | P-R Interval (sec) | | | | | | |
| | Q-T Interval (sec) | | | | | | |
| | T-R Interval (sec) | | | | | | |
| | R-Pulse Interval (sec) | | | | | | |
| | Pulse Wave Amplitude (mV) | | | | | | |

(*continues*)

**TABLE 1**: ECG Measurements for Pre- and Postexercise (Continued)

| | | Trial 1 | Trial 2 | Trial 3 | Trial 4 | Trial 5 | Mean |
|---|---|---|---|---|---|---|---|
| Post-exercise<br><br>60–90 sec | Beat period<br>(sec) | | | | | | |
| | R-wave amplitude<br>(mV) | | | | | | |
| | P-R Interval<br>(sec) | | | | | | |
| | Q-T Interval<br>(sec) | | | | | | |
| | T-R Interval<br>(sec) | | | | | | |
| | R-Pulse Interval<br>(sec) | | | | | | |
| | Pulse Wave Amplitude<br>(mV) | | | | | | |
| Post-exercise<br><br>120–150 sec | Beat period<br>(sec) | | | | | | |
| | R-wave amplitude<br>(mV) | | | | | | |
| | P-R Interval<br>(sec) | | | | | | |
| | Q-T Interval<br>(sec) | | | | | | |
| | T-R Interval<br>(sec) | | | | | | |
| | R-Pulse Interval<br>(sec) | | | | | | |
| | Pulse Wave Amplitude<br>(mV) | | | | | | |

# Exercise 17

# iWorx Blood Pressure

## Objectives

- Be able to define and describe *unidirectional flow* in the circulatory system.
- Identify the valves associated with the heart, their locations, and their functions.
- Be able to define the terms *systole* and *diastole* in reference to actions in the heart and to blood pressure.
- Be able to define the term *pulsatile* and relate it to blood flow and blood pressure in the arteries.
- Be able to identify blood pressures that fall into the normal range, prehypertension, and hypertension.
- Be able to define and calculate *pulse pressure*.
- Be able to relate pulse pressure to heart rate, stroke volume, and vascular resistance.
- Be able to define and calculate *mean arterial pressure*.
- Be able to relate *mean arterial pressure* to the equation $F = \Delta P/R$.
- Be able to identify the three methods of measuring blood pressure: *invasive measurement*, *oscillometric measurement*, and *auscultatory measurement*.
- Be able to identify the equipment used for the auscultatory method of blood pressure measurement.
- Be able to relate the *sounds of Korotkoff* and systolic and diastolic pressure.
- Be able to relate pressure in the sphygmomanometer cuff and blood pressure and blood flow in the restricted artery relative to the first and second sounds of Korotkoff.
- Be able to identify the proper conditions, methods, and position of the subject for taking an accurate blood pressure reading.

## Lab Safety Reminders

- Use caution with any electrical equipment.

## Hints for Success

All members of your lab group should feel comfortable operating the iWorx system.

**Never unplug any sensors from the iWorx box!** If you unplug a sensor during an experiment, you will lose all data. You will need to turn the iWorx box off, plug the sensor in and start the lab from the beginning.

On the iWorx recording screens, it is helpful to remember that the blue bars are for data that are being recorded. The green bars are for data that is being calculated based upon the data you have collected.

Be sure that you have the blood pressure cuff oriented correctly. It is important that you **do not overinflate the cuff or leave a high level of pressure in the cuff for an extended period of time (over 1 minute)**. Doing so is the same as placing a tourniquet on the subject's arm. Blood flow to and from the arm will be stopped, and tissue death will occur after an extended period of time.

Although blood pressure is more routinely taken on the left arm, blood pressure may be taken on either arm. The instructions for this lab are written for the left arm. Depending upon the space available in the lab, it may be more convenient to use the right arm. That is fine, but you should be consistent in which arm is used throughout the lab. Blood pressure may also be taken on the forearm, the thigh, and proximal to the ankle. However, blood pressure taken in these alternative areas is generally higher than blood pressure taken on the upper arm, with the difference in the systolic pressures being more pronounced than the difference in the diastolic pressures.

For one of the experimental components of this lab, the subject is asked to exercise, then once again record and measure blood pressure. To collect valid data, your subject must exercise to the point that the heart rate and blood pressure have truly increased and will remain elevated long enough to record the data. It is important that after the exercise you **quickly** replace the blood pressure cuff and the pulse plethysmograph and begin your recording. Delaying too long between exercise and the recording will allow your heart rate and blood pressure to return to normal, and you will not see the effects of exercise on heart rate and blood pressure.

While using the iWorx, it is important that you do not get so caught up with obtaining the correct numbers from your data that you forget the function or process you are trying to observe. As you are using the iWorx, keep in mind the principles from the lecture being illustrated in the lab exercise.

## Introduction

Normal blood flow is unidirectional, only flowing in one direction. This is the result of the action of the valves of the heart. Blood flow is also **pulsatile**, a result of the alternating diastolic and systolic actions of the heart. As a result of ventricular systole, a pulse of blood is ejected into the great arteries. The pulse of blood entering the arteries increases the pressure in the arteries. **Systolic blood pressure** is the force of the blood pushing outward against the walls of the arteries as a result of systole of the ventricles. **Diastolic blood pressure** is the force of the blood against the walls of the arteries as a result of the diastole of the ventricles. During ventricular diastole, no blood is ejected into the great arteries from the ventricles, and the great arteries recoil. The blood pressure in the arteries is expected to be lower during diastole. Blood pressures are written as systolic pressure/diastolic pressure (e.g., 120/80 mm Hg). **Pulse pressure** is the arithmetic difference between the systolic pressure and the diastolic pressure. The **mean arterial pressure** (MAP) is the average pressure the arteries are exposed to during a single cardiac cycle. This is not an arithmetic mean because the time spent in systole is usually (at rest) less than the time spent in diastole. MAP is calculated using either of the following two equations:

$$\text{MAP} = 1/3 \text{ pulse pressure} + \text{diastolic pressure}$$

or

$$\text{MAP} = (\text{systolic pressure} + 2 \bullet \text{diastolic pressure})/3$$

MAP is important because it is a measure of hemodynamic perfusion pressure of the body's organs. That means, it is a measure of how well blood is reaching individual organs. MAP in the range of 70 to 110 mm Hg is considered normal. If MAP falls below 60 mm Hg, there is a risk that organs, especially the heart supplied with blood through the coronary arteries, the brain, and the kidneys, will not be sufficiently perfused. This is critically important for the brain, although all tissues would suffer from a low MAP and associated low perfusion or lack of blood flow. The equation for MAP is not accurate for individuals with tachycardia or with bradycardia.

Blood pressure can be measured using three different methods. The **invasive method** of blood pressure measurement is the only direct measure of blood pressure. It is considered invasive because it requires the placement of a pressure transducer on the tip of a catheter inserted into the artery of interest. This introduces some risk as an artery has to be punctured. The radial artery at the wrist is the site often used because the ulnar artery provides an alternate path of blood flow to the hand. Although this method has the potential to be very accurate, there are two factors that can have a large influence on the value of the blood pressure measurements. The pressure in the radial artery can be dramatically changed by changing the position of the wrist relative to the heart. The pressure measured at the wrist can increase or decrease by 2 mm Hg for every 1 inch decrease or increase in elevation relative to the heart. When comparing the blood pressure in the radial artery at the wrist relative to the brachial artery, the systolic pressure tends to be higher and the diastolic pressure tends to be lower in the radial artery, although the MAP tends to be very similar. The distance from the heart as well as the difference in the stiffness of the arteries both have an impact on the difference in the blood pressure measurements in the radial artery at the wrist versus in the brachial artery. This is also not a method of blood pressure measurement that can be performed routinely, so it has a reduced value for the consistent monitoring of blood pressure.

The **oscillometric method** of blood pressure measurement is used with an automatic blood pressure device. In the oscillometric method, a cuff is placed on the arm and inflated to a pressure above the person's expected systolic pressure. The cuff pressure is slowly released and the pressure in the cuff is monitored for small pressure changes as blood begins to pulse through the artery. Note that some automatic devices for taking blood pressure contain a microphone and are actually using the auscultatory method of blood pressure measurement.

Blood in the vessels has **laminar flow**, a smooth, undisrupted flow of blood in concentric layers with the flow rate increasing as you move from the edge of a blood vessel to the center of the vessel. Laminar flow makes no noise, so normal blood flow cannot be heard. However, when laminar flow is disrupted, the turbulence that is created causes the flow to become noisy. This is the principle behind the **auscultatory method** of blood pressure measurement. When the sphygmomanometer is placed on the subject's arm and air is pumped into the cuff to place pressure on the subject's arm, the result is to squeeze the arteries closed. When the pressure applied to the arm is higher than systolic blood pressure, blood flow stops (see the precautions identified in the lab). As cuff pressure is released and it decreases to the point just below the systolic pressure of the subject, some blood flow returns. This initial flow is not laminar flow, so it is noisy and will be heard as a tapping sound. This is the **first Korotkoff sound** (K-1 or Phase 1). The cuff pressure at which K-1 is heard represents the subject's systolic pressure. As pressure in the cuff continues to drop, murmurs or swooshing sounds can be heard in the tapping sounds (K-2 or Phase 2). The sounds then become crisper and louder in K-3 (Phase 3). As the blood flow becomes less turbulent, the sounds are muffled and softer (K-4 or Phase 4). The **final Korotkoff sound** (K-5 or phase 5) is not a sound but rather silence. The cuff pressure at which K-5 is detected represents the subject's diastolic pressure. At this pressure, the arterial blood flow is no longer restricted or disrupted, and the blood flow returns to silent laminar flow.

There are four common errors associated with the auscultatory method. Incorrect cuff size can produce inaccurate measurements. A cuff that is too large will result in an inflated reading due to the large volume of air that must be pumped into the cuff to occlude the brachial artery. A cuff that is too small will not be able to be pumped up to a pressure that can adequately occlude the brachial artery. Incorrect placement of the cuff may not allow for proper placement of the stethoscope. Incorrect placement of the bell of the stethoscope so that it is not placed directly over the brachial artery often results in a low systolic pressure reading and a high diastolic pressure reading. Incorrect interpretation of the sounds of Korotkoff can occur if there are background noises that prevent proper hearing of the sounds. Additional factors include poor support of the arm during blood pressure measurement, palm not up during measurement, and too rapid release of the cuff pressure.

A person's blood pressure is not static. It is expected to change based upon the body's needs for blood supply. As a result, blood pressure is often related to heart rate. The systolic pressure can vary depending upon the person's activity level, increasing with an increase in physical activity as well as mental activity. The diastolic pressure does vary but not as much as systolic pressure.

Blood pressure falls into several categories:

| Category | Systolic Pressure mm Hg | And / Or | Diastolic Pressure mm Hg |
|---|---|---|---|
| Hypotensive | <90 | And | <60 |
| Normal | < 120 | And | < 80 |
| Elevated | 120–139 | And | < 80 |
| Hypertensive Stage 1 | 130–139 | Or | 80–89 |
| Hypertensive Stage 2 | 140 + | Or | 90 + |
| Hypertensive Stage 3 | 180 + | And / Or | 120 + |

The diagnostic category is determined if either systolic **or** diastolic pressure falls into the range of the category. The "worst" category is the diagnosis for the subject.

## Starting the iWorx Software

1. Be sure that the iWorx box is turned on. A small green light will be lit when the power is on. If you need to turn the power on, the on/off switch is on the back of the iWorx box.
2. Click on the **LabScribe3 shortcut** on the computer's desktop to open the program.
   You should see an information box that says "Hardware Found". Click **OK**; this will put you on the Main Window page. If the hardware is not found, check to be sure the iWorx box is turned on.
3. On the Main Window, pull down the **Settings** menu. Scroll to **Human Circulation,** then select the **BloodPressure-BodyPosition** settings file from the "Human Circulation" list.
   *Note: if "Human Circulation" is empty, then complete the following instructions. On the Main Window, pull down the "Settings" menu and select "Load Group." Scroll down to locate the folder that contains the settings group, "IPLMv6Complete.iwxgrp." Select this group and click "Open." At this point, you should be able to then complete step 3.*
4. Instructions for the BloodPressure-BodyPosition Setup will automatically open. These instructions are not necessary. Close this page by clicking on the close button in the upper right corner of this document.
5. LabScribe will appear on the computer screen as configured by the BloodPressure-BodyPosition settings file.
6. For this lab, you do not need any information from the following files. But as a reminder, once the settings file has been loaded, clicking the **Experiment** button on the toolbar will allow you to access any of the following documents:
   - Introduction—introductory material for iWorx Human Circulation lab exercises
   - Appendix—may have additional information and resources
   - Background—may have some background information regarding the lab topic
   - Labs—the lab exercise as written by iWorx
   - Setup (opens automatically)—instructions for setting up this particular iWorx lab. You already closed this out.
7. If you choose to save your data, it should be saved in the **iWorx Saved Files** folder on the **Desktop**. You may create your own folder within the iWorx Saved Files folder.

## Blood Pressure and Pulse Transducers Setup

The blood pressure cuff and pulse plethysmograph have already been connected to the iWorx box (see Figure 1). Do not unplug any of these cables. Doing so will cause you to lose your data.

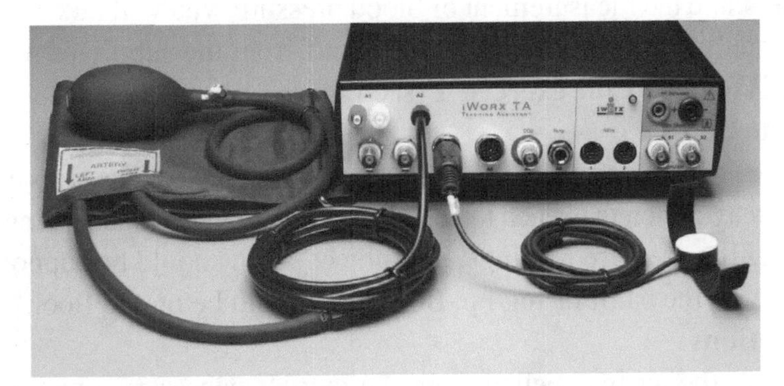


Figure 1 Equipment

Place the sensor surface of the plethysmograph on the volar (fingerprint) surface of the distal phalange of the subject's middle finger or thumb. Wrap the Velcro™ strap around the end of the finger to hold the sensor firmly in place, but not so tight as to restrict blood flow. **Do not squeeze the plethysmograph**

## Calibrating the Blood Pressure Transducer

- Lay the blood pressure cuff on the table.
- Click the "Record" button.
- Click on the "Autoscale" button in the Pulse and Blood Pressure channels.
- Record data for about 10 seconds.
- Click "Stop."
- Scroll to the beginning of the calibration data.
- Be sure that two cursors are present on your screen. If necessary, click the **Two Cursors** icon on the iWorx toolbar.
- Place the cursors on either side of a 10-second period of data collected, then click the **Zoom Display** icon in the iWorx toolbar. This will expand the data section to the width of the Main Window. Adjust your data section as necessary with the display time icons or the zoom icon.
- Click the arrow to the left of the "Blood Pressure" channel title. This will open the channel menu.
- In the channel menu select "Units," and within the submenu, select "Set Offset."
- In the unit conversion window, click to add a check to the "Apply units to all blocks" box. Change the value in the "Set Menu Value between Cursors to:" box to 0 (see Figure 2).
- This procedure is used to convert the voltages at the positions of the cursors to the correct pressure values.

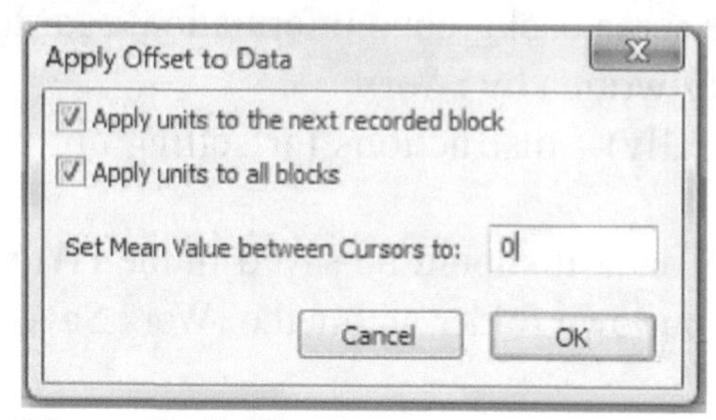


Figure 2 Calibration

## Lab Exercises

- **Exercise 1: Resting Blood Pressure—Manual Method**
  - To better understand the measurement of blood pressure, you will take blood pressure manually. Obtain a stethoscope and blood pressure cuff from the supply provided in the lab—**do NOT use the iWorx cuff for exercise 1** (see Figure 3).
  - **Subject Instructions**
    - The subject sit with the left arm easily accessible, body relaxed. The arms may rest in the lap, or the left arm may rest on the bench top or be supported by the person taking the blood pressure.
    - When taking the pressure readings, the subject's arm should be supported and roughly at heart level with the palm facing up. Both feet should be on the floor.
  - **Tester Instructions**
    - Make sure all the air is expelled from the cuff. If necessary, open the pressure release valve and squeeze any residual air out of the cuff. Close the pressure release valve.

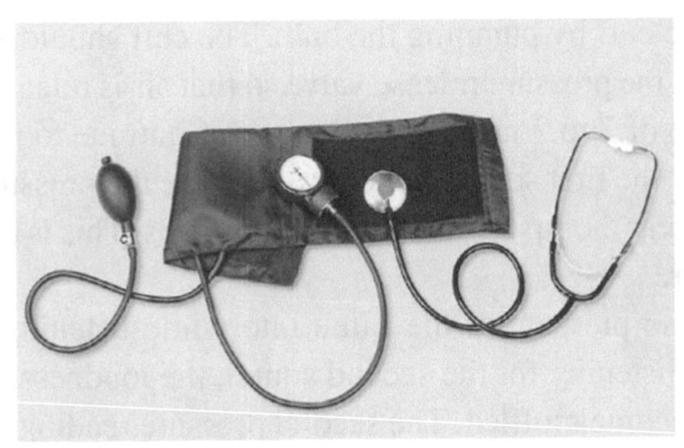

Figure 3 Equipment

- Place the blood pressure cuff around the upper portion of the left arm, with the lower edge of the cuff approximately 1 to 2 inches above the elbow. The "Artery" label should be positioned over the brachial artery and the arrow on the label should be pointing toward the elbow. There should be room between the cuff and the elbow to position the stethoscope over the brachial artery without it being covered by the cuff (see Figure 4).

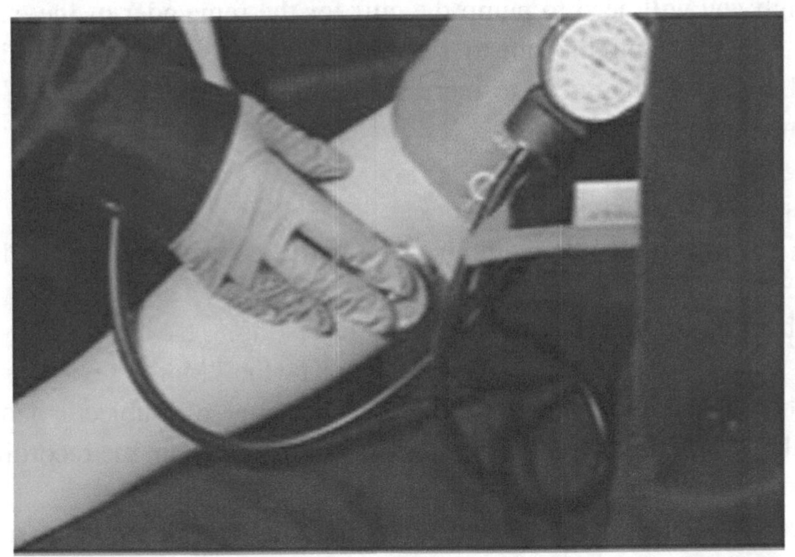

Figure 4 Manual blood pressure cuff and stethoscope placement

- Wrap the cuff snuggly and attach the cuff with the Velcro®. The cuff should stay in place on the arm without any additional support. A cuff that is too loose will result in an inflated reading due to the large volume of air that must be pumped into the cuff to occlude the brachial artery.
- Position the pressure gauge so that you can read the dial straight on and not at an angle. It should be positioned at eye level if possible (see Figure 4).
- The subject's arm should be supported at his or her heart level with the palm facing up.
- Be sure the tubing of the cuff is not pinched or obstructed in any way.
- Put the stethoscope ear pieces in the tester's ears. Position the stethoscope over the subject's brachial artery between the cuff and the elbow. The tester should hold the bell of the stethoscope in place with two fingers with a moderate amount of pressure so that the bell does not slip. Care should be taken to not apply too much pressure resulting in the impedance of blood flow in the brachial artery (see Figure 4).

- Rapidly inflate the cuff by pumping the bulb. The cuff should be inflated to 160 mm Hg. Immediately open the pressure release valve so that air is released and pressure in the cuff decreases at a rate of 2 to 3 mm Hg per second. Continue to release pressure at this rate while listening for the first sound of Korotkoff. Note the pressure reading on the pressure gauge when you hear the first soft "thumping" sounds. This is a very close approximation of systolic pressure.
- Continue to release pressure at the same rate while listening for the second sound of Korotkoff. While listening for the second sound, the loudness of the "thumping" will increase and then become muffled. The second pressure reading is taken when the "thumping" disappears and you hear silence again. Note the pressure reading on the pressure gauge when you hear silence. This is a very close approximation of diastolic pressure.
- You may now rapidly and completely release the remaining pressure in the cuff.
- Record the numbers here. Repeat for a second trial.

  Trial 1: Systolic pressure __________ Diastolic pressure __________

  Trial 2: Systolic pressure __________ Diastolic pressure __________
- Look at the subject's systolic pressure. This number will give you a good estimate as to how high you will need to pump the cuff for the remainder of the experiments. For exercises 2 and 3, pump the cuff to 20 mm Hg pressure over the systolic pressure you measured here. You may need to increase the pressure for the exercise 4 postexercise measurements.

## Exercise 2: Resting Blood Pressure—iWorx

- Be sure you have calibrated the equipment properly (see Calibrating the Blood Pressure Transducer).
- Read through all instructions before beginning the testing.
- To label the data for exercise 2, move the cursor to the "Mark" text box to the right of the "Mark" button and click to activate the text box. Type the subject's name followed by "supine" in the "Mark" text box. Insert the mark when you begin the recording period.
- **Subject Instructions**
  - The subject should rest in a supine position for 5 minutes prior to the first blood pressure reading.
  - The subject should lie quietly with both hands at the sides, palms facing upward. There should be no extraneous muscle activity during testing.
  - Between measurements, the subject may flex and extend the fingers to increase blood flow to the arm.
- **Tester Instructions**
  - Make sure all the air is expelled from the iWorx blood pressure cuff. If necessary, open the pressure release valve and squeeze any residual air out of the cuff. Close the pressure release valve.
  - Place the blood pressure cuff around the upper portion of the left arm, with the lower edge of the cuff approximately 2 inches above the elbow. The "Artery" label should be positioned over the brachial artery and the arrow on the label should be pointing toward the elbow. There should be room between the cuff and the elbow to position the stethoscope over the brachial artery without it being covered by the cuff.

- Wrap the cuff snuggly and attach the cuff with the Velcro®. The cuff should stay in place on the arm without any additional support. A cuff that is too loose will result in an inflated reading due to the large volume of air that must be pumped into the cuff to occlude the brachial artery.
- The subject's arm should be supported at his or her heart level with the palm facing up.
- Be sure the tubing of the cuff is not pinched or obstructed in any way.
- Place the sensor surface of the plethysmograph on the volar (fingerprint) surface of the distal phalange of the subject's middle finger or thumb of the left hand. Wrap the Velcro™ strap around the end of the finger to hold the sensor firmly in place, but not so tight as to restrict blood flow.
- At the end of the rest period, click the "Record" button in the upper right corner of your screen. Instruct the subject that the testing has begun and to remain still. Insert your new mark when you begin the recording.
- Inflate the blood pressure until the finger pulse wave on the "Pulse" channel disappears (see Figure 5). This should be close to the systolic pressure you measured manually.
- Release the pressure in the cuff at a rate of 3 to 5 mm Hg per second. Continue to release the pressure until the pressure is approximately 20 mm Hg.
- Click the "Stop" button.
- Release the rest of the pressure in the cuff. Be sure the blood pressure cuff is completely deflated and is not restricting blood flow in the subject's arm.
- The subject should rest for 2 to 3 minutes.
- Repeat the measurement.

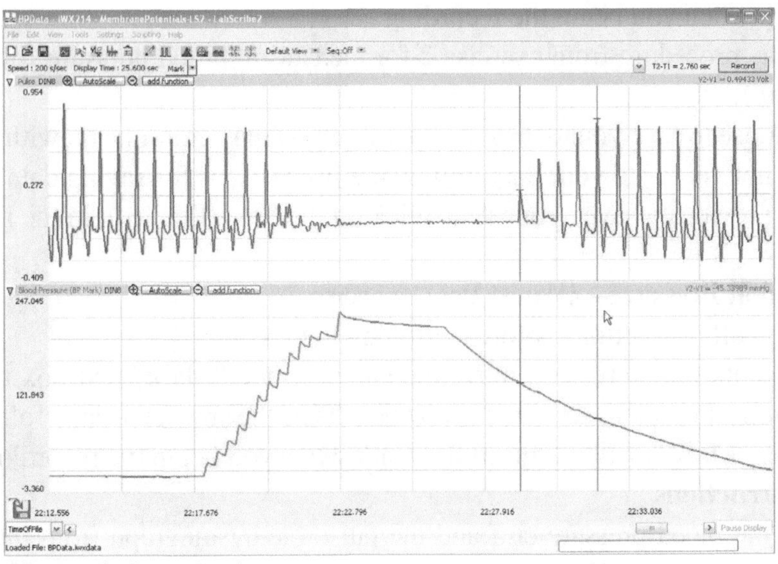

Figure 5 Sample data

- You may choose to save your data at this point or you may continue on with the experiment or data analysis. To save your data, on the Main Window, either
  1. Click on the **File** menu. Scroll to **Save As**, then save your data; it should be saved in the **iWorx Saved Files** folder on the **Desktop**. You may create your own folder within the

iWorx Saved Files folder. Give your file a name and designate the file type as .iwxdata, then click **Save**.

Or

2. Click on the **Save File** icon in the iWorx toolbar, then save your data; it should be saved in the **iWorx Saved Files** folder on the **Desktop**. You may create your own folder within the iWorx Saved Files folder. Give your file a name and designate the file type as .iwxdata, then click **Save**.

## Exercise 3: Blood Pressure—iWorx, Effect of Gravity

- Read through all instructions before beginning the testing.
- To label the data for exercise 3, move the cursor to the "Mark" text box to the right of the "Mark" button and click to activate the text box. Type the subject's name followed by "sitting" in the "Mark" text box. Insert the mark when you begin the recording period.
- **Subject Instructions**
  - The subject will move to a sitting position. It should not be necessary to remove the blood pressure cuff or pulse plethysmograph while changing position, but take care to not pull on the connections.
  - The subject should sit quietly with both hands in his/her lap, palms facing upward. There should be no extraneous muscle activity during testing.
  - Between measurements, the subject may flex and extend the fingers to increase blood flow to the arm.
- **Tester Instructions**
  - Once the subject is repositioned, click the "Record" button in the upper right corner of your screen. Instruct the subject that the testing has begun and to remain still. Insert your new mark when you begin the recording.
  - Repeat the procedure from exercise 2 for measuring blood pressure. Be sure to take two measurements.
- You may choose to save your data at this point or you may continue on with the data analysis. If you have not yet saved your data, follow the instructions for saving data under exercise 2. If you have previously saved your data, click on the **Save File** icon in the iWorx toolbar.

## Exercise 4: Blood Pressure—iWorx, Post-exercise

- Read through all instructions before beginning the testing.
- To label the data for exercise 4, move the cursor to the "Mark" text box to the right of the "Mark" button and click to activate the text box. Type the subject's name followed by "postexercise" in the "Mark" text box. Insert the mark when you begin the recording period.
- **Subject Instructions**
  - Remove the blood pressure cuff and the pulse plethysmograph from the subject. **Do not** unplug the wires from iWorx box; you will lose your data and must begin the exercise again.
  - The subject must exercise the leg muscles for a minimum of 3 minutes. Activities such as jogging in place or vigorously walking stairs are appropriate.
  - The subject should return to the supine position. There should be no extraneous muscle activity during testing.
  - Between measurements, the subject may flex and extend the fingers to increase blood flow to the arm.

- **Tester Instructions**
  - Replace the blood pressure cuff and pulse plethysmograph to the same arm used in the previous exercises.
  - As soon as possible after the exercise period and after replacing the blood pressure cuff and pulse plethysmograph, click the "Record" button in the upper right corner of your screen. Instruct the subject that the testing has begun and to remain still. Insert your new mark when you begin the recording.
  - Repeat the procedure from exercise 2 for measuring blood pressure. Be sure to take two measurements.
- You may choose to save your data at this point or you may continue on with the data analysis. If you have not yet saved your data, follow the instructions for saving data under exercise 2. If you have previously saved your data, click on the **Save File** icon in the iWorx toolbar.
- **Data Analysis Instructions**

Figure 6, a diagram of icons in the iWorx toolbar may be helpful.

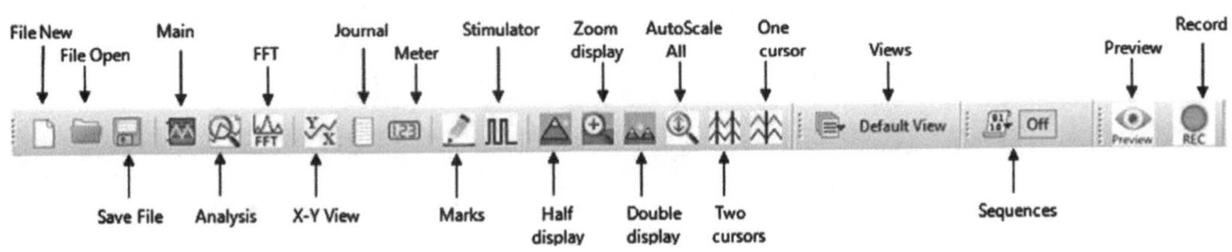


Figure 6 iWorx toolbar icons

- Click on the **Analysis** icon in the iWorx toolbar.
- Go to the beginning of your recordings. Click on the **Marks** icon to bring up the Mark Dialogue box with a table of the marks created in your lab. Once you have the table of marks, click on the number at the beginning of the row for the mark that identifies the beginning of the experimental data segment you are going to analyze. This will highlight that mark. Next, click on the **Go To Mark** button near the bottom of the Marks Dialogue box. Data at that mark will be displayed on your screen.
- Be sure that two cursors are present on your screen. If necessary, click the **Two Cursors** icon on the iWorx toolbar.
- Place the cursors on either side of the first measurement of blood pressure, one cursor on the baseline just before inflating the blood pressure cuff and the second cursor just after the cuff pressure was fully released, then click the **Zoom Display** icon in the iWorx toolbar. This will expand the data section to the width of the Main Window. Adjust your data section as necessary with the display time icons or the zoom icon (see Figure 7).
- Click on the **Analysis** icon in the iWorx toolbar.
- In the Analysis window, a "Function Table" will be displayed above the uppermost channel (ECG channel). The mathematical functions "V2−V1," "Value 1," "Value2," and "T2−T1" should be present. The values for each of these functions are displayed in the top margin of each channel. The "V2−V1" function measures the difference in amplitude between cursor 2 and cursor 1. The "Value 1" function is the value of the data point identified by cursor one. The "Value 2" function is the value of the data point identified by cursor two. The "T2−T1" function measures the difference in time between cursor 2 and cursor 1.

- To measure systolic blood pressure, find the area in the data after the blood pressure cuff has been inflated and blood flow to the finger has been occluded. Place the first cursor on the first pulse wave in the **Pulse** channel as the blood pressure cuff is being released (see Figure 7). Record the value for "Value1" from the **Blood Pressure** channel in Table 1. Leave this cursor in place.
- To measure diastolic pressure, place the second cursor on the first full-sized pulse pressure wave in the **Pulse** channel as the blood pressure cuff is being released (see Figure 7). Record the value for "Value 2" from the **Blood Pressure** channel in Table 1. Leave this cursor in place.
- The value for pulse pressure is the difference between systolic and diastolic pressures. This is the value measured by "V2−V1" from the **Blood Pressure** channel (see Figure 7). Record this value in Table 1.

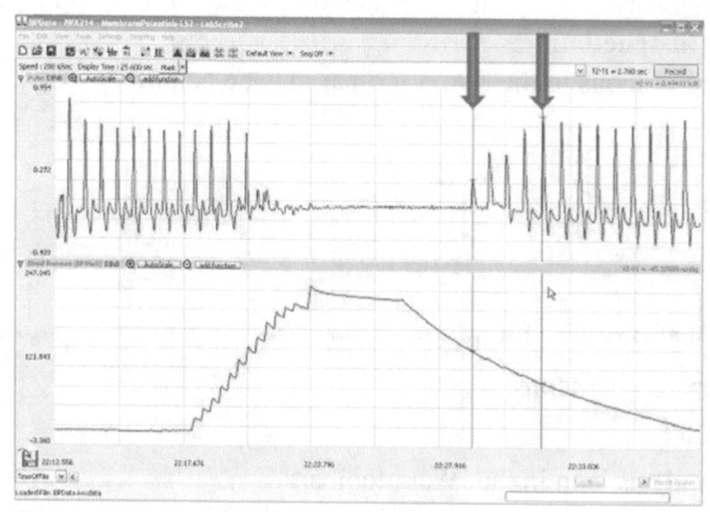

Figure 7 Data analysis blood pressure

- To measure peripheral pulse period, in the **Pulse** channel place one cursor at the peak of a pulse wave and the second cursor at the peak of the sequential pulse wave (see Figure 8). Record the "T2−T1" value in Table 1.
- Repeat the steps for data analysis for the second blood pressure trial in the current exercise. Record this value in Table 1. Calculate the mean value of the two trials.

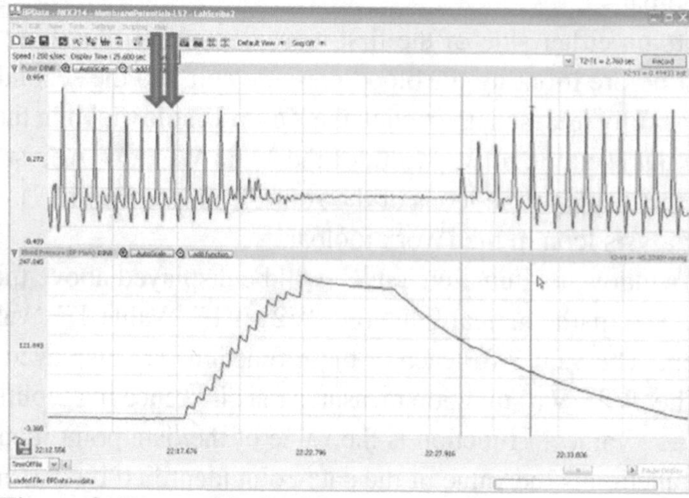

Figure 8 Data analysis pulse period

- Repeat the steps for data analysis for the blood pressure measurements in exercises 3 and 4.

**TABLE 1**: Blood Pressure Measurments and Calculations

| | | **Trial 1** | **Trial 2** | **Mean** |
|---|---|---|---|---|
| Supine | Systole (mm Hg) | | | |
| | Diastole (mm Hg) | | | |
| | Pulse Pressure (mm Hg) | | | |
| | Pulse Period (sec/pulse) | | | |
| Sitting | Systole (mm Hg) | | | |
| | Diastole (mm Hg) | | | |
| | Pulse Pressure (mm Hg) | | | |
| | Pulse Period (sec/pulse) | | | |
| Post-exercise | Systole (mm Hg) | | | |
| | Diastole (mm Hg) | | | |
| | Pulse Pressure (mm Hg) | | | |
| | Pulse Period (sec/pulse) | | | |

# Exercise 18

# Blood Cells and Blood Types

## Objectives

- Be able to identify the formed elements of the blood.
- Be able to identify the structure and characteristics of each of the formed elements.
- Be able to identify the whole blood concentrations of each of the formed elements.
- Be able to define the terms *antigen* and *antibody*.
- Be able to explain the effect of sensitization on antibody production.
- Be able to identify the antigens and antibodies associated with each blood type.
- Be able to identify blood type as a result of agglutination.
- Be able define the terms *major agglutination* and *minor agglutination*.
- Be able to explain the processes involved in major and minor agglutination.
- Understand the genetics of blood type.
- Be able to identify acceptable blood donors for any blood type and determine whether minor agglutination will occur.

## Lab Safety Reminders

- Handle only your own blood.
- Dispose of hazardous materials properly.
- Disinfect your work area.

## Introduction

Whole blood is a connective tissue that is composed of a fluid matrix and formed elements. **Plasma**, the fluid portion of the blood, is composed primarily of water, but also carries many chemicals including a variety of ions, dissolved gasses, proteins (enzymes, antibodies, and others), and hormones. The fluid component of blood allows the blood to flow. Changing the fluid composition of the plasma can have an effect on the ability of blood to flow. Of the approximately 5.5 L of whole blood in a person, approximately 3 L are plasma. The **formed elements** of the blood include **erythrocytes**, **leukocytes**, and **platelets**. Each of these began as a nucleated cell in the red bone marrow. The following table contains the average numbers or ranges of the formed elements in whole blood.

| Cellular Contents of Blood | |
|---|---|
| Formed element | Average whole blood concentration |
| Platelets | 150,000–450,000 per μL |
| Erythrocytes | 3.9–6.5 million per μL |
| Leukocytes | 4000–11,000 per μL |

## Leukocytes

**Leukocytes** are the white blood cells, often abbreviated **WBC**. The prefix *leuko-* means white or colorless, and the suffix *-cyte* means cell. The leukocytes have this name because the cytoplasm of cells is clear compared to the red color of the red blood cells. There are five different types of white blood cells. They can be divided into groups in a number of ways. One of the groupings is based on the presence or absence of granules in the cytoplasm. When stained, the cytoplasm of the **granulocytes** appears grainy. The granulocytes include the **neutrophils**, the **eosinophils**, and the **basophils**. The cytoplasm of the **agranulocytes** does not appear grainy when stained. The agranulocytes include the **monocytes** and the **lymphocytes**. The leukocytes can be identified not only by their cytoplasm but also by their size and the shape of the nucleus. Monocytes are the largest of the leukocytes and have an oval to kidney bean shaped nucleus. Lymphocytes are the smallest of the lymphocytes and have a round to slightly indented nucleus that may nearly fill the cell. Neutrophils are the most common leukocyte and have a multilobed or segmented nucleus. The eosinophils also have a multilobed nucleus, often bilobed. The basophils are the least common and have so many granules that the round nucleus is often not visible. See Figure 1, additional photos in your text, or the lab posters as references.

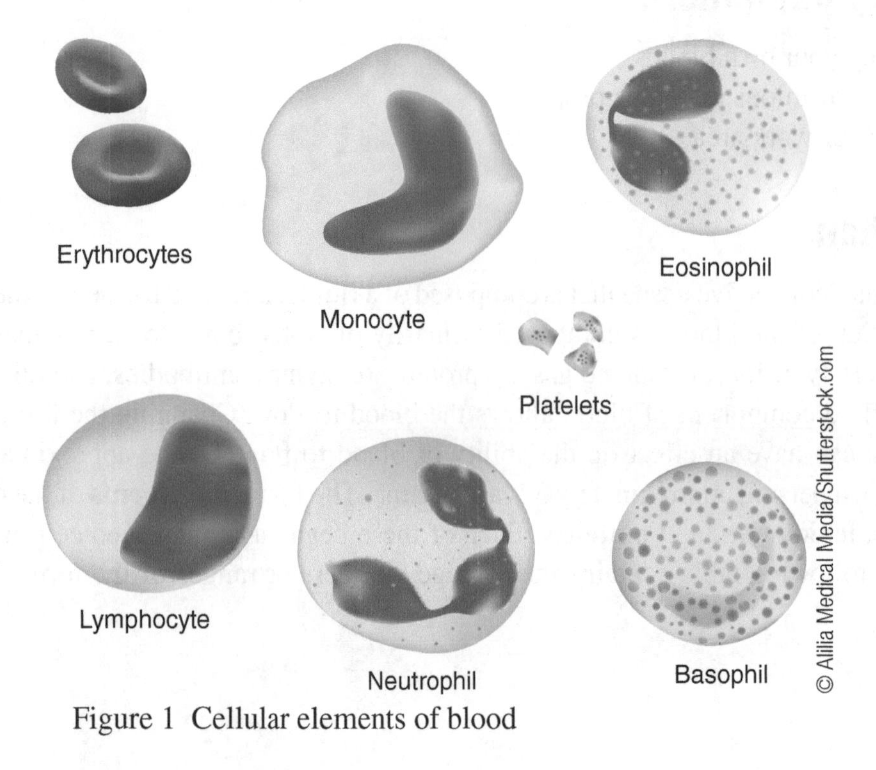


Figure 1 Cellular elements of blood

- **Leukocyte Activity**

  Leukocytes are the largest of the formed elements of the blood. They are also usually the only nucleated cells circulating in the plasma.
  - Using the prepared slide of the whole blood smear, identify each type of leukocyte. The basophils are the least common and may be more difficult to find. If you find one, let others know and see your specimen.
  - Refer to the chart in the lab or your textbook for pictures of the leukocytes.
  - Which of the leukocytes was most frequently observed? __________

## Platelets

**Platelets** are also referred to as **thrombocytes**. As functional components of blood, platelets are not cells but rather membrane-bound fragments of cytoplasm that came from **megakaryocytes**. Megakaryocytes are nucleated cells located in the red bone marrow while platelets function in the plasma.

- **Platelet Activity**

  Platelets are very small cell fragments. They are more common than leukocytes but much less common than erythrocytes. They are also much smaller than both the leukocytes and the erythrocytes.
  - Using the prepared slide of the whole blood smear, identify the platelets. Refer to the chart in the lab or your textbook for pictures of the platelets.

## Erythrocytes

**Erythrocytes** are the red blood cells, often abbreviated **RBC**. The prefix *erythro-* means red, and the suffix *-cyte* means cell. Erythrocytes have this name because of the red color of the cells. This color is the result of the presence of hemoglobin (Hb) in the cells. Hemoglobin contains iron. It is the iron of the hemoglobin that is capable of binding oxygen. When hemoglobin is exposed to oxygen, the iron binds oxygen, is oxidized, and turns bright red. Mature, circulating erythrocytes generally do not have a nucleus. Occasionally, an immature reticulocyte may be identified, but this is very rare. Erythrocytes also have a shape unique among the formed elements of the blood. They are biconcave. Think of them as looking like a donut but without the middle completely cut out, so they are thick around the edges and thin in the center. They are also very flexible, so they can be deformed as they flow through the small capillaries of the body.

- **Erythrocyte Activity**

  Erythrocytes are smaller than the leukocytes but larger than the platelets, and they have a biconcave shape, so they often look as though they have more color around the edges than in the middle. Also, circulating erythrocytes are not nucleated. They are much more common than either leukocytes or platelets.
  - Using the prepared slide of the whole blood smear, identify the erythrocytes. Refer to the chart in the lab or your textbook for pictures of the erythrocytes.

## Blood Type

Most people are familiar with the terminology of blood type, one or two letters followed by a + (positive) or a – (negative) sign. A person's basic blood type is referred to as the ABO blood type. This is followed by the Rh factor, the positive or negative notation. The major blood types are listed in practice tables later in this lab exercise. Blood type identification is based on the reaction between the **antigens, Ag**, (or agglutinogens) found on the cell membrane of the erythrocytes and free **antibodies, Ab**, (or agglutinins) that are produced by lymphocytes and added to the blood. The antigens serve as markers to identify your cells as belonging to you. Some cells of your immune system produce antibodies (specific types of proteins) that can react with antigens. However, your immune system should not produce antibodies that can react with your own antigens. So, if you have the A antigen on your erythrocytes (type A blood), you will not produce antibody a (anti-a). However, if you have been **sensitized** or exposed to the B antigen, you will be able to produce antibody b (anti-b). If you have the B antigen on your erythrocytes (type B blood), you will not produce anti-b, but you will produce anti-a after sensitization. If you have both the A and B antigens on your erythrocytes (type AB blood), you will not produce either anti-a or anti-b. If you have no antigens on your erythrocytes (type O blood), you will produce both anti-a and anti-b after sensitization. Sensitization to the A or B antigen generally occurs early in life and occurs easily and does not require direct exposure to erythrocytes. It may even occur as a result of digesting materials, including some bacteria, which contain the antigen. Infants aged 3 to 6 months may already be sensitized the ABO blood type antigens and have the appropriate antibodies circulating in their blood. All individuals should be assumed to be sensitized to the A and/or B antigens. For the Rh factor, if you have the Rh positive blood (+), you have the Rh or D antigen on your erythrocytes, you will not produce antibody d (anti-d). If you have the Rh negative blood (−), you do not have the Rh or D antigen on your erythrocytes, you will produce anti-d. However, you must have been **sensitized** or exposed to the D antigen, before you will be able to produce anti-d. Sensitization to the Rh antigen requires a direct exposure to erythrocytes with the D antigen.

| | **Transfusion Donor Blood Type** | | | |
|---|---|---|---|---|
| **Transfusion Recipient Blood Type** | **A** | **B** | **AB** | **O** |
| A | | Major agglutination & Minor agglutination | Major agglutination | |
| B | Major agglutination & Minor agglutination | | Major agglutination | |
| AB | Minor agglutination | Minor agglutination | | |
| O | Major agglutination | Major agglutination | Major agglutination | |

For the Rh factor, if you have the Rh positive blood (+), you have the Rh (D) antigen on your erythrocytes, you will not produce antibody d (anti-d). If you have the Rh negative blood (−), you do not have the Rh (D) antigen on your erythrocytes, you will produce anti-d. However, you must have been **sensitized** or exposed to the D antigen, before you will be able to produce anti-d. Sensitization to the Rh antigen requires a direct exposure to erythrocytes with the D antigen.

| | **Transfusion donor blood type** | |
|---|---|---|
| **Transfusion Recipient Blood Type** | **Rh +** | **Rh −** |
| Rh + | | Minor agglutination |
| Rh − | Major agglutination | |

When an antibody interacts with an antigen on the surface of your erythrocytes, it forms an **immune complex**. Because one antibody can bind to antigens on two erythrocytes and because one erythrocyte has the ability to bind with many antibodies, the formation of the immune complexes forms **antibody bridges** between erythrocytes and causes the erythrocytes to clump together and become nonfunctional. The clumping of the erythrocytes due to the interaction between the antibodies and erythrocyte antigens is called **agglutination**.

When a person receives blood in a transfusion, the blood must be properly matched to avoid the formation of antibody bridges and agglutination. There are two possible types of agglutination, **minor agglutination** and **major agglutination**. A minor agglutination occurs when the antibodies in the donated blood react with the antigens on the erythrocytes in the recipient blood and cause a small amount of agglutination. It is considered minor because there are usually very few antibodies in a transfusion of 1 or 2 pints of blood, and as a result, there is only a small amount of agglutination. A major agglutination occurs when the recipient receives the incorrect antigens on the erythrocytes in the donated blood. As a result, the recipient produces many antibodies toward the foreign antigen, and the result is a potentially large amount of agglutination. This can result in **transfusion shock** and potential death for the recipient. It is important to note that the formation of an antigen–antibody complex and the resulting agglutination of erythrocytes is NOT the same as blood coagulation or clotting.

Whole blood is not always given to patients in need of a transfusion. In many cases, whole blood is separated into the various components, erythrocytes, leukocytes, platelets, plasma proteins, and plasma or serum. This makes a variety of blood products available for separate use. Packed erythrocytes have only a residual amount of plasma and may be transfused when functional red blood cells are needed without additional blood components. Washed erythrocytes have had all plasma and plasma contents removed. The erythrocytes are then resuspended in isotonic saline solution. When washed erythrocytes are transfused, the antibodies in the plasma are not transfused. An additional advantage of packed and washed erythrocytes is that the plasma may undergo fractionation and the various components or fractions may be used for specific purposes. Plasma, the fluid component of blood, may be used in some transfusions. Serum, plasma without the clotting factors, may also be used in some cases. The advantage of transfusing plasma or serum is that no erythrocytes are being transfused. As a result, no antigens are introduced into the recipient's blood. However, the donor's antibodies will still be transfused into the recipient.

There are a variety of variations of blood transfusions. Each type of transfusion has its own specific composition and its specific use. The following table identifies some of the transfusions and their purposes.

| Type of Transfusion | Contents of the Transfusion | Use |
|---|---|---|
| Whole blood | Erythrocytes, leukocytes, platelets, plasma | Anemia due to a variety of causes, including renal failure, malignancies, internal bleeding, acute blood loss due to trauma |
| Packed red blood cells | Erythrocytes, small amount of plasma | Anemia, see above |
| Washed red blood cells | Erythrocytes, resuspended in saline solution | Anemia, see above<br>Used in cases of multiple transfusion and in patients with hypersensitivity to plasma proteins |
| Platelet concentrates | Platelets | Thrombocytopenia |
| Plasma | Water, plasma proteins, ions | Low blood volumes and some bleeding disorders that have no specific factor concentrate available |
| Plasma protein fraction | Plasma proteins | Replacement of all plasma proteins |
| Albumin | Albumin proteins | Correct colloid or oncotic pressure in cases of edema, burn victims, liver failure, hemorrhagic or surgery-induced shock |
| Gamma globulins | IgG antibodies | Used in intravenous and intramuscular vaccinations |
| Cryoprecipitated antihemophiliac globulin | Clotting factors, especially factors VIII and IX | Hemophilia A and hemophilia B |
| Lyophilized factor VIII | Factor VIII | Hemophilia A |

Blood type is an inherited trait. You inherited half of your chromosomes and thus genes in the egg produced by your mother and the other half in the sperm produced by your father. A **gene** is a segment of DNA that codes for a trait (a gene specifically codes for a polypeptide chain). There are often different forms of genes. These alternate gene forms are called **alleles**. For ex a gene that controls the production of the surface antigens that determine your blood is the specific form of that gene, and the allele determines whether you will produc the B antigen, or no antigen. So, for each trait, such as blood type, you have two co (two alleles) that control that trait, one copy from each parent. To determine wh an individual has, you must look at the individual's combination of the two alleles. If an allele is **dominant**, it will be expressed no matter what the other allele in the pair is. If an allele is **recessive**, it will be hidden if the other allele of the pair is a dominant allele. The recessive allele will be expressed only if both alleles of the pair are recessive. In blood type, the allele for type O blood (no antigen) is recessive, and both the allele for the A antigen and the allele for the B antigen are dominant. Because both the A and B alleles are dominant, they are called **codominant**. Both the A and the B alleles will be expressed when both are present. The Rh factor is a second gene that regulates the production of another antigen on the surface of erythrocytes. For the Rh factor, often referred to as antigen D, the allele that codes for the production of the Rh antigen is dominant, and the allele that codes for no Rh antigen production is recessive. A person with the Rh antigen is referred to as Rh+ (Rh positive), and a person without the Rh antigen is referred to as Rh– (Rh negative). Rh antibody production follows the same pattern as for a and b antibody production in the ABO blood types. An Rh+ person has the antigen and will not produce the Rh antibody. An Rh– person does not have the Rh antigen and after sensitization will produce the Rh antibody. The two antigen traits, ABO blood type and the Rh factor, are written together for an individual. For example, a person with the A antigen and the Rh antigen would be A+ (A positive). A person with the A antigen but not the Rh antigen would be A– (A negative).

## Blood Type Practice: Genetics

- What blood type would be the result of each of the following genetic combinations?

| Individual's allele combination | Individual's blood type |
|---|---|
| ABO alleles: | |
| A & A | ________ |
| A & O | ________ |
| B & B | ________ |
| B & O | ________ |
| A & B | ________ |
| O & O | ________ |
| Rh alleles: | |
| + & + | ________ |
| + & – | ________ |
| – & – | ________ |
| Both ABO & Rh alleles: | |
| A & A with + & + | ________ |

A & A
with
+ & – ________
A & A
with
– & – ________

## Blood Type Practice: ABO Blood Types

- Identify the antigens present in each of the following recipients.
- Assuming sensitization, identify the antibodies expected to be circulating in the plasma of the recipient.
- Identify the blood types that could be donated to the recipient without the risk of major agglutination. Which would result in minor agglutination?
- Identify the blood types that would result in major agglutination and should never be donated to that recipient.

| Blood Type of Recipient | Antigen Present on Erythrocyte | Antibody Present in Serum | Blood Type (s) of Possible Donors Will there be Minor Agglutination? | Blood Type (s) that Should Never be Given to this Recipient |
|---|---|---|---|---|
| A | | | | |
| B | | | | |
| AB | | | | |
| O | | | | |

## Blood Type Practice: ABO Blood Types and the Rh Factor (+ or –)

- Identify the antigens present in each of the following recipients.
- Assuming sensitization, identify the antibodies expected to be circulating in the plasma of the recipient.
- Identify the blood types that could be donated without the risk of major agglutination. Which would result in minor agglutination?
- Identify the blood types that would result in major agglutination and should never be donated to that recipient.

| Blood Type of Recipient | Antigen Present on Erythrocyte | Antibody Present in Serum | Blood Type (s) of Possible Donors Will there be Minor Agglutination? | Blood Type (s) that Should Never be Given to this Recipient |
|---|---|---|---|---|
| A+ | | | | |
| A– | | | | |

| B+ | | | | |
|---|---|---|---|---|
| B− | | | | |
| AB+ | | | | |
| AB− | | | | |
| O+ | | | | |
| O− | | | | |

## Blood Type Practice: Major and minor agglutination

- In the following hypothetical whole blood transfusion, identify the source and type of antigens and antibodies that would result in a major agglutination in the recipient. Assume all individuals have been sensitized to all antigens.

| | Antigen Involved in Major Agglutination | Antibody Involved in Major Agglutination |
|---|---|---|
| Donor: type A whole blood | | Not applicable |
| Recipient: type B whole blood | Not applicable | |

- In the following hypothetical whole blood transfusion, identify the source and type of antigens and antibodies that would result in a minor agglutination in the recipient. Assume all individuals have been sensitized to all antigens.

| | Antigen Involved in Minor Agglutination | Antibody Involved in Minor Agglutination |
|---|---|---|
| Donor: type A whole blood | Not applicable | |
| Recipient: type B whole blood | | Not applicable |

- In the following hypothetical whole blood transfusion, identify the source and type of antigens and antibodies that would result in a major agglutination in the recipient. Assume all individuals have been sensitized to all antigens.

| | Antigen (s) Involved in Major Agglutination | Antibody (ies) Involved in Major Agglutination |
|---|---|---|
| Donor: type A+ whole blood | | Not applicable |
| Recipient: type A− whole blood | Not applicable | |

- In the following hypothetical whole blood transfusion, identify the source and type of antigens and antibodies that would result in a minor agglutination in the recipient. Assume all individuals have been sensitized to all antigens.

| | Antigen (s) Involved in Minor Agglutination | Antibody (ies) Involved in Minor Agglutination |
|---|---|---|
| Donor: type A− whole blood | Not applicable | |
| Recipient: type A+ whole blood | | Not applicable |

- What blood type could serve as a universal donor for packed erythrocytes with no risk of either major or minor agglutination?
- In the following hypothetical plasma transfusion, identify the source and type of antigens and antibodies that would result in a minor agglutination in the recipient. Assume all individuals have been sensitized to all antigens.

| | Antigen (s) Involved in Minor Agglutination | Antibody (ies) involved in Minor Agglutination |
|---|---|---|
| Donor: type A+ whole blood | Not applicable | |
| Recipient: type B− whole blood | | Not applicable |

- Would major agglutination ever be a problem in plasma transfusions?

- What blood type could serve as a universal donor for plasma or serum transfusions with no risk of minor agglutination?

## Blood Type Activity

Blood type is determined by adding a drop of **antiserum** containing a specific antibody to a drop of blood and looking for agglutination. If agglutination can occur, the antibodies in the antiserum will bind to the antigens on the erythrocytes, form antibody bridges, and cause erythrocytes to agglutinate. Because of the positive reaction between that particular antibody and antigen, you know the blood contains that antigen and, thus, is that blood type.

- You will identify the types of four different blood samples.
- Obtain four clean microscope slides. Label each slide with the number or letter of the unknown blood you will be testing.
- Obtain at least 12 toothpicks.
- Using a wax pencil, draw three circles on each of the four slides.

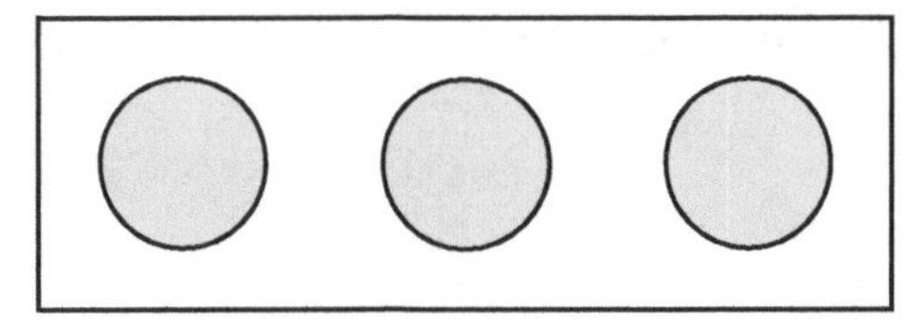

Blood Typing Step 1

- For each of your unknowns, place one drop of blood in each of the three circles.
- For each of your unknowns, add one drop of anti-a serum to the first circle of the slide. Mix the blood and anti-a serum with a clean toothpick. After mixing, discard the toothpick in the biohazards box. You **must** use a clean toothpick for mixing **each** drop of blood and antiserum.

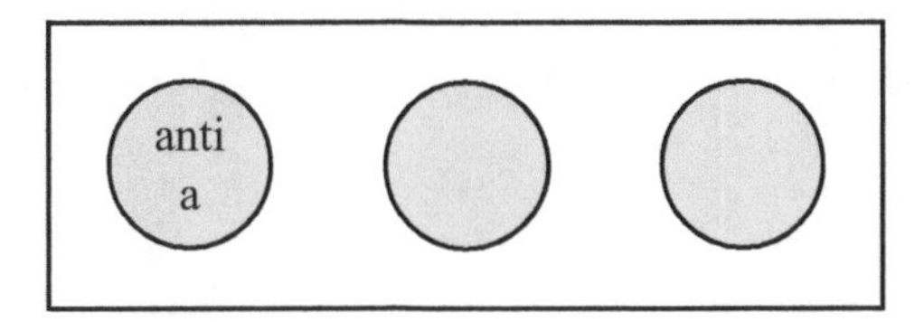


Blood Typing Step 2

- For each of your unknowns, add one drop of anti-b serum to the second circle of the slide. Mix the blood and anti-b serum with a clean toothpick. After mixing, discard the toothpick in the biohazards box.

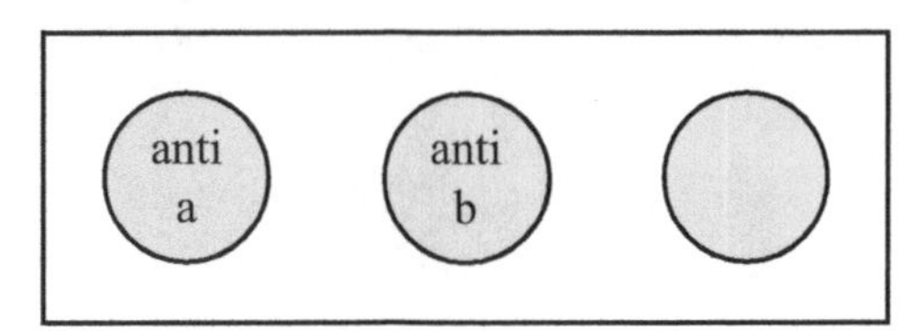


Blood Typing Step 3

- For each of your unknowns, add one drop of anti-d serum to the third circle of the slide. Mix the blood and anti-d serum with a clean toothpick. After mixing, discard the toothpick in the biohazards box.

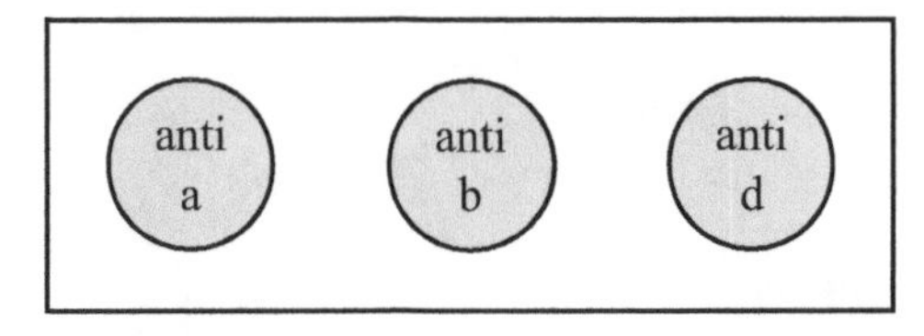


Blood Typing Step 4

- Observe each circle for agglutination.
- Record your results in the Review Questions section at the end of the lab.

- Why is it important to add only one type of antiserum to each drop of blood?

- Why is it important that you mix each antiserum and blood drop with a clean toothpick?

- Why is there no anti-o serum?

# Exercise 19

# Erythrocyte Count and Other Indices and Hemoglobin Content

## Objectives

- Be able to discuss the components of hemoglobin and their functions.
- Be able to define each of the following terms and indicate where in the body you would be most likely to find each: *oxyhemoglobin*, *deoxyhemoglobin*, *carboxyhemoglobin*, *reduced hemoglobin*, *carbaminohemoglobin*, and *glycosylated hemoglobin*.
- Be able to identify the cause of each of the terms listed above.
- Define the term *anemia*.
- Be able to identify the hemocytometer and procedures used to count erythrocytes.
- Given a red blood cell count from a hemocytometer, be able to calculate the number of red blood cells per μL.
- Be able to identify when an erythrocyte count is appropriate, low, or high for an adult of either sex.
- Be able to define and use the term *erythrocytopenia*.
- Be able to define and use each of the following terms: *relative polycythemia*, *polycythemia vera* (*erythemia*), and *secondary polycythemia* (*erythrocytosis*).
- Be able to explain the procedure used in a hematocrit.
- Be able to determine the value of an apparent hematocrit.
- Be able to identify when a hematocrit value (%) is appropriate, low, or high for an adult of either sex.
- Be able to define the following terms describing erythrocyte size: *normocytic*, *microcytic*, and *macrocytic*.
- Be able to explain the effect each of the terms listed above could have on a hematocrit.
- Be able to determine hemoglobin content from a Tallquist test.
- Be able to identify whether a hemoglobin value (g/dL) is appropriate, low, or high for an adult of either sex.
- Be able to define the following terms describing erythrocyte color: *normochromic*, *hypochromic*, and *hyperchromic*.
- Be able to identify the effect each of the terms listed above could have on a person's ability to transport oxygen.
- Be able to calculate and interpret an MCV, MCH, and MCHC.
- Given values for a red blood cell count, a hematocrit, and a Tallquist test, determine whether or not abnormalities exist and their possible causes.

## Lab Safety Reminders

- Handle only your own blood.
- Dispose of hazardous materials properly.
- Disinfect your work area.

## Introduction

If you have ever donated blood, you may have noticed that the blood that fills the donation bag has a dark maroon color as compared to the bright red blood you see if you suffer from a minor cut or scrape. In the case of the donated blood, the blood is being taken from a vein and is not exposed to atmospheric oxygen ($O_2$). In the case of a minor cut, the blood is coming from capillaries or small arterioles or venules, and the blood is exposed to oxygen in the air. A molecule of **hemoglobin (Hb)** is a multimeric protein composed of four polypeptide chains that make up the globin portion of the hemoglobin, two alpha chains and two beta chains. Each of these polypeptide chains is bound to a heme group made up of a porphyrin ring containing an atom of iron. It is the iron atom that binds the oxygen. Because each molecule of hemoglobin contains four heme groups and thus four iron atoms, each molecule of hemoglobin is capable of binding a total of four oxygen atoms. When exposed to oxygen, the hemoglobin molecules in the erythrocytes (RBC) become bright red. Hemoglobin with oxygen bound is call **oxyhemoglobin**. When the oxygen is released from the hemoglobin, it is called **deoxyhemoglobin**. Carbon monoxide (CO) can also bind to and oxidize the iron in the hemoglobin, resulting in a bright red color. When carbon monoxide is bound to a hemoglobin molecule, oxygen cannot bind and the person is at risk of carbon monoxide poisoning that may lead to hypoxia and asphyxiation. Hemoglobin with carbon monoxide bound is called **carboxyhemoglobin**. In addition to binding oxygen and carbon monoxide, a molecule of hemoglobin can also bind hydrogen (**reduced hemoglobin**), carbon dioxide (**carbaminohemoglobin**), and nitric oxide (NO), but these bind to the protein portion not to the iron and do not turn the hemoglobin bright red. As hemoglobin releases oxygen in the systemic tissues, nitric oxide is also released from hemoglobin. The presence of the nitric oxide in the systemic tissues may act as a vasodilator. Some hemoglobin molecules may have a glucose molecule bound to an amino acid in the protein portion of the hemoglobin. This form of hemoglobin is called **glycosylated hemoglobin** (Hb A1C). In most individuals, the percent of glycosylated hemoglobin is low, usually 3–5%. In individuals who tend to routinely experience hyperglycemia, as in uncontrolled diabetes mellitus, the percent of glycosylated hemoglobin may be 2–3 times higher and is used as an indicator of how well blood glucose levels have been controlled over the past 2–3 months.

It is the ability of hemoglobin to bind oxygen as well as to release the oxygen that is essential to the functioning of your cells. When red blood cells are in the pulmonary circulation and pass through the capillaries of the lungs, they release carbon dioxide and hydrogen and pick up oxygen. When the red blood cells are in the systemic circulation and pass through the capillaries they release oxygen and pick up carbon dioxide and hydrogen. The oxygen released from the hemoglobin is used for aerobic cellular respiration and ATP production. The carbon dioxide and hydrogen picked up are the byproducts of aerobic cellular respiration. Thus, it is essential that the blood be capable of transporting enough oxygen to meet the demands of the cells. **Anemia** is a condition that limits the ability to transport oxygen in the blood. There are many causes of anemia, but the end result is the same; anemia occurs when the blood is unable to transport more than 13.5 mL oxygen per dL of whole blood. Because both red blood cells and hemoglobin are essential to the transport of oxygen, you can conclude that altering the number of circulating erythrocytes, the amount of hemoglobin, the ability of hemoglobin to function properly, or any combination of these could alter the volume of oxygen being transported. The activities in this lab are designed to give you the opportunity to see some common blood assays and consider the potential effects of abnormal results from these tests.

## Red Blood Cell Count

Because the hemoglobin is contained within the erythrocytes, the number of erythrocytes plays a key role in the ability of the blood to transport oxygen. The normal range for a red blood cell count in adult females is 3.9–5.6 million cells per μL of whole blood. The normal range for a red blood cell count in adult males is 4.5–6.5 million cells per μL of whole blood (note: these values will vary slightly among sources and depending upon the method of testing). Counts falling below these ranges produce a condition called erythrocytopenia. Counts above these ranges produce a condition called polycythemia. There are three forms of polycythemia. **Polycythemia vera**, or erythemia, occurs when there has been no arterial hypoxia. Some unknown reason has caused the body to make an excess of erythrocytes. One possibility is the overproduction of erythropoietin by the kidneys. Another possibility is a malfunction of the red bone marrow. **Secondary polycythemia** (erythrocytosis) occurs when a person has experienced systemic arterial hypoxia. The decreased transport of oxygen during systemic arterial hypoxia stimulates the release of erythropoietin from the kidney and thus the increased production of erythrocytes. Diseases of the lung such as chronic obstructive pulmonary disease (COPD) and restrictive pulmonary diseases (RPD) reduce the ability of the blood to be oxygenated. The increased oxygen demand of pregnancy will reduce oxygen availability to the systemic tissues. A person who is at a high altitude experiences a lower concentration of oxygen in the air. This will reduce the ability of the blood to be oxygenated. In all these cases, the person will experience arterial hypoxia and polycythemia will result. Both polycythemia vera and secondary polycythemia will exhibit not only a higher than normal red blood cell count but also a higher than normal total red blood cell mass. **Relative polycythemia** also results in a higher than normal red blood cell count but not a higher than normal total red blood cell mass. In this case, the red blood cells have become more concentrated, usually due to water loss such as in dehydration or loss of fluids from the plasma as might be experienced in shock or in more severe burns.

Red cell counts can be performed using a special slide called a **hemocytometer**. The hemocytometer is a precise slide that contains a ruled scale ground into the surface (see Figure 1). This produces a counting chamber of precise volume. Each of the smallest of squares of the slide counted for a red blood cell count covers an area of 0.05 mm x 0.05 mm or 0.0025 $mm^2$. Because there are 16 small squares in each of the sections counted, the total area of each of the sections is 0.04 $mm^2$. Because 5 sections are counted, the total area counted in a red blood cell count is 0.2 $mm^2$. The depth of the space between the counting chamber and the coverslip is 0.1 mm. Therefore, the total volume of the areas being counted in a red blood cell count is 0.02 $mm^3$, or 0.02 μL. To perform a red blood cell count, a known amount of whole blood must be drawn and diluted with a known amount of isotonic solution. If the blood is not diluted, the red blood cells would be too thick to count. Care must be taken not to count cells more than once. Before you begin counting, determine some rules for counting that will help you avoid counting a cell more than once (see Figure 2). For example, count the cells that touch the line to the left of the square but do not count the cells that touch the line to the right of the square (those will be counted in the square to the right or not at all); count the cells that touch the line at the top of the square but do not count the cells that touch the line at the bottom of the square (those will be counted in the square below or not at all).

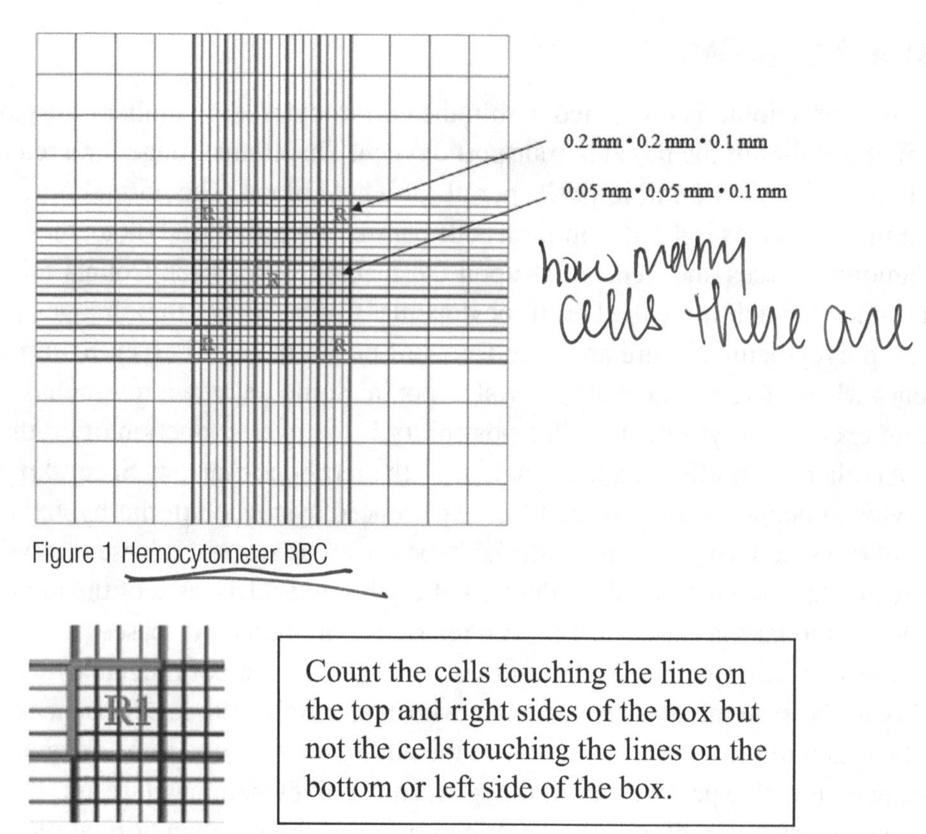
0.2 mm • 0.2 mm • 0.1 mm
0.05 mm • 0.05 mm • 0.1 mm
R1
Count the cells touching the line on the top and right sides of the box but not the cells touching the lines on the bottom or left side of the box.

Figure 1 Hemocytometer RBC

Figure 2 Counting rules RBC

### Hemocytometer Activity

The hemocytometer grid and process of slide preparation (Becton-Dickinson Unopette 5851, dilution factor of 200) will be explained by your lab instructor. Photographs have been taken of a hemocytometer prepared for a red blood cell count. There is a photograph of the entire counting surface as well as individual photographs of each of the five red blood cell counting sections.

- Using the photographs, count each of the five red blood cell sections of the counting surface. Each section contains 16 small squares.
- Record the number of red blood cells counted in each of the five red blood cell sections in the chart.
- To calculate the total red blood cell count, simply add the counts from all five sections and multiply by 10,000.

$$\text{Total \# of red blood cells counted} \cdot 10{,}000 = \text{RBC per } \mu\text{L}$$

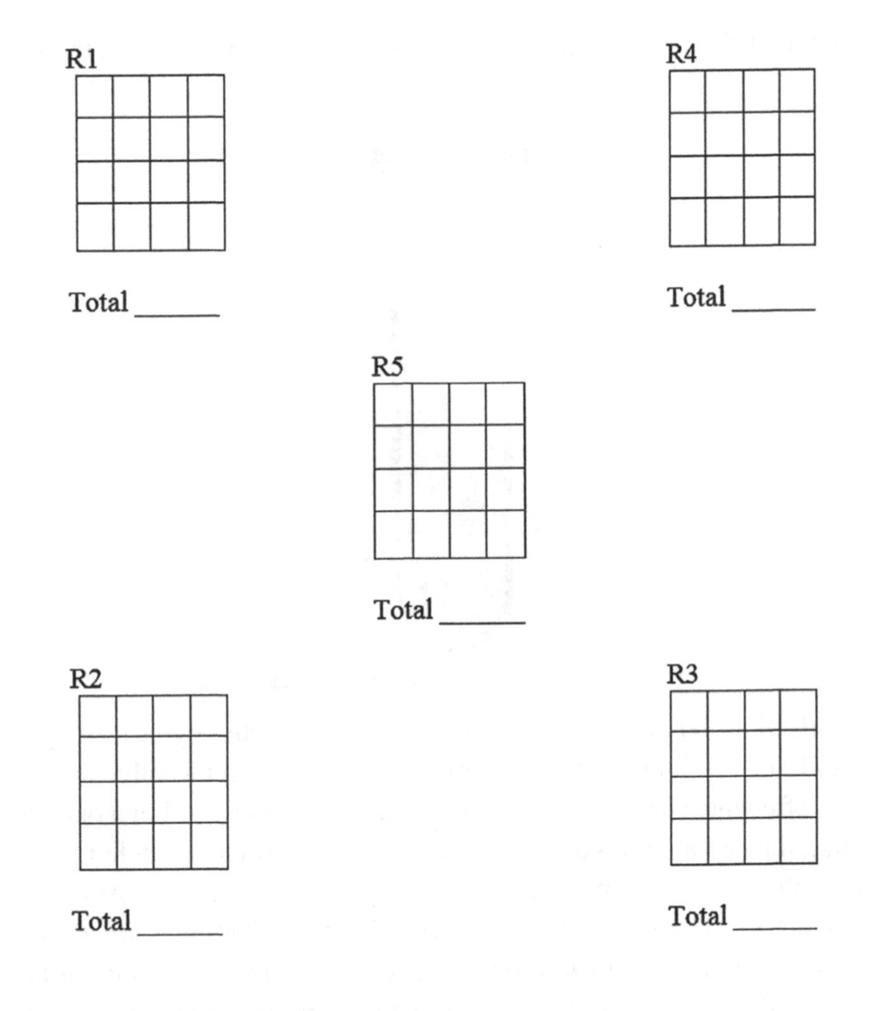


Calculation: total red blood cell count = ________________

- How does this count compare to the acceptable range for an adult female?

- How does this count compare to the acceptable range for an adult male?

## Hematocrit

A **hematocrit** determines the % volume of packed red blood cells in a sample of whole blood. The normal range for a hematocrit in adult females is 42 ± 5%. The normal range for a hematocrit in adult males is 47 ± 7% (note: these values will vary slightly among sources and depending upon the method of testing). A sample of blood is drawn into a heparinized capillary tube (heparin is a chemical that prevents clot formation in blood). The end of the capillary tube is then sealed, and the capillary tube is placed in a centrifuge and spun. Centrifugal force causes the heavier red blood cells to be compacted at the bottom of the tube. A small buffy layer containing the leukocytes and platelets settles on top of the erythrocytes. The lightest of blood components, the plasma is on top (see Figure 3).

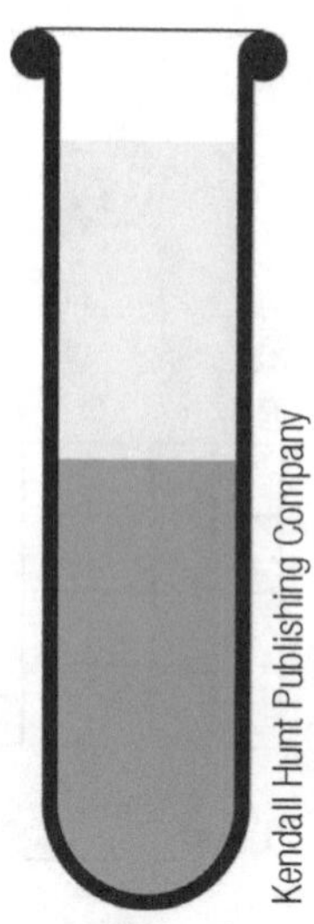


Figure 3 Centrifuged whole blood

If a sample of 20 mL of blood is centrifuged, and the packed erythrocytes take up 9 mL, then the hematocrit is 9 mL/20 mL = 0.45 or 45%. Several factors can affect the value of a hematocrit. The most obvious factor affecting a hematocrit is the number of erythrocytes. For a person with a normal red blood cell count, the hematocrit would be expected to be normal. An individual with an abnormal red blood cell count would be expected to have an abnormal hematocrit. Another factor that can affect a person's hematocrit is erythrocyte size. Normal-sized erythrocytes are termed **normocytic**. Erythrocytes smaller than normal are **microcytic**, and those larger than normal are **macrocytic**. How could an abnormality in erythrocyte size alter the hematocrit?

- **Hematocrit Activity**

The hematocrits have been prepared for you and have been properly placed on the hematocrit scale (see Figure 4). There are three sets of hematocrits. Set A is below normal; set B is within the normal range; and set C is above the normal range.

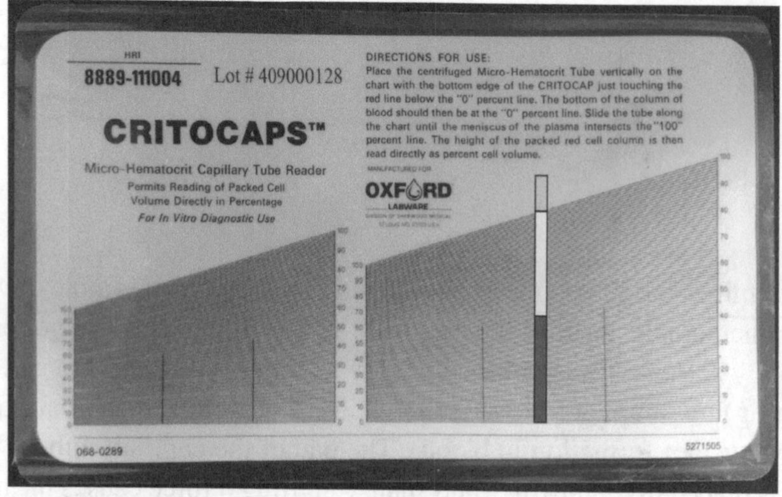


Figure 4 Hematocrit scale with example tube

Volume

- Using the prepared micro-hematocrits, determine the value of each of the prepared hematocrits in the set. The micro-hematocrit capillary tubes have been properly place on the Critocaps ™ hematocrit scale with the base of the packed erythrocytes placed on the 0% line and the top of the plasma placed on the 100% line. The hematocrit value is read by finding the value for the line at the intersection of the packed erythrocytes and the plasma (see Figure 4). Record the hematocrit values for each of the samples, A, B, and C, below.
- Set A hematocrit value: __________
  How would this value affect the person's ability to transport oxygen?

- Set B hematocrit value: __________
  How would this value affect the person's ability to transport oxygen?

- Set C hematocrit value: __________
  How would this value affect the person's ability to transport oxygen?

## Tallquist Test

The **Tallquist test** uses a colored hemoglobin scale to estimate the amount of hemoglobin in the blood. Because the amount of hemoglobin determines the amount of oxygen that can be transported by the blood, this is a very informative test. The normal range for hemoglobin content in adult females is 12–16 g/dL whole blood and 14–17 g/dL whole blood in adult males (average for adults is approximately 15 g/dL) (note: these values will vary slightly among sources and depending upon the method of testing). The standard for the Tallquist scale is 15.6 g/dL (100%). By comparing fresh blood color to the colors of the chart, the hemoglobin content as well as the % of normal (15.6 g/dL) can be determined. The value obtained can then be used to assess an individual's ability to transport oxygen. Because each gram of hemoglobin can transport 1.34 mL of $O_2$, the normal person is expected to transport 20.1 mL $O_2$ per dL of whole blood (g Hb • 1.34 mL $O_2$). The normal range of the ability of blood to transport oxygen is 16–25 mL $O_2$/dL whole blood. These values are well above 13.5 mL $O_2$ per dL whole blood that defines anemia.

Keep in mind that the hemoglobin is contained within the erythrocytes, so the amount of hemoglobin per erythrocyte determines the color of the cells and thus the color of the whole blood. Erythrocytes that are normal in color are called **normochromic**. Those that are lighter in color are called **hypochromic**, and those that are darker than normal are called **hyperchromic**.

- **Tallquist Activity**

You will determine the hemoglobin content of three strips of simulated blood. Note that the color of the "blood" on the strips may not perfectly match one of the colors on the Tallquist scale (see Figure 5). This could be true of real blood, too. For colors that don't match perfectly, find the closest color match or estimate the hemoglobin value.

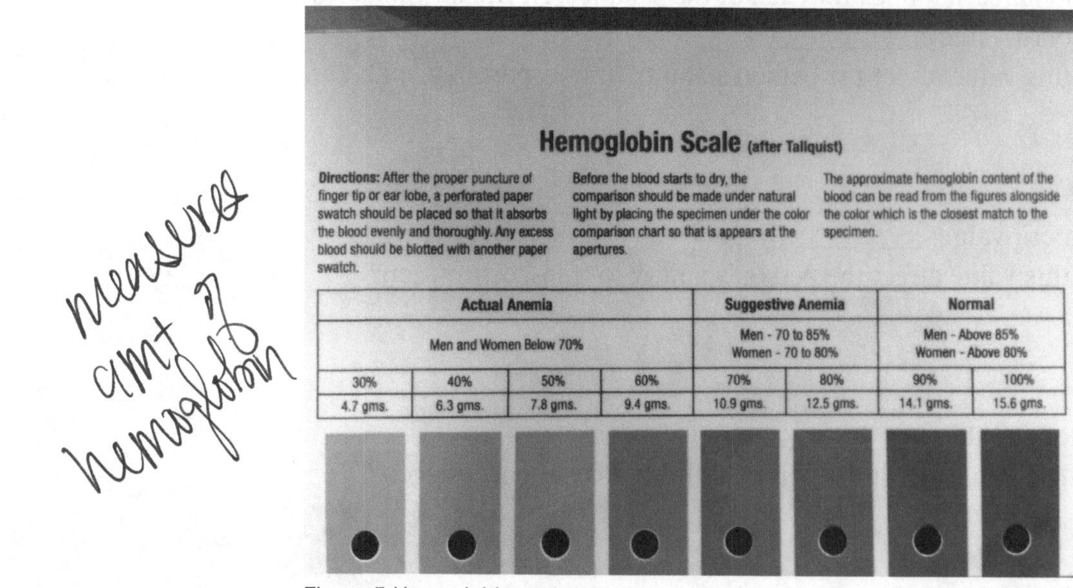

**Hemoglobin Scale** (after Tallquist)

**Directions:** After the proper puncture of finger tip or ear lobe, a perforated paper swatch should be placed so that it absorbs the blood evenly and thoroughly. Any excess blood should be blotted with another paper swatch.

Before the blood starts to dry, the comparison should be made under natural light by placing the specimen under the color comparison chart so that is appears at the apertures.

The approximate hemoglobin content of the blood can be read from the figures alongside the color which is the closest match to the specimen.

| Actual Anemia | | | | Suggestive Anemia | | Normal | |
|---|---|---|---|---|---|---|---|
| Men and Women Below 70% | | | | Men - 70 to 85%<br>Women - 70 to 80% | | Men - Above 85%<br>Women - Above 80% | |
| 30% | 40% | 50% | 60% | 70% | 80% | 90% | 100% |
| 4.7 gms. | 6.3 gms. | 7.8 gms. | 9.4 gms. | 10.9 gms. | 12.5 gms. | 14.1 gms. | 15.6 gms. |

Figure 5 Hemoglobin scale

- Paper with three strips of color (labeled A, B, and C) simulating blood have been prepared for you.
- Place each strip of "blood" under the Tallquist Hemoglobin Scale to find the color on the scale that most closely matches the color of the "blood." Record the amount of hemoglobin in the sample (g/dL).
- Determine the % of normal hemoglobin content from the Tallquist Hemoglobin Scale for each strip of "blood." Record the amounts.
- Categorize each strip of "blood" with regards to anemia by calculating the ability of each "blood" to transport oxygen.
  Oxygen carrying capacity calculation:

$$\text{Hb g/dL} \cdot 1.34 \text{ mL } O_2 \text{ per g Hb} = \text{mL } O_2 \text{ per dL whole blood}$$

- Set A hemoglobin content: __________ g/dL % of normal __________
  Ability to transport $O_2$: __________ mL $O_2$ per dL whole blood
  How does the calculated value of $O_2$/dL whole blood compare to the normal value (16–25 mL $O_2$/dL whole blood)?

How does the calculated value of $O_2$/dL whole blood compare to the value that defines anemia?

- Set B hemoglobin content: __________ g/dL % of normal __________
  Ability to transport $O_2$: __________ mL $O_2$ per dL whole blood
  How does this value compare to the value that defines anemia?

- Set C hemoglobin content: __________ g/dL % of normal __________
  Ability to transport $O_2$: __________ mL $O_2$ per dL whole blood
  How does this value compare to the value that defines anemia?

## Red Cell Indices—Combining Test Data

As you may have concluded from the information, although each test performed today gives important information, the whole picture comes from combining the information from all these tests. For example, information from a Tallquist test may indicate that a person's hemoglobin content is low enough that it suggests anemia. This information alone is not enough to determine the cause of the problem. You still need to determine why the hemoglobin content is below normal. It could be because there is not much hemoglobin in each cell (hypochromic erythrocytes), but it could also be the result of a red blood cell count that is below normal.

Red cell indices combine the data from these three tests (RBC count, hematocrit, and Tallquist test) to give a more complete picture for a more accurate assessment of the individual's condition. The following are the red cell indices you will calculate and use to assess your data. The electronic diagnostic red cell counts automatically also calculate the red cell indices. The values given for each of the following indices are within the normal range. However, the normal ranges may vary by lab and specific testing procedures.

**MCV: Mean corpuscular volume** is the average volume of an erythrocyte measured in femtoliters (fL) ($10^{-15}$ L). The normal adult value is 87 ± 5 fL. It is calculated as:

$$\text{MCV (fL)} = \text{hematocrit ratio} \cdot 10^3 / \text{red blood cell count in millions per } \mu\text{L}$$

$$\text{Example: } 0.45 \cdot 10^3 / 4.9 \text{ million per } \mu\text{L}$$

Note: The hematocrit ratio is the raw number (e.g., 0.45) and not the percent value (45%).

**MCH: Mean corpuscular hemoglobin** is the average amount of hemoglobin in an erythrocyte measured in picograms (pg). The normal adult value is 29 ± 5 pg. It is calculated as:

$$\text{MCH (pg)} = \text{g/dL Hb} \bullet 10 \,/\, \text{red blood cell count in millions per } \mu\text{L}$$
$$\text{Example: } 15 \text{ g/dL Hb} \bullet 10 \,/\, 4.9 \text{ million per } \mu\text{L}$$

**MCHC: Mean corpuscular hemoglobin concentration** is the proportion of hemoglobin per erythrocyte measured as a percent (%). The normal adult value is 34 ± 2%. It is calculated as:

$$\text{MCHC (\%)} = \text{g/dL Hb} \,/\, \text{hematocrit ratio (raw number, see above)}$$
$$\text{Example: } 15 \text{ g/dL Hb} \,/\, 0.45$$

■ **Activity**

Using the one value obtained from the red blood cell count along with the three values obtained from the hematocrits and the three values obtained from the Tallquist tests, you will make several combinations. For each of the combinations, you will calculate MCV, MCH, and MCHC. With these values, you will then characterize the cells using the following terms: normocytic, microcytic, macrocytic, normochromic, hypochromic, and hyperchromic.

- Using the red blood cell count and sample A from both the hematocrit and the Tallquist test, calculate and record the following values. Indicate whether the value is lower than normal, normal, or higher than normal (circle the correct answer). From this information, characterize the red blood cells by size and color (circle the correct answer).

  MCV __________ Low / Normal / High

  MCH __________ Low / Normal / High

  MCHC __________ Low / Normal / High

  Characterization by size: Microcytic / Normocytic / Macrocytic

  Characterization by color: Hypochromic / Normochromic / Hyperchromic

- Using the red blood cell count and sample B from both the hematocrit and the Tallquist test, calculate and record the following values.

  MCV __________ Low / Normal / High

  MCH __________ Low / Normal / High

  MCHC __________ Low / Normal / High

  Characterization by size: Microcytic / Normocytic / Macrocytic

  Characterization by color: Hypochromic / Normochromic / Hyperchromic

- Using the red blood cell count and sample C from both the hematocrit and the Tallquist test, calculate and record the following values.

  MCV __________ Low / Normal / High

  MCH __________ Low / Normal / High

  MCHC __________ Low / Normal / High

  Characterization by size: Microcytic / Normocytic / Macrocytic

  Characterization by color: Hypochromic / Normochromic / Hyperchromic

- Using the red blood cell count and sample A from the hematocrit and sample B from the Tallquist test, calculate and record the following values.

  MCV __________ Low / Normal / High

  MCH __________ Low / Normal / High

  MCHC __________ Low / Normal / High

  Characterization by size: Microcytic / Normocytic / Macrocytic

  Characterization by color: Hypochromic / Normochromic / Hyperchromic

- Using the red blood cell count and sample A from the hematocrit and sample C from the Tallquist test, calculate and record the following values.

  MCV __________ Low / Normal / High

  MCH __________ Low / Normal / High

  MCHC __________ Low / Normal / High

  Characterization by size: Microcytic / Normocytic / Macrocytic

  Characterization by color: Hypochromic / Normochromic / Hyperchromic

- Using the red blood cell count and sample B from the hematocrit and sample C from the Tallquist test, calculate and record the following values.

  MCV __________ Low / Normal / High

  MCH __________ Low / Normal / High

  MCHC __________ Low / Normal / High

  Characterization by size: Microcytic / Normocytic / Macrocytic

  Characterization by color: Hypochromic / Normochromic / Hyperchromic

Normocytic/normochromic anemia is generally the result of a loss of blood either through hemorrhage or sepsis, long-term illnesses, aplastic anemia, or lack of erythropoietin production often associated with kidney disease or failure. Microcytic/hypochromic anemia may be caused by iron deficiency, inflammation, lead poisoning, and the genetic condition of thalassemia. Macrocytic/normochromic anemia may be caused by vitamin $B_{12}$ deficiency, folate deficiency, or some forms of chemotherapy.

# Exercise 20

# Leukocyte and Platelet Counts

## Objectives

- Be able to identify the hemocytometer and procedures used to count leukocytes and platelets.
- Given a white blood cell count from a hemocytometer, be able to calculate the number of white blood cells per μL.
- Be able to identify when a leukocyte count is appropriate, low, or high for an adult.
- Be able to define and use the terms *leucopenia* and *leukocytosis*.
- Given the values from a differential white cell count, be able to identify whether the values for the specific types of leukocytes are too high or too low.
- Be able to define and use each of the following terms: *neutropenia*, *neutrophilia*, *lymphocytosis*, *monocytosis*, *eosinophilia*, and *basophilia*.
- Given a platelet count from a hemocytometer, be able to calculate the number of platelets per μL.
- Be able to identify when a platelet count is appropriate, low, or high for an adult.
- Be able to define and use the terms *thrombocytopenia* and *thrombocytosis*.
- Be able to define the term *hemostasis*.
- Be able to identify the stages involved in coagulation and the major reactions that occur in each stage.
- Be able to explain the difference between coagulation and agglutination.
- Be able to explain the difference between plasma and serum.

## Lab Safety Reminders

- Handle only your own blood.
- Dispose of hazardous materials properly.
- Disinfect your work area.

## Introduction

Along with erythrocytes, the leukocytes and platelets make up the cellular components of the blood. The leukocytes are the active cells of the immune system. They provide the body with defense against foreign pathogens and also protect the body from cancer cells. They are also responsible for removing damaged and dead cells as well as antigen–antibody complexes. Although we expect to find leukocytes in the circulatory system, they may also be found in other tissues of the body, especially the primary and secondary lymph organs, including the lymph nodes. Platelets are responsible in part for the process of hemostasis or the reduction or stop of blood loss.

The activities in this lab are designed to give you the opportunity to see some common blood assays and consider the potential effects of abnormal results from these tests.

## White Blood Cell Count

The normal range for a white blood cell count in adults is 4000 to 11,000 cells per μL of whole blood (note: these values will vary slightly among sources and depending upon the method of testing). When the total white blood cell count is lower than normal, the person has **leukopenia**. When the total white blood cell count is higher than normal, the person has **leukocytosis**.

White cell counts can be performed using a special slide called a hemocytometer. The hemocytometer is a precise slide that contains a ruled scale ground into the surface (see Figure 1). This produces a counting chamber of precise volume. Each of the largest of squares of the slide counted for a white blood cell count covers an area of 0.25 mm × 0.25 mm or 0.0625 $mm^2$. Because there are 16 small squares in each of the sections counted, the total area of each of the sections is 1 $mm^2$. Because 4 sections are counted, the total area counted in a white blood cell count is 4 $mm^2$. The depth of the space between the counting chamber and the coverslip is 0.1 mm. Therefore, the total volume of the areas being counted in a red blood cell count is 0.4 $mm^3$ or 0.4 μL. To perform a white blood cell count, a known amount of whole blood must be drawn and diluted with a known amount of slightly hypotonic solution. If the blood is not diluted, the white blood cells are too concentrated and would be difficult to count. The solution is hypotonic so that the erythrocytes will lyse and the white blood cells will be much more visible. Care must be taken not to count cells more than once. Before you begin counting, determine some rules for counting that will help you avoid counting a cell more than once (see Figure 2). For example, count the cells that touch the line to the left of the square but do not count the cells that touch the line to the right of the square (those will be counted in the square to the right or not at all); count the cells that touch the line at the top of the square but do not count the cells that touch the line at the bottom of the square (those will be counted in the square below or not at all).

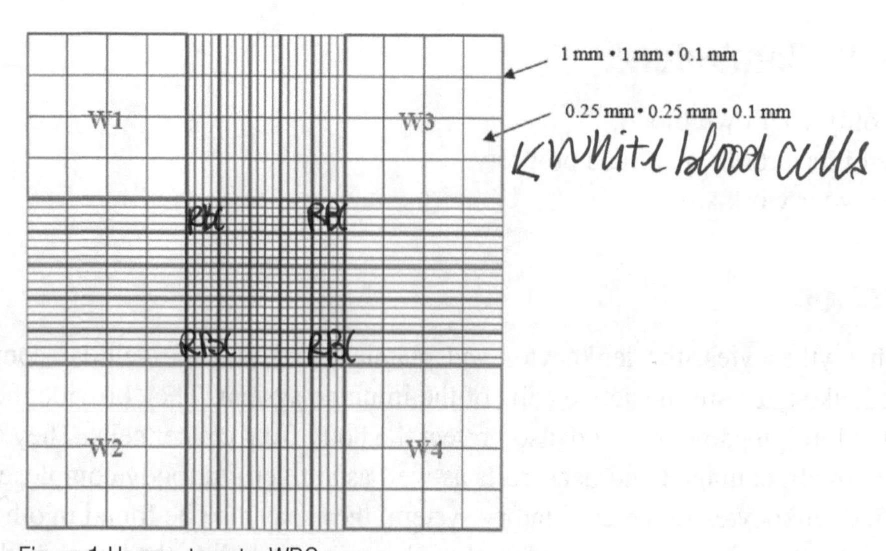


Figure 1 Hemocytometer WBC

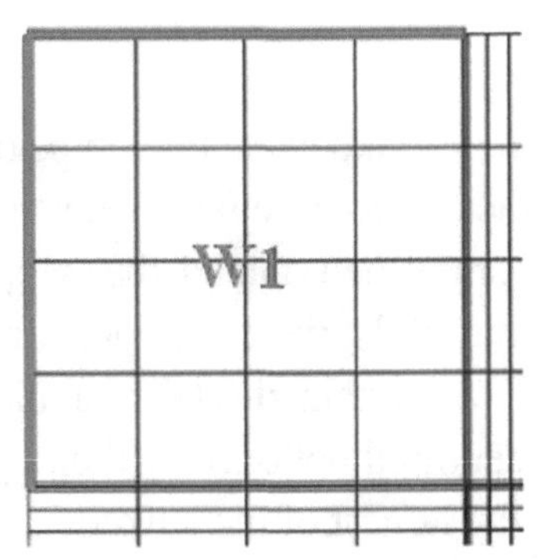


Count the cells touching the line on the top and right sides of the box but not the cells touching the lines on the bottom or left side of the box.

Figure 2 Counting rules WBC

- **Hemocytometer Activity**

The hemocytometer grid and process of slide preparation (Becton-Dickinson Unopette 5853, dilution factor of 100) will be explained by your lab instructor. Photographs have been taken of a hemocytometer prepared for a white blood cell count. There is a photograph of each of the four white blood cell counting sections.

- Using the photographs, count each of the four white blood cell sections of the counting surface. Each section contains 16 small squares.
- Record the number of white blood cells counted in each of the four white blood cell sections in the chart.
- To calculate the total white blood cell count, simply add the counts from all four sections and multiply by 250.

Total # of white blood cells counted • 250 = WBC per μL

W1

**Total** _______

W3

**Total** _______

W2

**Total** _______

W4

**Total** _______

- Calculation: total white blood cell count = _______________
- How does this count compare to the acceptable range for an adult?

# Differential White Cell Count

As you observed in a previous lab, there are different types of leukocytes. Each type of leukocyte has a unique function and appearance (see Figure 3). One characteristic of leukocytes is presence (granulocytes) or absence (agranulocytes) of granules in the cytoplasm. When stained, the cytoplasm of the granulocytes appears grainy. The granulocytes include the neutrophils, the eosinophils, and the basophils. The cytoplasm of the agranulocytes does not appear grainy when stained. The agranulocytes include the monocytes and the lymphocytes. The leukocytes can be identified not only by their cytoplasm but also by their size and the shape of the nucleus. Monocytes are the largest of the leukocytes and have an oval to kidney bean shaped nucleus. Lymphocytes are the smallest of the lymphocytes and have a round to slightly indented nucleus that nearly fills the cell. Neutrophils are the most common leukocyte and have a multilobed or segmented nucleus. The eosinophils also have a multilobed nucleus, often bilobed. The basophils are the least common and have so many granules that the round nucleus is often not visible.

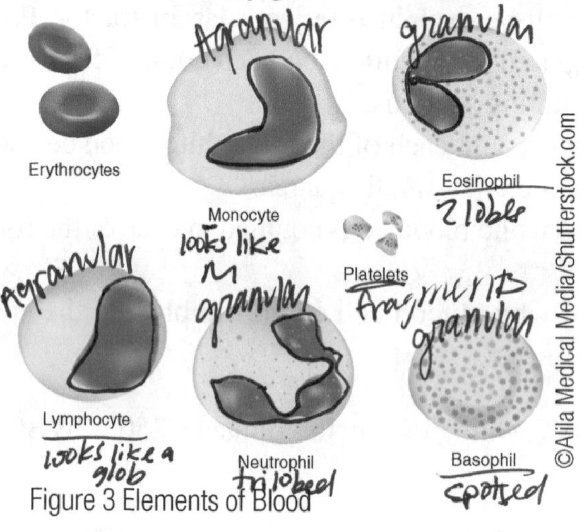


Figure 3 Elements of Blood

In the case of leukopenia, the cause is often a decrease in the number of neutrophils—**neutropenia**. Some common causes of leukopenia are chemotherapy, radiation, aplastic anemia, and HIV. In the case of leukocytosis, the cause is often an increase in the number of neutrophils—**neutrophilia**. However, an increase in any of the types of leukocytes may be seen under certain conditions. A common cause of leukocytosis is activation of the immune system due to invasion by a pathogen or toxin. Different types of pathogens, diseases, or physiological conditions will affect different types of leukocytes.

| Leukocyte Type | Normal % of Leukocytes | High Count | Some Potential Causes |
|---|---|---|---|
| Neutrophil | 50–70% | Neutrophilia | Chronic infections, appendicitis, meningitis, late pregnancy, heavy exercise, prolonged vomiting |
| Lymphocyte | 20–40% | Lymphocytosis | Infectious diseases, lymphatic leukemia |
| Monocyte | 2–8% | Monocytosis | Tuberculosis, mononucleosis, malaria, monocytic leukemia |
| Eosinophil | 1–4% | Eosinophilia | Parasitic infection, allergies |
| Basophil | 0.1–1% | Basophilia | Inflammatory diseases |

- **Differential White Cell Count Activity**

You will use prepared slides of whole blood that have been stained so the nuclei of the white blood cells are visible. Use the figures in your text or the wall charts for assistance in the identification of leukocytes.

- Identify each type of white blood cell.
- Count 100 leukocytes. Record how many of each type of leukocyte you observed in the following table.

| | Neutrophils | Lymphocytes | Monocytes | Eosinophils | Basophils | Total |
|---|---|---|---|---|---|---|
| Number counted | | | | | | |
| % of total counted | | | | | | NA |

## Platelet Count and Action

The normal range for a platelet count in adults is 150,000 to 450,000 platelets per μL of whole blood (note: these values will vary slightly among sources and depending upon the method of testing). When the platelet count is lower than normal, the person has **thrombocytopenia**. When the platelet count is higher than normal, the person has **thrombocytosis** (thrombocythemia). Platelets are essential for the process of hemostasis. It is the action of platelets and some of the blood proteins that reduces blood loss when an injury occurs to the blood vessels. The platelets play several roles in this process. Platelets are responsible for aggregating and forming a platelet plug at the site of injury. The platelets also secrete several platelet factors including thromboxane 2, which act as vasoconstrictors and also provide positive feedback to activate additional platelets. Once platelets are activated, fibrinogen can bind to fibrinogen receptors on the platelets. Platelet factors also play a role in the process of coagulation resulting in the conversion of fibrinogen to fibrin. The fibrinogen bound to the platelets of the platelet plug can then form fibrin bridges between the platelets.

Platelet counts can be performed using a special slide called a hemocytometer. The hemocytometer is a precise slide that contains a ruled scale ground into the surface. This produces a counting chamber of precise volume. The same areas are used for a platelet count as were used for the RBC count. Each of the smallest of squares of the slide counted for a red blood cell count covers an area of 0.05 mm × 0.05 mm or 0.0025 $mm^2$. Because there are 16 small squares in each of the sections counted, the total area of each of the sections is 0.04 $mm^2$. Because 5 sections are counted, the total area counted in a platelet count is 0.2 $mm^2$. The depth of the space between the counting chamber and the coverslip is 0.1 mm. Therefore, the total volume of the areas being counted in a platelet count is 0.02 $mm^3$, or 0.02 μL. To perform a platelet count, a known amount of whole blood must be drawn and diluted with a known amount of slightly hypotonic solution. If the blood is not diluted, the platelets are too concentrated and would be difficult to count. The solution is hypotonic so that the erythrocytes will lyse and the platelets will be much more visible. Care must be taken not to count platelets more than once. Before you begin counting, determine some rules for counting that will help you avoid counting a platelet more than once. For example, count the platelets that touch the line to the left of the square but do not count the platelets that touch the line to the right of the square (those will be counted in the square to the right or not at all); count the platelets that touch the line at the top of the square but do not count the platelets that touch the line at the bottom of the square (those will be counted in the square below or not at all).

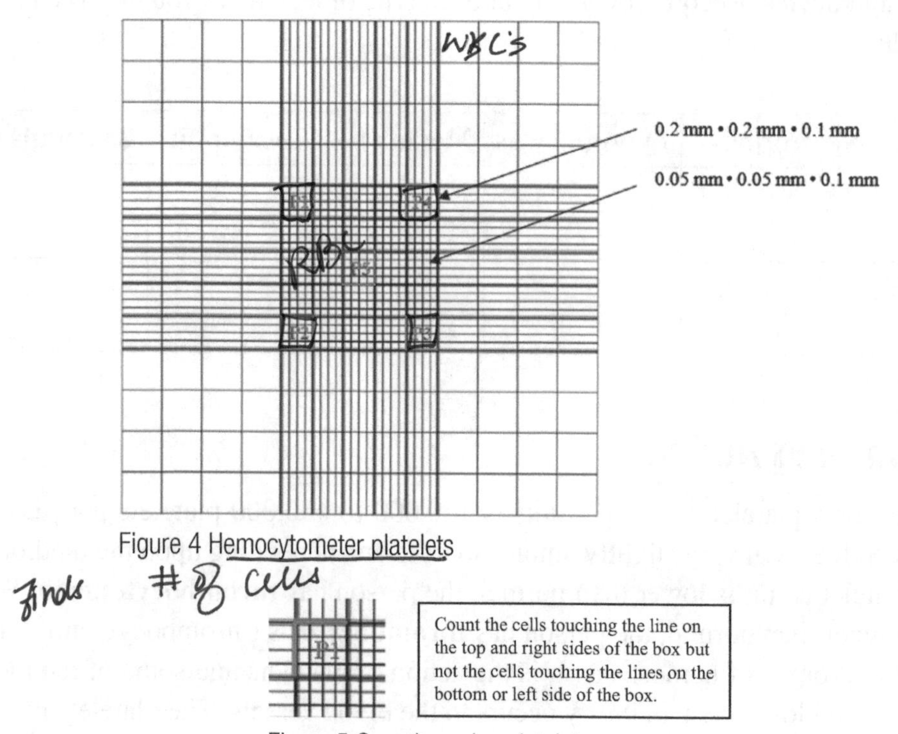
0.2 mm • 0.2 mm • 0.1 mm
0.05 mm • 0.05 mm • 0.1 mm
Count the cells touching the line on the top and right sides of the box but not the cells touching the lines on the bottom or left side of the box.

Figure 4 Hemocytometer platelets

Figure 5 Counting rules platelets

## Hemocytometer Activity

The hemocytometer grid and process of slide preparation (Becton-Dickinson Unopette 5855, dilution factor of 100) will be explained by your lab instructor. Photographs have been taken of a hemocytometer prepared for a platelet count. There is a photograph of the entire counting surface as well as individual photographs of each of the five platelet counting sections (same as the red blood cell sections).

- Using the photographs, count each of the five platelet sections of the counting surface. Each section contains 16 small squares.
- Record the number of platelets counted in each of the five platelet sections in the chart.
- To calculate the total platelet count, simply add the counts from all five sections and multiply by 5000.

Total # platelets counted • 5000 = platelets per μL

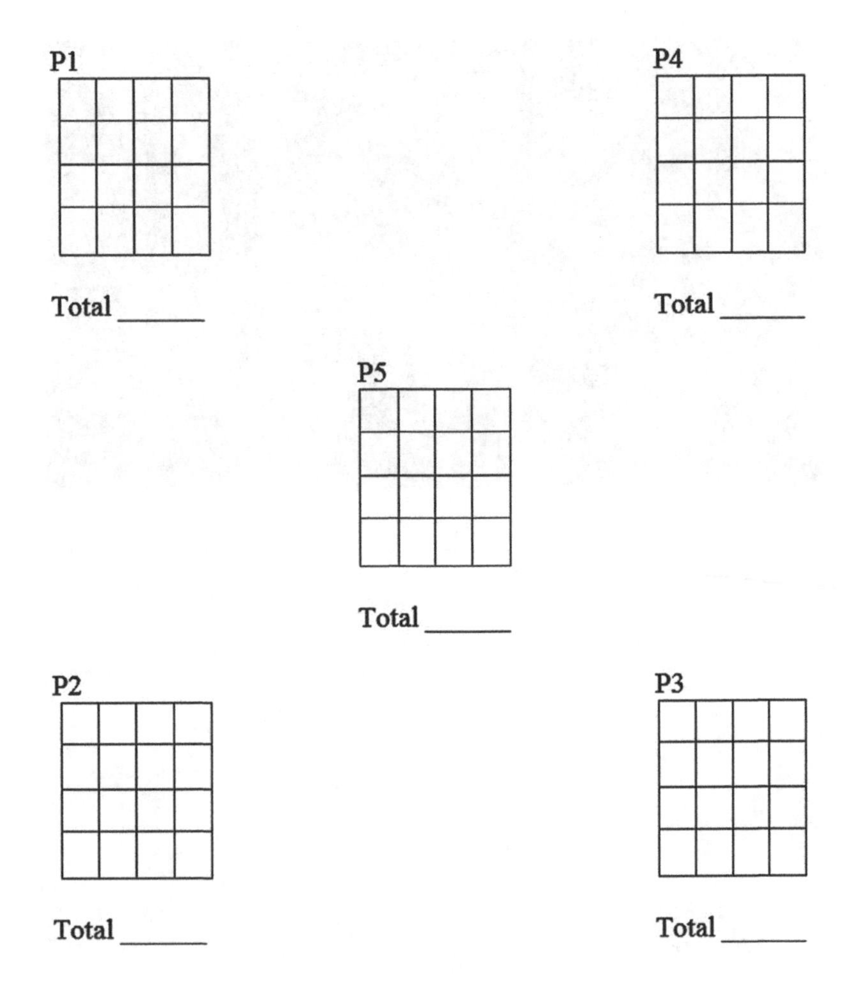


- Calculation: total platelet count = _______________
- How does this count compare to the acceptable range for an adult?

## Coagulation Activity

Although the process of coagulation is a very complex process, in this lab we will focus on the key event in each of the three stages of the process. For the purpose of this lab, we will not differentiate between intrinsic or extrinsic initiation of the process. In stage I, prothrombin-converting factor is produced. In stage II, the use of prothrombin-converting factor results in the conversion of prothrombin to thrombin. In stage III, the use of thrombin results in the conversion of soluble fibrinogen to insoluble fibrin. This is what then forms the fibrin bridges between platelet of the platelet plug, resulting in a clot. The proteins (or their precursors) necessary for coagulation are present in the plasma. When the clotting factors and clotting proteins are removed from the plasma, the remaining fluid and its contents are called serum.

coagulation = platelets clotting

agglutination = antigen-antibody complexes

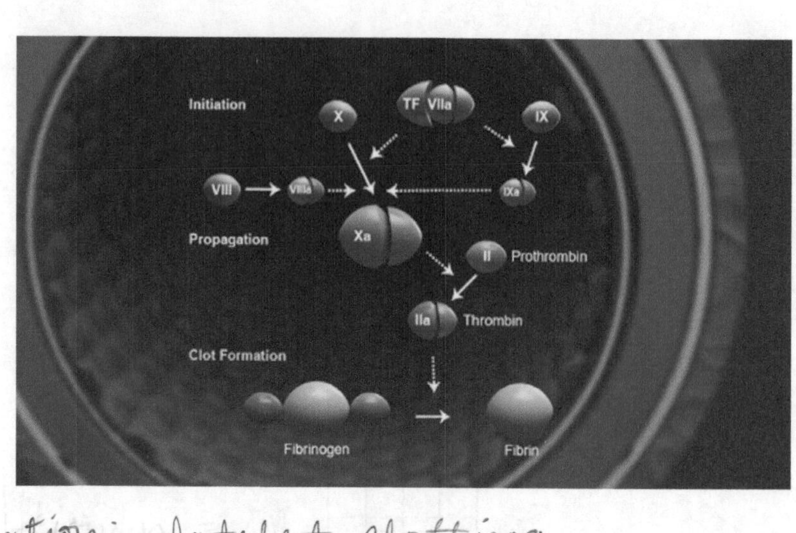


Coagulation: platelet clotting

agglutination: antigen-antibody complexes

| Plasma | Serum |
|---|---|
| blood liquid before coagulation · contains clotting factors | blood liquid after coagulation · doesn't contain any clotting factors |

blood vessel

healthy

damaged

→ clotting factor cascade: series of events taking place rapidly ↓ activation

Prothrombin → thrombin

vasoconstriction. starting to form the platelet plug

★ fibrinogen "soluble" → fibrin "insoluble" →

blood clot formation "hemostasis"

# Exercise 21

# iWorx Lung Volumes and Capacities

## Objectives

- Be able to define the term *respiratory cycle.*
- Be able to identify the components of the respiratory cycle.
- Be able to define each of the respiratory volumes: *tidal volume, inspiratory reserve volume, expiratory reserve volume,* and *residual volume.*
- Be able to define each of the respiratory capacities: *inspiratory capacity, vital capacity, functional residual capacity,* and *total lung capacity.*
- Be able to define the terms *eupnea, hyperpnea, hyperventilation, hypoventilation, tachypnea, dyspnea, voluntary apnea,* and *apnea vera.*
- Be able to define and calculate *minute ventilation* or *pulmonary ventilation.*
- Be able to define the term *anatomic dead space.*
- Be able to define and calculate *alveolar ventilation.*

## Lab Safety Reminders

- Use caution with any electrical equipment.

## Hints for Success

All members of your lab group should feel comfortable operating the iWorx system.

Considerations for selecting your subject: Your subject should be healthy, free of respiratory infections, and have no history of respiratory or cardiovascular problems.

Although the subject may be consciously breathing through the mouth, it is highly likely that some air will also be moving through the nose. The use of a nose clip will prevent any air from escaping or entering the subject's respiratory system through the nose. Using a nose clip will provide a more accurate set of data. Please be sure to clean the nose clips with alcohol both before and after use, and return the nose clips when done with the exercise.

Be sure to follow the instructions for calibration and warm-up.

**Never unplug any sensors from the iWorx box!** If you unplug a sensor during an experiment, you will lose all data. You will need to turn the iWorx box off, plug the sensor in, and start the lab from the beginning.

On the iWorx recording screens, it is helpful to remember that the blue bars are for data that are being recorded. The green bars are for data that is being calculated based upon the data you have collected.

While using the iWorx, it is important that you do not get so caught up with obtaining the correct numbers from your data that you forget the function or process you are trying to observe. As you are using the iWorx, keep in mind the principles from the lecture being illustrated in the lab exercise.

## Introduction

A **respiratory cycle** is a complete cycle of lung ventilation that includes inspiration and expiration. A respiratory cycle starts at the beginning of an inspiration and ends at the completion of the expiration. Respiratory rate is the number of respiratory cycles or breaths per minute. **Pulmonary** or **minute ventilation** is the total volume of air exchanged between the lungs and the atmosphere per minute. It is the product of the volume of gas exchanged in ventilation (tidal volume) multiplied by the respiratory rate. This is not a perfect calculation of the volume of air or gasses available for exchange between the alveoli and capillaries of the lungs. When considering ventilation, you must consider that the respiratory system is composed of the conduction zone and the respiratory zone. The conduction zone is a passageway for air. It does not allow the exchange of gasses between the respiratory system and the cardiovascular system and is considered **anatomic dead space**. Only the respiratory zone is involved in the exchange of gasses. For every inspiration, approximately 150 mL of the total volume of the breath (approximately 500 mL for tidal volume) is contained in the conduction zone and is not available for exchange. As a result, pulmonary ventilation is an overestimate of the volume of gas available for exchange. **Alveolar ventilation** is the volume of air available in the respiratory zone and thus available for gas exchange. It is calculated as (tidal volume – conduction zone volume) multiplied by the respiratory rate.

Lung volumes are volumes of air, most of which can be moved during the process of lung ventilation. Lung volumes include the following:

- **Tidal Volume (TV)**: the volume of air exchanged per respiratory cycle, generally measured during normal quiet breathing.
- **Inspiratory Reserve Volume (IRV)**: the volume of air that can be inhaled above the tidal volume. Inspiration of this volume in addition to the normal tidal volume would completely fill your lungs.
- **Expiratory Reserve Volume (ERV)**: the volume of air you can exhale after you have exhaled your tidal volume.
- **Residual Volume (RV)**: the volume of air that remains in your lungs after a maximal expiration. The residual volume is the one volume that is not exchanged.

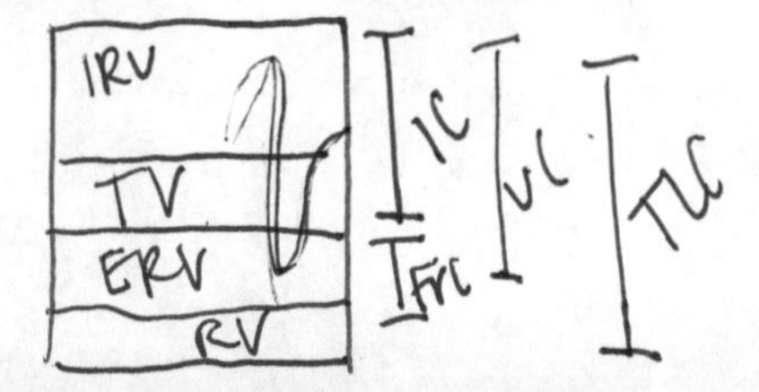

Lung capacities are the sum of two or more lung volumes. Lung Capacities include the following:

- **Inspiratory Capacity** (IC): This is the sum of TV + IRV.
- **Vital Capacity** (VC): This is the sum of TV + IRV + ERV.
- **Functional Residual Capacity** (FRC): This is the sum of ERV + RV. FRC is the starting volume of a breathing cycle.
- **Total Lung Capacity** (TLC): This is the sum of TV + IRV + ERV + RV.

The following table of lung volumes and capacities is based upon the average values for the standard 70-kg male and 56-kg female, at rest.

| Lung Volume or Capacity | Volume Male | Volume Female |
|---|---|---|
| Tidal Volume (TV) | 500 mL | 500 mL |
| Inspiratory Reserve Volume (IRV) | 3000 mL | 1900 mL |
| Expiratory Reserve Volume (ERV) | 1100 mL | 700 mL |
| Residual Volume (RV) | 1200 mL | 1100 mL |
| Inspiratory Capacity (IC) | 3500 mL | 2400 mL |
| Vital Capacity (VC) | 4600 mL | 3100 mL |
| Functional Residual Capacity (FRC) | 2300 mL | 1800 mL |
| Total Lung Capacity (TLC) | 5800 mL | 4200 mL |

Figure 1 and figures from your text may help you understand the relationships among the respiratory volumes and capacities.

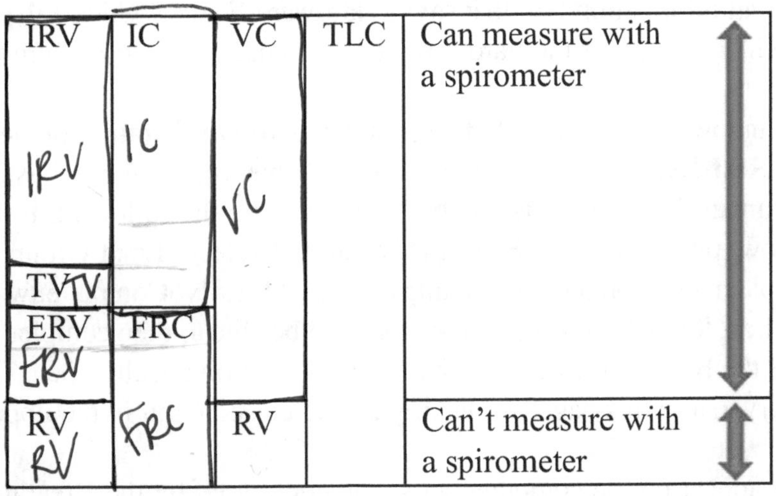

Figure 1 Lung volumes and capacities

FRC = ERV + RV
IC = IRV + TV
VC = IRV + TV + ERV
TLC = IRV + TV + ERV + RV

There are several patterns of ventilation that can occur under normal conditions as well as during respiratory distress.

- **Eupnea**: normal quiet breathing that maintains a normal level of systemic arterial blood gasses.
- **Hyperpnea**: an increase in respiratory rate and/or volumes as a result of an increase in metabolic rate.
- **Hyperventilation**: an increase in respiratory rate and/or volumes without an increase in metabolic rate. This usually results in a decrease in the partial pressure of systemic arterial carbon dioxide ($CO_2$).
- **Hypoventilation**: a decrease in respiratory rate and/or volumes without a decrease in metabolic rate. This usually results in an increase in the partial pressure of systemic arterial carbon dioxide ($CO_2$).
- **Tachypnea**: rapid ventilation generally accompanied by an increase in ventilation volume.
- **Dyspnea**: difficulty in breathing, often described as an inability to catch one's breath or feeling starved for oxygen.
- **Voluntary Apnea**: voluntary cessation of breathing. This will usually begin with a maximal inspiration.
- **Apnea Vera**: involuntary cessation of breathing. This is expected to begin at the end of an expiration when no stimulus to breathe is given.

The goal of this lab is to identify and measure the basic lung volumes involved in lung ventilation as well as to calculate lung capacities and minute and alveolar ventilation in eupnea.

## Starting the iWorx Software

1. Be sure that the iWorx box is turned on. A small green light will be lit when the power is on. If you need to turn the power on, the on/off switch is on the back of the iWorx box.
2. Click on the **LabScribe3 shortcut** on the computer's desktop to open the program. You should see an information box that says "Hardware Found." Click **OK**; this will put you on the Main Window page. If the hardware is not found, check to be sure the iWorx box is turned on.
3. On the Main Window, pull down the **Settings** menu. Scroll to **Human Spirometry**, then select the **Breathing-Rest-Exercise** settings file from the "Human Spirometry" list.
   - Note: If "Human Spirometry" is empty, then complete the following instructions. On the Main Window, pull down the "Settings" menu and select "Load Group." Scroll down to locate the folder that contains the settings group, "IPLMv6Complete.iwxgrp." Select this group and click "Open." At this point, you should be able to then complete step 3.
4. Instructions for the Breathing-Rest-Exercise Setup will automatically open. These instructions are not necessary. Close this page by clicking on the close button in the upper right corner of this document.
5. LabScribe will appear on the computer screen as configured by the Breathing-Rest-Exercise settings file.

6. For this lab, you do not need any information from the following files. But as a reminder, once the settings file has been loaded, clicking the **Experiment** button on the toolbar will allow you to access any of the following documents:
   - Introduction—introductory material for iWorx Human Spirometry lab exercises
   - Appendix—may have additional information and resources
   - Background—may have some background information regarding the lab topic
   - Labs—the lab exercise as written by iWorx
   - Setup (opens automatically)—instructions for setting up this particular iWorx lab. You already closed this out.
7. If you choose to save your data, it should be saved in the **iWorx Saved Files** folder on the **Desktop**. You may create your own folder within the "iWorx Saved Files" folder.

## Spirometer Setup

The spirometer has already been connected to the iWorx box (see Figure 2). Do not unplug the cable. Doing so will cause you to lose your data.

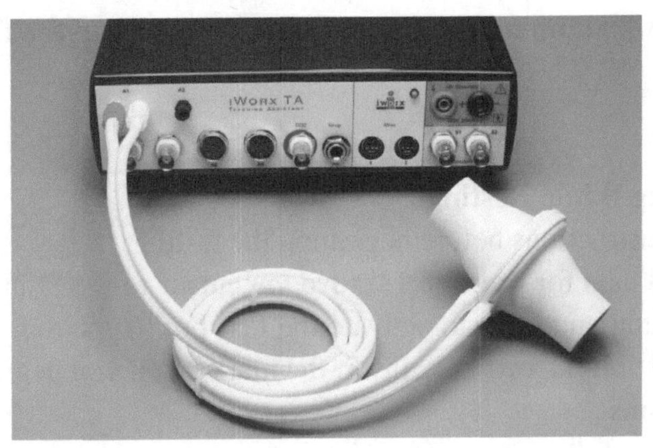


Figure 2 Spirometer flowhead

You will be using a biological filter to breathe through. Simply attach the biological filter to the inlet of the spirometer (see Figure 3). The inlet side has been marked with red on the spirometer flowhead as well as on the tubing. The subject is to breathe through the biological filter, *not* the open end of the spirometer flowhead. You will be using your same biological filter for all iWorx respiratory labs conducted, so you will need to label your filter and turn it in for storage at the end of each lab.

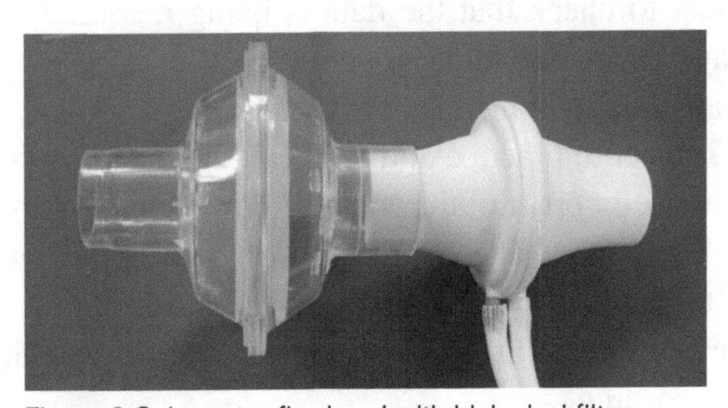

Figure 3 Spirometer flowhead with biological filter

## Spirometer Calibration

You will need to check the calibration of the spirometer.

- In the Main Window, locate the "Lung Volumes" channel.
- Click on the words **Vol.Human(AirFlow)** in the heading line of the "Lung Volumes" channel. This will open a drop down menu.
- Click **Setup Function** to open the "Spirometer Calibration Dialogue."
- Check to be sure the internal spirometer being used is the **IX-TA**. If it is not, open the drop down menu associated with the spirometer and select **IX-TA**.
- Check to be sure the type of flowhead being used is the **300L**. If it is not, open the drop down menu associated with the flowhead and select **300L**.
- Check to be sure the Reset option being used is the **No Reset**. If it is not, open the drop down menu associated with the reset and select **No Reset**.
- Enter the room air temperature in degrees Celsius for the "Temperature of Inhaled Air." The "Temperature of Exhaled Air" should be **29**°C. The "Atmospheric Relative Humidity" should be **60**%.
- Click the **OK** button at the bottom of the "Spirometer Calibration Dialogue" box.
- **Allow the IX-TA to warm up for 10 minutes before recording the first data.**

## Lab Exercises

- **Exercise 1: Breathing While Resting**
  - Read through all instructions before beginning the testing.
  - To label the actual data for exercise 1, move the cursor to the "Mark" text box to the right of the "Mark" button and click to activate the text box. Type the subject's name followed by "resting" in the "Mark" text box. Insert the mark when you begin the recording period.
  - **Subject Instructions**
    - The subject should sit quietly and maintain good posture through the testing period.
    - It is very important for the subject to become comfortable breathing through the apparatus and that the subject breathes normally. Initially the subject will have the tendency to exaggerate tidal breathing. The subject may also fail to exhale the normal tidal volume. To become comfortable, the subject should take some practice inspirations and expirations through the apparatus. Practice ventilations should include both tidal volume ventilations and maximal volume ventilations.
    - It is also important to check that the data is being recorded correctly. Inspiration data should be displayed as an upward deflection on the "Air Flow" channel (see Figure 4; note the arrows at a tidal volume inspiration).
      - To allow the subject to practice and to check the data display, click the **Save to Disk** icon in the lower left corner of the Main Window screen. A red "X" will appear across the "Save to Disk" icon. The LabScribe software is put into preview mode and the recording system works **without saving any data**.
      - Click the **Preview** button in the upper right corner of the Main Window screen. This will start the recording.
      - Have the subject begin by inhaling. Click the **Autoscale** button in both the "Air Flow" channel and the "Lung Volume" channel.

- If the flowhead is oriented correctly, the inspiration will be recorded as an upward deflection in both the "Air Flow" and the "Lung Volume" channels. If the inspiration is recorded as a downward deflection on these channels, click on the arrow to the left of the "Air Flow" channel. In the drop down menu, click "Invert."
- If it was necessary to invert the data, test the inspiration again.
- Have the subject complete as many respiratory cycles as necessary to become comfortable breathing through the apparatus.
- Click the **Stop** button.
- Click the **Save to Disk** icon in the lower left corner of the Main Window screen. The red "X" across the "Save to Disk" icon will be replaced by a green arrow. The LabScribe software is now in recording mode and the system will save your data.

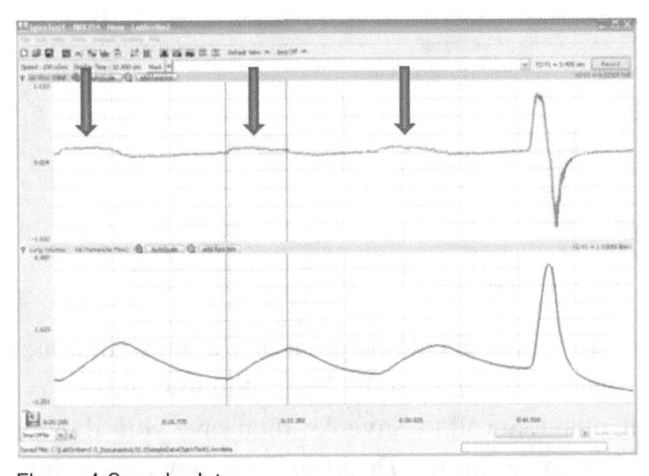

Figure 4 Sample data

- **Tester Instructions**
  - Instruct the subject to sit quietly and maintain good posture while conducting the test. Explain the testing procedure to the subject.
  - Begin the test with the biological filter and spirometer flowhead out of the subject's mouth. The apparatus should be held at mouth level in a position that does not allow air flow into the apparatus. **The flowhead outlets and tubing should be pointing down**. Click the "Record" button and allow the testing to run for 10 seconds. These 10 seconds of recording are necessary to zero the "Lung Volumes" channel.
  - After the 10 seconds required for zeroing, instruct the subject to place the open end of the biological filter into the mouth and begin breathing.
  - Click the "Autoscale" buttons in both the "Air Flow" and "Lung Volumes" channels as necessary to be able to see the data as it is being collected.
  - Record eight resting tidal volume breathing cycles (inspiration and expiration). This should take approximately one minute. More breaths may be recorded if the breaths do not look consistent.
  - Immediately following the last respiratory cycle (do not interrupt the recording process), instruct the subject to **inhale as DEEPLY as possible** then **IMMEDIATELY exhale as QUICKLY and as DEEPLY as possible**. Even though the subject thinks they have exhaled all the air possible, ask them to continue exhaling until no more air can be moved (see Figure 5, note the arrow at the maximal inspiration and expiration).

- Instruct the subject to return to resting tidal volume breathing for eight tidal volume breathing cycles (or until tidal volume breathing has returned to normal for at least two breathing cycles).
- Repeat the maximal inspiration and expiration and an additional four resting tidal volume breathing cycles.
- Click "Stop."

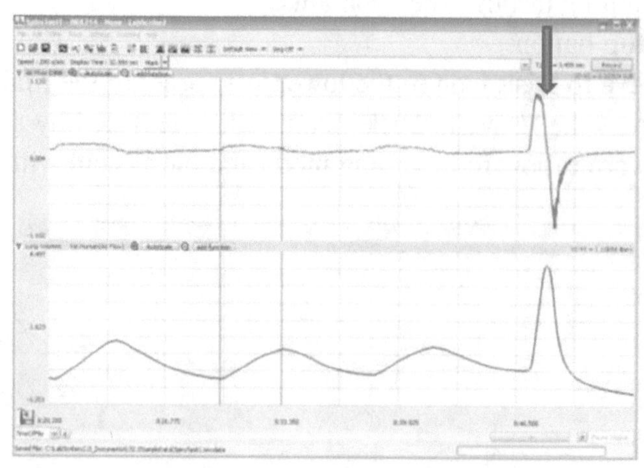

Figure 5 Sample data

- You may choose to save your data at this point or you may continue on with the experiment or data analysis. To save your data, on the Main Window, either
  1. click on the **File** menu. Scroll to **Save As,** then save your data; it should be saved in the **iWorx Saved Files** folder on the **Desktop**. You may create your own folder within the iWorx Saved Files folder. Give your file a name and designate the file type as .iwxdata, then click **Save.**

     Or
  2. click on the **Save File** icon in the iWorx toolbar, then save your data; it should be saved in the **iWorx Saved Files** folder on the **Desktop**. You may create your own folder within the iWorx Saved Files folder. Give your file a name and designate the file type as .iwxdata, then click **Save.**
- **Data Analysis Instructions**

  Figure 6, a diagram of icons in the iWorx toolbar may be helpful.

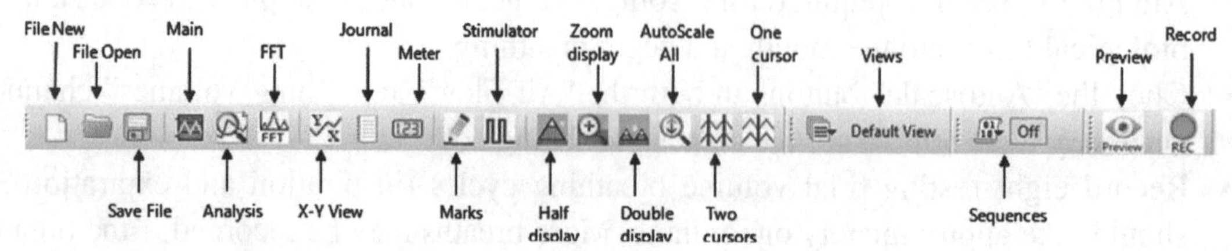


Figure 6 iWorx toolbar icons

- Click on the **Analysis** icon in the iWorx toolbar.
- Go to the beginning of your recordings. Click on the **Marks** icon to bring up the Mark Dialogue box with a table of the marks created in your lab. Once you have the table of marks, click on the number at the beginning of the row for the mark that identifies the beginning of

the experimental data segment you are going to analyze. This will highlight that mark. Next, click on the **Go To Mark** button near the bottom of the Marks Dialogue box. Data at that mark will be displayed on your screen.

- Be sure that two cursors are present on your screen. If necessary, click the **Two Cursors** icon on the iWorx toolbar.
- Place the cursors on either side of four consistent breathing cycles, one cursor on the baseline just before inspiration and the second cursor just after the final expiration, then click the **Zoom Display** icon in the iWorx toolbar. This will expand the data section to the width of the Main Window. Adjust your data section as necessary with the display time icons or the zoom icon.
- Click on the **Analysis** icon in the iWorx toolbar.
- In the Analysis window, a "Function Table" will be displayed above the uppermost channel (Air Flow channel). The mathematical functions "V2–V1," Max_dv/dt, Min_dv/dt, and "T2–T1" should be present. The values for each of these functions are displayed in the top margin of each channel. The "V2–V1" function measures the difference in amplitude between cursor 2 and cursor 1. The "Max_dv/dt" function is the maximum rate of air movement during inspiration in the time period between cursors. The "Min_dv/dt" function is the maximum rate of air movement during expiration in the time period between cursors. The "T2–T1" function measures the difference in time between cursor 2 and cursor 1.
- Click on the arrow to the left of the "Air Flow" channel. In the drop down menu, click "Minimize" to reduce the "Air Flow" channel to a bar across the top of the data screen.
- Click on the arrow to the left of the "Lung Volume" channel. In the drop down menu, move the cursor to "Scale" and click "Autoscale" from the submenu. This will maximize the data displayed in the "Lung Volume" channel.
- **All measurements will be made in the "Lung Volume" channel. All data for the measurements of air flow and volume should be taken from the "Lung Volume" channel data.**
- To measure tidal volume (TV) and maximum inspiratory flow of a breathing cycle, place the first cursor at the lowest point of the breathing cycle, the beginning of the breathing cycle just prior to an inspiration. Place the second cursor at the highest point of the same breathing cycle, the end of the inspiration (see Figure 7).

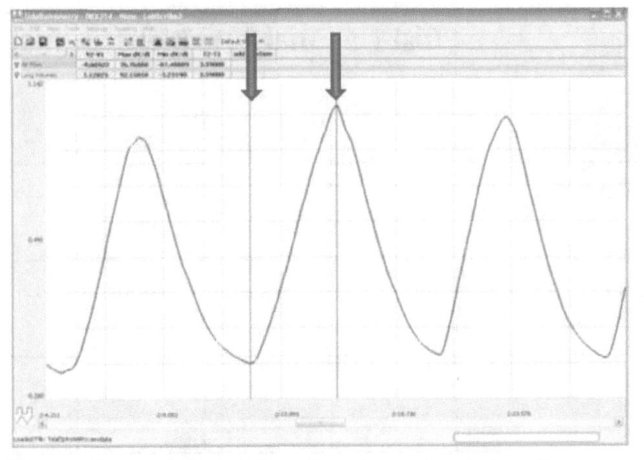

Figure 7 Data analysis TV and maximum inspiratory flow

- Record the "V2–V1" data for the tidal volume in Table 1.
- Record the "Max_dv/dt" data in Table 1 for the maximum inspiratory flow rate.
- To measure the maximum expiratory flow rate, place the first cursor at the highest point of the same breathing cycle, at the end of the inspiration. Place the second cursor at the lowest point of the same breathing cycle; the end of the expiration and the end of the breathing cycle (see Figure 8).

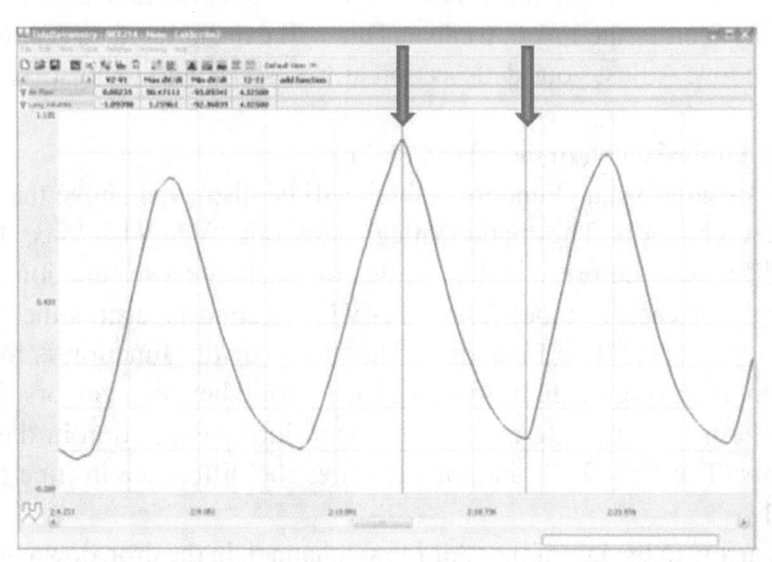

Figure 8 Data analysis maximum expiratory flow

- Record the "Min_dv/dt" data for the maximum expiratory flow rate in Table 1.
- To measure the breath period of a breathing cycle, place the first cursor at the lowest point of the breathing cycle, the beginning of the breathing cycle just prior to inspiration (as in measuring TV), and place the second cursor at the lowest point of the breathing cycle, at the end of expiration (as in measuring minimum flow rate) (see Figure 9). If the troughs are not distinct, you may measure from peak to peak (at the end of inspiration of two sequential breathing cycles).

**TABLE 1:** Tidal Volume Breathing Cycle Measurements
(record absolute values for all numbers)

| | Trial 1 | Trial 2 | Trial 3 | Trial 4 | Trial 5 | Mean |
|---|---|---|---|---|---|---|
| Tidal volume (L) | | | | | | |
| Maximum flow rate, Inspiration (L/sec) | | | | | | |
| Maximum flow rate, Expiration (L/sec) | | | | | | |
| Breath Period (sec) | | | | | | |

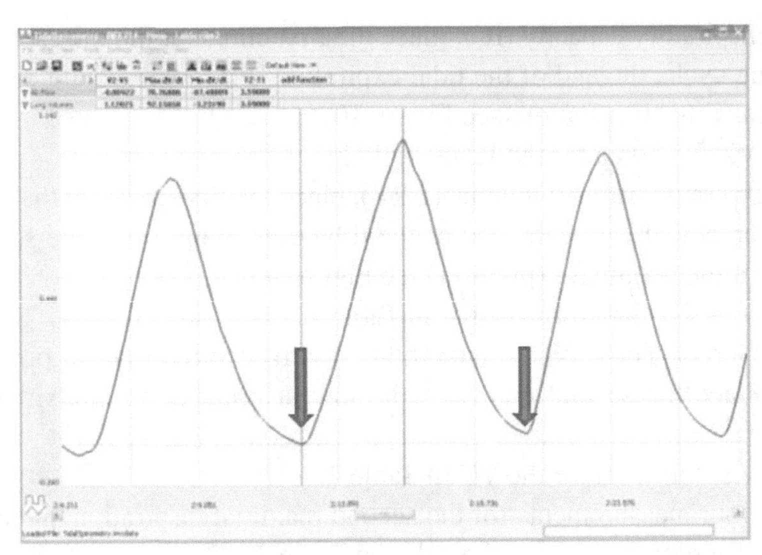

Figure 9 Data analysis breath period

- Record the "T2–T1" data for breath period in Table 1.
- Repeat these four measurements for four more breathing cycles. Calculate the mean for each and record the value in Table 1.
- Go to the data segment that includes the maximal inspiration and expiration to measure the other respiratory volumes and capacities.
- Adjust the data display so that two resting tidal volume breathing cycles, the first maximal inspiration and expiration, and one subsequent resting tidal volume breathing cycle are all visible on the data screen.
- To measure inspiratory reserve volume (IRV), place the first cursor at the peak of the resting tidal breathing cycle, the end of the tidal inspiration, just prior to the maximal inspiration. Place the second cursor at the highest point of the maximal inspiration (see Figure 10).
- Record the "V2–V1" data for the IRV in Table 2.

**TABLE 2:** Respiratory Volumes and Capacities Measurements (record absolute values for all numbers)

| | Trial 1 | Trial 2 | Mean |
|---|---|---|---|
| Inspiratory Reserve Volume (L) | | | |
| Inspiratory Capacity (L) | | | |
| Expiratory Reserve Volume (L) | | | |
| Vital Capacity (L) | | | |

- To measure inspiratory capacity (IC), place the first cursor at the lowest point of the resting tidal breathing cycle, the end of the tidal expiration. Place the second cursor at the highest point of the maximal inspiration (see Figure 10).
- Record the "V2–V1" data for the IC in Table 2.
- To measure expiratory reserve volume (ERV), place the first cursor at the lowest point of the resting tidal breathing cycle, the end of the tidal expiration. Place the second cursor at the lowest point of the maximal expiration (see Figure 10).
- Record the "V2–V1" data for the ERV in Table 2.
- To measure vital capacity (VC), place the first cursor at the highest point of the maximal inspiration. Place the second cursor at the lowest point of the maximal expiration (see Figure 10).
- Record the "V2–V1" data for the VC in Table 2.
- Repeat these four measurements for the second maximal inspiration and expiration. Calculate the mean for each and record the value in Table 2.

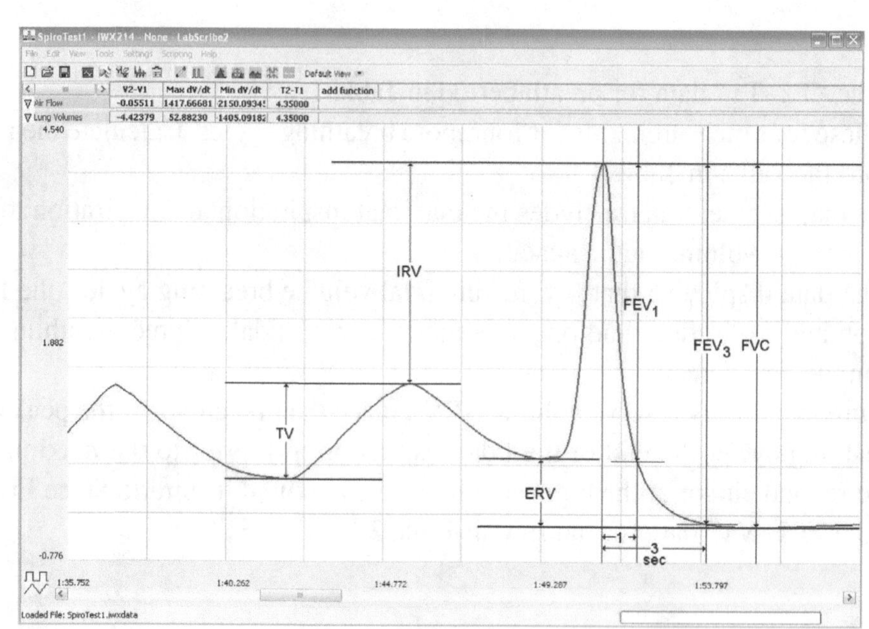

Figure 10 Data analysis respiratory volumes and capacities

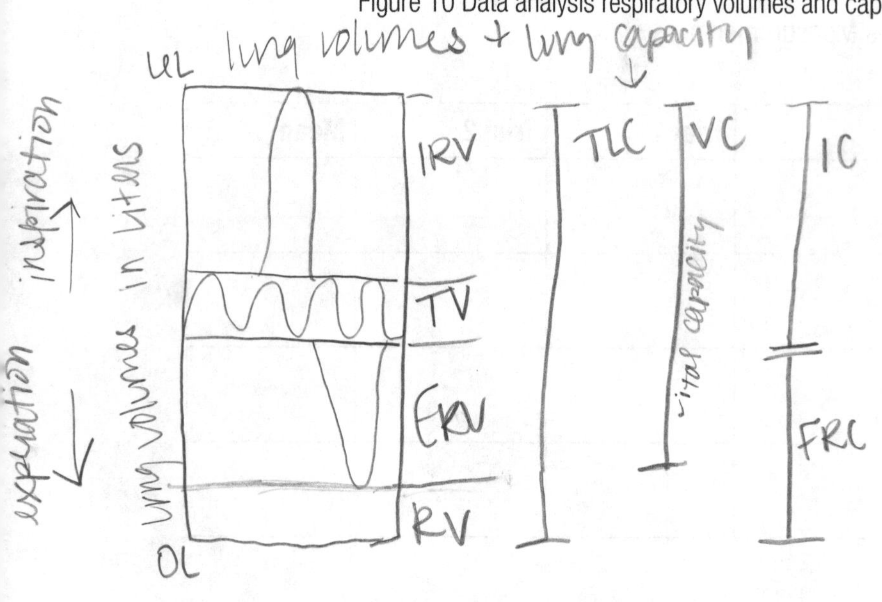

# Exercise 22

# iWorx Changes in Breathing Patterns

## Objectives

- Be able to define the term *respiratory cycle*.
- Be able to identify the components of the respiratory cycle.
- Be able to define each of the respiratory volumes: *tidal volume, inspiratory reserve volume, expiratory reserve volume,* and *residual volume*.
- Be able to define each of the respiratory capacities: *inspiratory capacity, vital capacity, functional residual capacity,* and *total lung capacity*.
- Be able to define the terms *eupnea*, *hyperpnea*, *hyperventilation, hypoventilation, dyspnea, voluntary apnea*, and *apnea vera*.
- Be able to define and calculate *minute ventilation* or *pulmonary ventilation*.
- Be able to define the term *anatomic dead space*.
- Be able to define and calculate *alveolar ventilation*.
- Be able to explain the roles of the primary respiratory centers of the medulla in the process of breathing.
- Be able to explain the action of chemoreceptors and somatic sensory receptors in the regulation of the respiratory system.
- Be able to predict the changes in alveolar ventilation due to changes in the partial pressure of carbon dioxide ($P_{CO_2}$) and oxygen ($P_{O_2}$), and changes in the pH (concentration of hydrogen, $H^+$) in the blood.
- Be able to define the following terms: *respiratory alkalosis* or *alkalemia* and *respiratory acidosis* or *acidemia*.
- Be able to explain the change in breathing patterns during hyperpnea.
- Be able to explain the changes in the partial pressure of carbon dioxide ($CO_2$) and oxygen ($O_2$) and blood pH associated with hyperpnea.
- Be able to explain the change in breathing patterns during alveolar hyperventilation.
- Be able to explain the changes in the partial pressure of carbon dioxide ($CO_2$) and oxygen ($O_2$) and blood pH associated with hyperventilation and how this would be expected to alter breathing patterns during hyperventilation recovery.
- Be able to explain the change in breathing patterns during alveolar hypoventilation.
- Be able to explain the changes in the partial pressure of carbon dioxide ($CO_2$) and blood pH associated with hypoventilation and how this would be expected to alter breathing patterns during hypoventilation recovery.

## Lab Safety Reminders

- Use caution with any electrical equipment.

## Hints for Success

All members of your lab group should feel comfortable operating the iWorx system.

Considerations for selecting your subject: Your subject should be healthy, free of respiratory infections, and have no history of respiratory or cardiovascular problems.

Although the subject may be consciously breathing through the mouth, it is highly likely that some air will also be moving through the nose. The use of a nose clip will prevent any air from escaping or entering the subject's respiratory system through the nose. Using a nose clip will provide a more accurate set of data. Please be sure to clean the nose clips with alcohol both before and after use, and return the nose clips when done with the exercise.

Be sure to follow the instructions for calibration and warm-up.

**Never unplug any sensors from the iWorx box!** If you unplug a sensor during an experiment, you will lose all data. You will need to turn the iWorx box off, plug the sensor in, and start the lab from the beginning.

On the iWorx recording screens, it is helpful to remember that the blue bars are for data that are being recorded. The green bars are for data that is being calculated based upon the data you have collected.

While using the iWorx, it is important that you do not get so caught up with obtaining the correct numbers from your data that you forget the function or process you are trying to observe. As you are using the iWorx, keep in mind the principles from the lecture being illustrated in the lab exercise.

## Introduction

Respiration rate and depth will be compared under the following conditions: (1) during eupnea (the control), (2) immediately after exercise, (3) during hyperventilation, and (4) during hypoventilation. Although you do have some conscious control over your breathing, ventilation is a reflex regulated by the primary **respiratory centers** of the medulla oblongata, the dorsal respiratory group (DRG) and the ventral respiratory group (VRG). These centers work cooperatively and also with input from the pontine respiratory group (PRG) in the pons to rhythmically stimulate inspiratory muscles (inspiration) and block stimulation of inspiratory muscles (expiration). They also regulate the changes in respiratory patterns for forced inspiration and expiration. Peripheral chemoreceptors in the carotid bodies and in the aortic arch are sensitive to partial pressures of carbon dioxide ($P_{CO_2}$) and oxygen ($P_{O_2}$) and pH (concentration of hydrogen, $H_2$) in the systemic arterial blood. Of these, $P_{O_2}$ usually requires the greatest change before it becomes an effective stimulus. Information regarding the systemic arterial blood is sent from these chemoreceptors to the DRG. An increase in the $P_{CO_2}$, a decrease in blood pH, and a decrease in the $P_{O_2}$ would all be expected to lead to an increase in **minute ventilation** ($V_E = V_t \bullet f$), the volume of air exchanged between the respiratory system and the atmosphere per minute, and **alveolar ventilation** ($V_A = (V_T - V_D) \bullet f$), the volume of air moved into and out of the respiratory zone per minute. A decrease in the $P_{CO_2}$, an increase in blood pH, and an increase in the $P_{O_2}$ would all be expected to lead to a decrease in ventilation. In addition to the peripheral chemoreceptors, there are also central chemoreceptors in the DRG. These receptors respond to the pH of the cerebrospinal fluid. However, $H_2$ does not cross the blood–brain barrier well, so the detection of changes in the pH of the cerebrospinal fluid is due to the diffusion of $CO_2$ from the blood into the cerebrospinal fluid. Once $CO_2$ is in the cerebrospinal fluid, it is converted to bicarbonate ($HCO_3^-$) and $H_2$ ($CO_2 + H_2O \leftrightarrow H_2CO_3 \leftrightarrow H^+ + HCO_3^-$).

In this lab, you will compare modified breathing patterns to the normal breathing pattern. Normal quiet breathing is referred to as **eupnea**. This breathing pattern maintains systemic arterial blood gasses ($P_{CO_2}$ and $P_{O_2}$, and also blood pH) within their respective homeostatic ranges ($P_{CO_2} = 40 \pm 5$ mm Hg; $P_{O_2} = 75$ to 100 mm Hg; pH $= 7.4 \pm 0.05$). During and for some time after exercise, a person is expected to alter the breathing pattern from eupnea to hyperpnea. **Hyperpnea** is an increase in respiratory rate and/or volumes as a result of an increase in metabolic rate. In the case of hyperpnea, systemic arterial $P_{O_2}$ is maintained within the normal homeostatic range. Due to the increase in ventilation, systemic arterial blood $P_{CO_2}$ may decrease to the lower end of the normal range. Systemic arterial blood pH may decrease as a result of an increase in metabolic acid production during exercise but not due to the increased ventilation. **Hyperventilation** occurs when alveolar ventilation exceeds the metabolic demand, the result of increasing both the rate and depth of the tidal volume without increasing metabolic activity. Hyperventilation would be expected to result in the following systemic arterial blood changes: a decrease in the $P_{CO_2}$, "blowing off $CO_2$," an increase in pH, and an increase in the $P_{O_2}$. Prolonged hyperventilation can lead to the process of **respiratory alkalosis** resulting in **alkalemia**, a condition of higher than acceptable blood pH (pH $> 7.45$). **Hypoventilation** occurs when metabolic demand exceeds alveolar ventilation, the result of decreasing both the rate and depth of the tidal volume without decreasing metabolic activity. Hypoventilation would be expected to result in the following systemic arterial blood changes: an increase in the $P_{CO_2}$, a decrease in pH, and a decrease in the $P_{O_2}$. Prolonged hypoventilation can lead to the process of **respiratory acidosis** resulting in **acidemia**, a condition of lower than acceptable blood pH (pH $< 7.35$).

**Dyspnea** is the disruption of normal breathing patterns often associated with a shortness of breath or a difficulty in breathing. There can be several causes of dyspnea, including chronic obstructive pulmonary disease (COPD), restrictive respiratory diseases, chronic congestive heart failure, and other temporary causes such as broken ribs. When breathing stops, the person is experiencing **apnea**. When a person voluntarily stops breathing, the apnea is referred to as **voluntary apnea**. Hold your breath. Did you notice what you did right before you held your breath? **Apnea vera** occurs when breathing stops involuntarily. Because inspiration requires stimulation and normal expiration is passive, when a person's respiratory center does not stimulate breathing, a person stops before inspiration, at functional residual capacity (FRC). Although damage to the respiratory center is one obvious cause for apnea vera, low partial pressure of $CO_2$ and high blood pH can also lead to temporary apnea vera.

A number of other factors can also influence the depth and rate of respiration, feedforward from the neurons of the motor cortex in the central command center, stimulation as a result of increased stimuli from mechanoreceptors and proprioceptors, and suppression as a result of some narcotics and alcohol.

## Starting the iWorx Software

1. Be sure that the iWorx box is turned on. A small green light will be lit when the power is on. If you need to turn the power on, the on/off switch is on the back of the iWorx box.
2. Click on the **LabScribe3 shortcut** on the computer's desktop to open the program. You should see an information box that says "Hardware Found". Click **OK**; this will put you on the Main Window page. If the hardware is not found, check to be sure the iWorx box is turned on.
3. On the Main Window, pull down the **Settings** menu. Scroll to **Human Spirometry**, then select the **Breathing-Rest-Exercise** settings file from the "Human Spirometry" list.
   - Note: If "Human Spirometry" is empty, then complete the following instructions. On the Main Window, pull down the "Settings" menu and select "Load Group." Scroll down to locate the folder that contains the settings group, "IPLMv6Complete.iwxgrp." Select this group and click "Open." At this point, you should be able to then complete step 3.
4. Instructions for the Breathing-Rest-Exercise Setup will automatically open. These instructions are not necessary. Close this page by clicking on the close button in the upper right corner of this document.
5. LabScribe will appear on the computer screen as configured by the Breathing-Rest-Exercise settings file.
6. For this lab, you do not need any information from the following files. But as a reminder, once the settings file has been loaded, clicking the **Experiment** button on the toolbar will allow you to access any of the following documents:
   - Introduction—introductory material for iWorx Human Spirometry lab exercises
   - Appendix—may have additional information and resources
   - Background—may have some background information regarding the lab topic
   - Labs—the lab exercise as written by iWorx
   - Setup (opens automatically)—instructions for setting up this particular iWorx lab. You already closed this out.
7. If you choose to save your data, it should be saved in the **iWorx Saved Files** folder on the **Desktop**. You may create your own folder within the iWorx Saved Files folder.

## Spirometer Setup

The spirometer has already been connected to the iWorx box (see Figure 1). Do not unplug the cable. Doing so will cause you to lose your data.

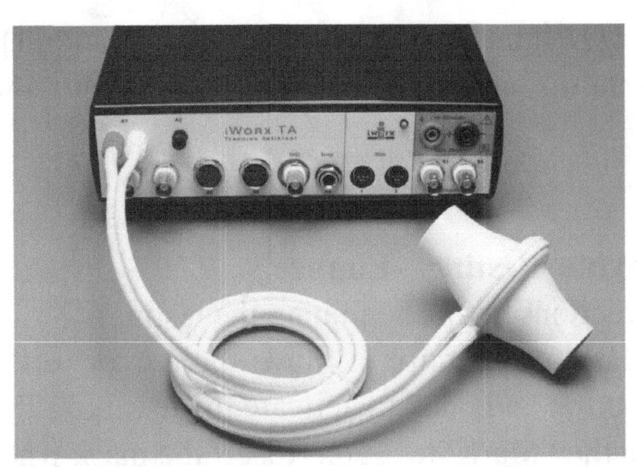


Figure 1 Spirometer flowhead

You will be using a biological filter to breathe through. Simply attach the biological filter to the inlet of the spirometer (see Figure 2). The subject is to breathe through the biological filter, *not* the open end of the spirometer flowhead. You will be using your same biological filter for all iWorx respiratory labs conducted, so you will need to label your filter and turn it in for storage at the end of lab.

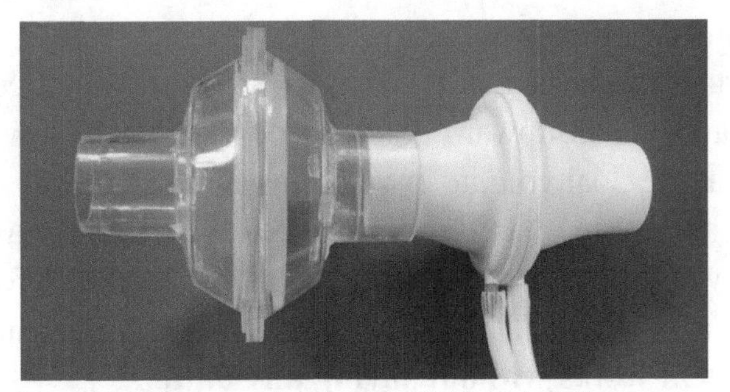

Figure 2 Spirometer flowhead with biological filter

## Spirometer Calibration

You will need to check the calibration of the spirometer.

- In the Main Window, locate the "Lung Volumes" channel.
- Click on the words **Vol.Human(AirFlow)** in the heading line of the "Lung Volumes" channel. This will open a drop down menu.
- Click **Setup Function** to open the "Spirometer Calibration Dialogue."
- Check to be sure the internal spirometer being used is the **IX-TA**. If it is not, open the drop down menu associated with the spirometer and select **IX-TA**.
- Check to be sure the type of flowhead being used is the **300L**. If it is not, open the drop down menu associated with the flowhead and select **300L**.
- Check to be sure the Reset option being used is the **No Reset**. If it is not, open the drop down menu associated with the reset and select **No Reset**.
- Enter the room air temperature in degrees Celsius for the "Temperature of Inhaled Air." The "Temperature of Exhaled Air" should be **29**°C. The "Atmospheric Relative Humidity" should be **60**%.

- Click the **OK** button at the bottom of the "Spirometer Calibration Dialogue" box.
- **Allow the IX-TA to warm up for 10 minutes before recording the first data.**

## Lab Exercises

- **Exercise 1: Breathing While Resting—Eupnea**
  - Read through all instructions before beginning the testing.
  - To label the actual data for exercise 1, move the cursor to the "Mark" text box to the right of the "Mark" button and click to activate the text box. Type the subject's name followed by "resting" in the "Mark" text box. Insert the mark when you begin the recording period.
  - **Subject Instructions**
    - The subject should sit quietly and maintain good posture through the testing period.
    - It is very important for the subject to become comfortable breathing through the apparatus and that the subject breathes normally. Initially the subject will have the tendency to exaggerate tidal breathing. The subject may also fail to exhale the normal tidal volume. To become comfortable, the subject should take some practice inspirations and expirations through the apparatus. Practice ventilations should include both tidal volume ventilations and maximal volume ventilations.
    - It is also important to check that the data is being recorded correctly. Inspiration data should be displayed as an upward deflection on the "Air Flow" channel (see Figure 3; note the arrows at the tidal volume inspirations).
      - To allow the subject to practice and to check the data display, click the **Save to Disk** icon in the lower left corner of the Main Window screen. A red "X" will appear across the "Save to Disk" icon. The LabScribe software is put into preview mode and the recording system works **without saving any data**.
      - Click the **Preview** button in the upper right corner of the Main Window screen. This will start the recording.
      - Have the subject begin by inhaling. Click the **Autoscale** button in both the "Air Flow" channel and the "Lung Volume" channel.
      - If the flowhead is oriented correctly, the inspiration will be recorded as an upward deflection in both the "Air Flow" and the "Lung Volume" channels. If the inspiration is recorded as a downward deflection on these channels, click on the arrow to the left of the "Air Flow" channel. In the drop down menu, click "Invert."
      - If it was necessary to invert the data, test the inspiration again.
      - Have the subject complete as many respiratory cycles as necessary to become comfortable breathing through the apparatus.
      - Click the **Stop** button.
      - Click the **Save to Disk** icon in the lower left corner of the Main Window screen. The red "X" across the "Save to Disk" icon will be replaced by a green arrow. The LabScribe software is now in recording mode and the system will save your data.

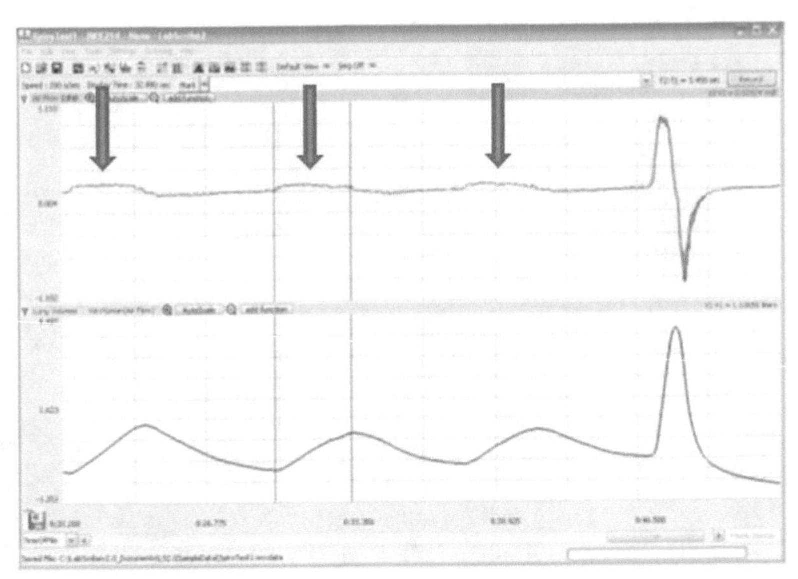

Figure 3 Sample data

- **Tester Instructions**
  - Instruct the subject to sit quietly and maintain good posture while conducting the test. Explain the testing procedure to the subject.
  - Begin the test with the biological filter and spirometer flowhead out of the subject's mouth. The apparatus should be held at mouth level in a position that does not allow air flow into the apparatus. **The flowhead outlets and tubing should be pointing down.** Click the "Record" button and allow the testing to run for 10 seconds. These 10 seconds of recording are necessary to zero the "Lung Volumes" channel.
  - After the 10 seconds required for zeroing, instruct the subject to place the open end of the biological filter into the mouth and begin breathing.
  - Click the "Autoscale" buttons in both the "Air Flow" and "Lung Volumes" channels as necessary to be able to see the data as it is being collected.
  - Record eight resting tidal volume breathing cycles (inspiration and expiration). This should take approximately 1 minute. More breaths may be recorded if the breaths do not look consistent.
  - Immediately following the last respiratory cycle (do not interrupt the recording process), instruct the subject to **inhale as DEEPLY as possible,** then **IMMEDIATELY exhale as QUICKLY and as DEEPLY as possible**. Even though the subject thinks he or she has exhaled all the air possible, ask him or her to continue exhaling until no more air can be moved (see Figure 4, note the arrow at the maximal inspiration and expiration).
  - Instruct the subject to return to resting tidal volume breathing for eight tidal volume breathing cycles (or until tidal volume breathing has returned to normal for at least two breathing cycles).
  - Click "Stop."

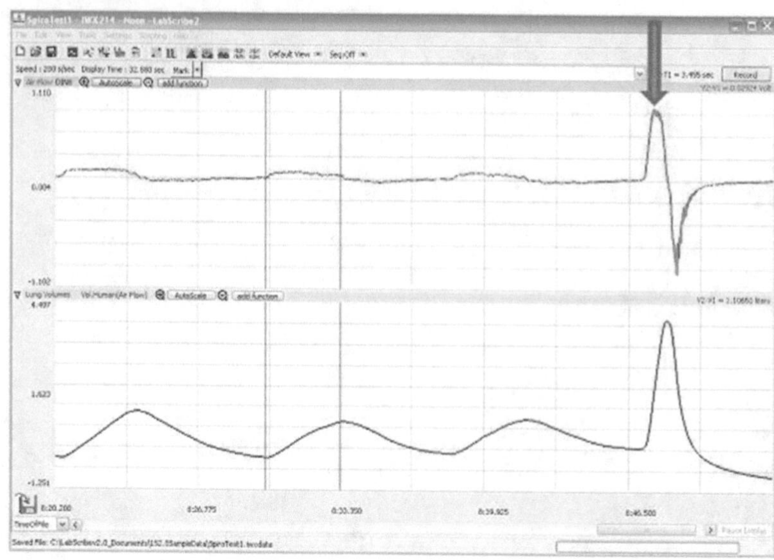

Figure 4 Sample data

- You may choose to save your data at this point or you may continue on with the experiment or data analysis. To save your data, on the Main Window, either
  1. click on the **File** menu. Scroll to **Save As,** then save your data; it should be saved in the **iWorx Saved Files** folder on the **Desktop**. You may create your own folder within the iWorx Saved Files folder. Give your file a name and designate the file type as .iwxdata, then click **Save**.

     Or
  2. click on the **Save File** icon in the iWorx toolbar, then save your data; it should be saved in the **iWorx Saved Files** folder on the **Desktop**. You may create your own folder within the iWorx Saved Files folder. Give your file a name and designate the file type as .iwxdata, then click **Save**.

**Exercise 2: Breathing Immediately After Exercise—Hyperpnea**

- Read through all instructions before beginning the testing.
- To label the actual data for exercise 2, move the cursor to the "Mark" text box to the right of the "Mark" button and click to activate the text box. Type the subject's name followed by "postexercise" in the "Mark" text box. Insert the mark when you begin the recording period.
- **Subject Instructions**
  - The subject must exercise the leg muscles for a minimum of 3 minutes. Activities such as jogging in place or vigorously walking stairs are appropriate. It is essential that the subject exercise enough to increase both respiratory depth and rate along with metabolic rate. These increased levels must be maintained through the data collection period.
  - Immediately after the exercise period, the subject must begin data collection. The subject should sit quietly and maintain good posture through the testing period.
  - It is very important for the subject to become comfortable breathing through the apparatus and that the subject breathes normally. Use the same precautions used in exercise 1.
- **Tester Instructions**
  - Instruct the subject to sit quietly and maintain good posture while conducting the test. Explain the testing procedure to the subject.

  - Begin the test with the biological filter and spirometer flowhead out of the subject's mouth. The apparatus should be held at mouth level in a position that does not allow air flow into the apparatus. **The flowhead outlets and tubing should be pointing down.** Click the "Record" button and allow the testing to run for 10 seconds. These 10 seconds of recording are necessary to zero the "Lung Volumes" channel.
  - After the 10 seconds required for zeroing, instruct the subject to place the open end of the biological filter into the mouth and begin breathing.
  - Click the "Autoscale" buttons in both the "Air Flow" and "Lung Volumes" channels as necessary to be able to see the data as it is being collected.
  - Record at least eight tidal volume breathing cycles (inspiration and expiration). This should take approximately 1 minute. More breaths may be recorded if the breaths do not look consistent.
  - Immediately following the last postexercise respiratory cycle (do not interrupt the recording process), instruct the subject to **inhale as DEEPLY as possible,** then **IMMEDIATELY exhale as QUICKLY and as DEEPLY as possible**. Even though the subject thinks he or she has exhaled all the air possible, ask him or her to continue exhaling until no more air can be moved (see Figure 4, note the arrow at the maximal inspiration and expiration).
  - Instruct the subject to return to tidal volume breathing for eight tidal volume breathing cycles (or until tidal volume breathing has returned to normal for at least two breathing cycles).
  - Click "Stop."
- You may choose to save your data at this point or you may continue on with the data analysis. If you have not yet saved your data, follow the instructions for saving data under exercise 1. If you have previously saved your data, click on the **Save File** icon in the iWorx toolbar.

■ **Exercise 3: Hypoventilation Breathing**

- Be sure the subject's breathing has returned to normal.
- Read through all instructions before beginning the testing.
- To label the actual data for exercise 3, move the cursor to the "Mark" text box to the right of the "Mark" button and click to activate the text box. Type the subject's name followed by "hypoventilation" in the "Mark" text box. Insert the mark when you begin the recording period.
- **Subject Instructions**
  - Hypoventilation requires that the subject decreases primarily the depth of the tidal volume. Breathe very shallowly so that you are exchanging air primarily in the conduction zone and not in the respiratory zone. Rate of breathing is not so important, but you may find that you increase your rate involuntarily as your breathing center is stimulated to increase your external gas exchange due to the increased partial pressure of $CO_2$ and decreased pH and partial pressure of $O_2$.
  - The subject should sit quietly and maintain good posture through the testing period.
  - It is very important for the subject to become comfortable breathing through the apparatus and that the subject alters his or her breathing cycle appropriately. Use the same precautions used in exercise 1.

- **Tester Instructions**
  - Be sure the subject's breathing has returned to normal.
  - Instruct the subject to sit quietly and maintain good posture while conducting the test. Explain the testing procedure to the subject.
  - Begin the test with the biological filter and spirometer flowhead out of the subject's mouth. The apparatus should be held at mouth level in a position that does not allow air flow into the apparatus. **The flowhead outlets and tubing should be pointing down.** Click the "Record" button and allow the testing to run for 10 seconds. These 10 seconds of recording are necessary to zero the "Lung Volumes" channel.
  - After the 10 seconds required for zeroing, instruct the subject to place the open end of the biological filter into the mouth and begin breathing.
  - Click the "Autoscale" buttons in both the "Air Flow" and "Lung Volumes" channels as necessary to be able to see the data as it is being collected.
  - Record hypoventilation breathing cycles (inspiration and expiration) for 30 seconds. More breaths may be recorded if the breaths do not look consistent.
  - Immediately following the last hypoventilation respiratory cycle (do not interrupt the recording process), instruct the subject to **inhale as DEEPLY as possible,** then **IMMEDIATELY exhale as QUICKLY and as DEEPLY as possible**. Even though the subject thinks he or she has exhaled all the air possible, ask him or her to continue exhaling until no more air can be moved (see Figure 4, note the arrow at the maximal inspiration and expiration).
  - Instruct the subject to return to normal tidal volume breathing for eight tidal volume breathing cycles (or until tidal volume breathing has returned to normal for at least two breathing cycles).
  - Click "Stop."
- You may choose to save your data at this point or you may continue on with the data analysis. If you have not yet saved your data, follow the instructions for saving data under exercise 1. If you have previously saved your data, click on the **Save File** icon in the iWorx toolbar.
- **Data Analysis Instructions**

Figure 5, a diagram of icons in the iWorx toolbar may be helpful.

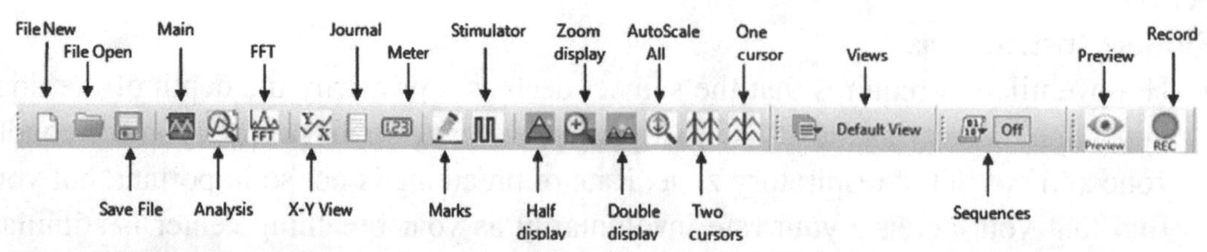

Figure 5 iWorx toolbar icons

- Click on the **Analysis** icon in the iWorx toolbar.
- Go to the beginning of your recordings. Click on the **Marks** icon to bring up the Marks Dialogue box with a table of the marks created in your lab. Once you have the table of marks, click on the number at the beginning of the row for the mark that identifies the beginning of the experimental data segment you are going to analyze. This will highlight that mark. Next,

click on the **Go To Mark** button near the bottom of the Marks Dialogue box. Data at that mark will be displayed on your screen.

- Be sure that two cursors are present on your screen. If necessary, click the **Two Cursors** icon on the iWorx toolbar.
- Place the cursors on either side of four consistent breathing cycles, one cursor on the baseline just before inspiration and the second cursor just after the final expiration, then click the **Zoom Display** icon in the iWorx toolbar. This will expand the data section to the width of the Main Window. Adjust your data section as necessary with the display time icons or the zoom icon.
- Click on the **Analysis** icon in the iWorx toolbar.
- In the Analysis window, a "Function Table" will be displayed above the uppermost channel (Air Flow channel). The mathematical functions "V2–V1," Max_dv/dt, Min_dv/dt, and "T2–T1" should be present. The values for each of these functions are displayed in the top margin of each channel. The "V2–V1" function measures the difference in amplitude between cursor 2 and cursor 1. The "Max_dv/dt" function is the maximum rate of air movement during inspiration in the time period between cursors. The "Min_dv/dt" function is the maximum rate of air movement during expiration in the time period between cursors. The "T2–T1" function measures the difference in time between cursor 2 and cursor 1.
- Click on the arrow to the left of the "Air Flow" channel. In the drop down menu, click "Minimize" to reduce the "Air Flow" channel to a bar across the top of the data screen.
- Click on the arrow to the left of the "Lung Volume" channel. In the drop down menu, move the cursor to "Scale" and click "Autoscale" from the submenu. This will maximize the data displayed in the "Lung Volume" channel.
- **All measurements will be made in the "Lung Volume" channel. All data for the measurements of air flow and volume should be taken from the "Lung Volume" channel data.**
- To measure tidal volume (TV) and maximum inspiratory flow of a breathing cycle, place the first cursor at the lowest point of the breathing cycle, the beginning of the breathing cycle just prior to an inspiration. Place the second cursor at the highest point of the same breathing cycle, the end of the inspiration (see Figure 6).

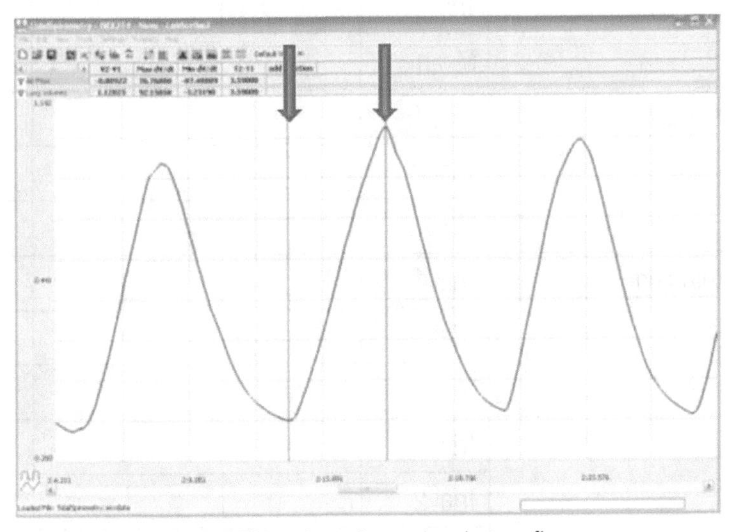

Figure 6 Data analysis TV and maximum inspiratory flow

- Record the "V2–V1" data for the tidal volume in Table 1.
- Record the "Max_dv/dt" data for the maximum inspiratory flow rate in Table 1.
- To measure the maximum expiratory flow rate, place the first cursor at the highest point of the same breathing cycle, at the end of the inspiration. Place the second cursor at the lowest point of the same breathing cycle; the end of the expiration and the end of the breathing cycle (see Figure 7).

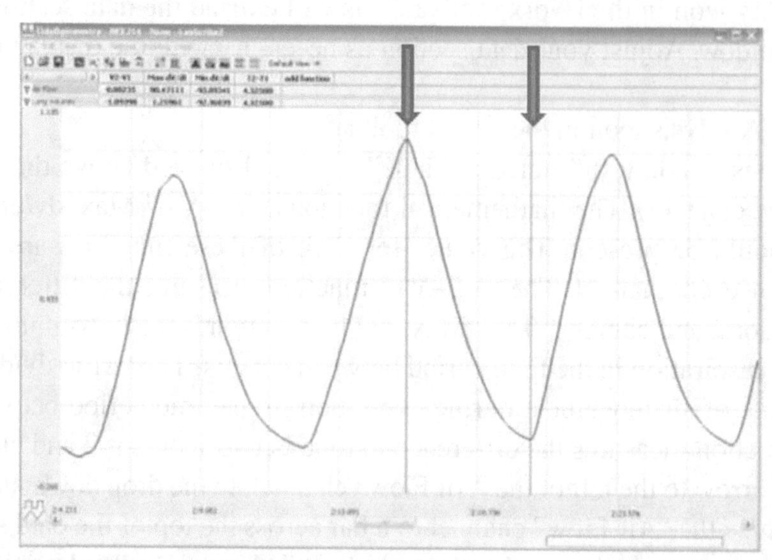

Figure 7 Data analysis maximum expiratory flow

- Record the "Min_dv/dt" flow data for the maximum expiratory flow rate in Table 1.

**TABLE 1:** Tidal Volume Breathing Cycle Measurements
(record absolute values for all numbers)

| | | **Eupnea** | **Hyperpnea** | **Hypoventilation** |
|---|---|---|---|---|
| Tidal volume (L) | Trial 1 | | | |
| | Trial 2 | | | |
| | Mean | | | |
| Maximum flow rate, Inspiration (L/sec) | Trial 1 | | | |
| | Trial 2 | | | |
| | Mean | | | |
| Maximum flow rate, Expiration (L/sec) | Trial 1 | | | |
| | Trial 2 | | | |
| | Mean | | | |
| Breath Period (sec) | Trial 1 | | | |
| | Trial 2 | | | |
| | Mean | | | |

- To measure the breath period of a breathing cycle, place the first cursor at the lowest point of the breathing cycle, the beginning of the breathing cycle just prior to inspiration (as in measuring TV), and place the second cursor at the lowest point of the breathing cycle, at the end of expiration (as in measuring minimum flow rate) (see Figure 8). If the troughs are not distinct, you may measure from peak to peak (at the end of inspiration of two sequential breathing cycles).

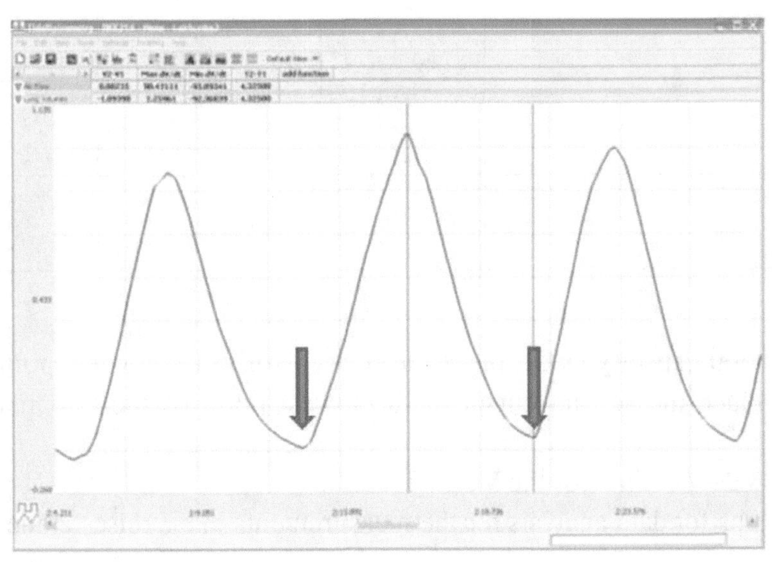

Figure 8 Data analysis breath period

- Record the "T2–T1" data for the breath period in Table 1.
- Repeat these four measurements for one more breathing cycle. Calculate the mean for each and record the value in Table 1.
- Go to the data segment that includes the maximal inspiration and expiration to measure the other respiratory volumes and capacities.
- Adjust the data display so that two resting tidal volume breathing cycles, the first maximal inspiration and expiration, and one subsequent resting tidal volume breathing cycle are all visible on the data screen.
- To measure inspiratory reserve volume (IRV), place the first cursor at the peak of the resting tidal breathing cycle, the end of the tidal inspiration, just prior to the maximal inspiration. Place the second cursor at the highest point of the maximal inspiration (see Figure 9).
- Record the "V2–V1" data for the IRV in Table 2.
- To measure inspiratory capacity (IC), place the first cursor at the lowest point of the resting tidal breathing cycle, the end of the tidal expiration. Place the second cursor at the highest point of the maximal inspiration (see Figure 9).
- Record the "V2–V1" data for the IC in Table 2.
- To measure expiratory reserve volume (ERV), place the first cursor at the lowest point of the resting tidal breathing cycle, the end of the tidal expiration. Place the second cursor at the lowest point of the maximal expiration (see Figure 9).
- Record the "V2–V1" data for the ERV in Table 2.

**TABLE 2:** Respiratory Volumes and Capacities Measurements
(record absolute values for all numbers)

| | Eupnea | Hyperpnea | Hypoventilation |
|---|---|---|---|
| Inspiratory Reserve Volume (L) | | | |
| Inspiratory Capacity (L) | | | |
| Expiratory Reserve Volume (L) | | | |
| Vital Capacity (L) | | | |

- To measure vital capacity (VC), place the first cursor at the highest point of the maximal inspiration. Place the second cursor at the lowest point of the maximal expiration (see Figure 9).
- Record the "V2–V1" data for the VC in Table 2.
- Repeat the steps for data analysis for the measurements in exercises 2 and 3.

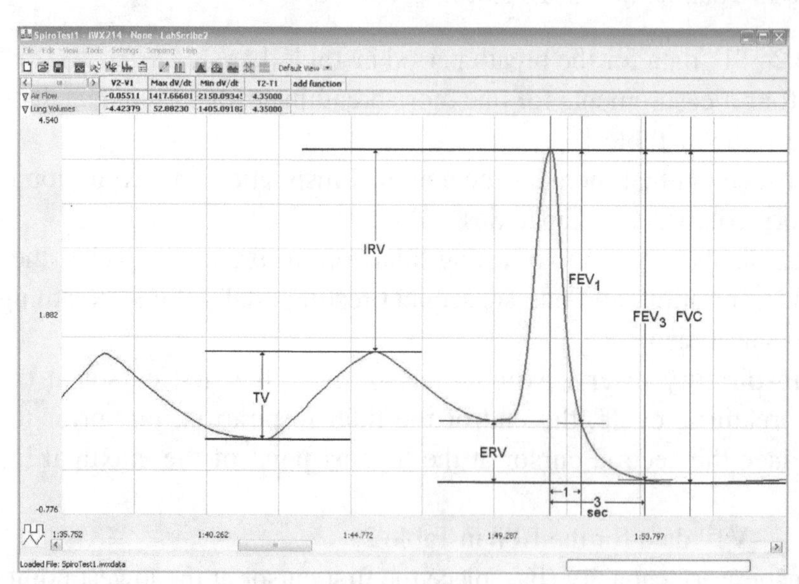


Figure 9 Data analysis lung volumes and capacities

# Exercise 23

# iWorx Forced Vital Capacity and Forced Expiratory Volumes

## Objectives

- Be able to define the term *forced vital capacity* (FVC) or *timed vital capacity*.
- Be able to define the terms *forced expiratory volume*, $FEV_1$, $FEV_2$, and $FEV_3$.
- Be able to calculate FEV for a given time.
- Be able to determine the normal values for $FEV_1$, $FEV_2$, and $FEV_3$.
- Be able to define the term *restrictive pulmonary disease* (RPD) and identify relatively common respiratory diseases that are classified as RPDs: *interstitial pulmonary fibrosis* (IPF) and *extrapulmonary RPD*.
- Be able to explain why FEV is not a tool for diagnosing RPD.
- Be able to define the term *chronic obstructive pulmonary disease* (COPD) and identify relatively common respiratory diseases that are classified as COPDs.
- Be able to explain how FEV would change for a person with COPD.

## Lab Safety Reminders

- Use caution with any electrical equipment.

## Hints for Success

All members of your lab group should feel comfortable operating the iWorx system. Considerations for selecting your subject: Your subject should be healthy, free of respiratory infections, and have no history of respiratory or cardiovascular problems.

Although the subject may be consciously breathing through the mouth, it is highly likely that some air will also be moving through the nose. The use of a nose clip will prevent any air from escaping or entering the subject's respiratory system through the nose. Using a nose clip will provide a more accurate set of data. Please be sure to clean the nose clips with alcohol both before and after use, and return the nose clips when done with the exercise.

Be sure to follow the instructions for calibration and warm-up.

**Never unplug any sensors from the iWorx box!** If you unplug a sensor during an experiment, you will lose all data. You will need to turn the iWorx box off, plug the sensor in and start the lab from the beginning.

On the iWorx recording screens, it is helpful to remember that the blue bars are for data that are being recorded. The green bars are for data that is being calculated based upon the data you have collected.

While using the iWorx, it is important that you do not get so caught up with obtaining the correct numbers from your data that you forget the function or process you are trying to observe. As you are using the iWorx, keep in mind the principles from the lecture being illustrated in the lab exercise.

## Introduction

In this lab, you will be measuring the forced vital capacity (FVC) and determine $FEV_1$, $FEV_2$, and $FEV_3$. These values are then used to calculate the ratios of $FEV_1/FVC$, $FEV_2/FVC$, and $FEV_3/FVC$. They will first be measured with no restrictions of the breathing cycle. The second measurement will mimic the effect of restrictive pulmonary disease (RPD). The third measurement will mimic the effect of chronic obstructive pulmonary disease (COPD). In general, respiratory diseases tend to restrict air flow and the ability to exchanges gases between the lungs and the environment. As a result, this limits the ability to exchange gases between the alveoli and the pulmonary capillaries and ultimately reduces the availability of oxygen to the cells of the body and removal of carbon dioxide from the body. In both cases, gas exchange does not meet the metabolic demands of the body. Sometimes this is a chronic condition that is experienced relatively constantly. Sometimes this is a condition that occurs under certain conditions such as increased metabolic demands, stress, or exposure to specific triggers. Two large classes of pulmonary diseases are restrictive pulmonary disease (RPD) and COPD. Both of these diseases alter lung function, but they do so in two very different ways. Individuals with RPD have lung tissue that is often described as stiff and resistant to expansion. They have decreased lung compliance but an increase in lung recoil. As a result, the lung tissue has a reduced ability to expand properly and the respiratory volumes that are exchanged are usually reduced. Because they have good recoil, their ability to exhale is not reduced. Individuals with COPD have a reduced ability to quickly exchange air between the lungs and the environment, but the respiratory volumes that are exchanged are usually normal. In some cases, it is the airway that restricts the flow of air in and out of the lungs and ultimately the ability to supply oxygen to the tissues and remove carbon dioxide from the tissues. In other cases, they have a high compliance with a reduced ability to recoil the lung tissue. Although these individuals can expand their lungs, they have a reduced flow rate.

The most common restrictive pulmonary diseases (RPDs) are: **interstitial pulmonary fibrosis (IPF)** and **extrapulmonary RPD**. IPF affects the functional cells of the lung and the associated connective tissue. In IPF, the lung tissue is damaged and chronic inflammation occurs. The chronic inflammation results in the conversion of functional and connective tissues to stiff fibrous tissues with a decrease in compliance. In advanced IPF, the walls of the alveoli and the walls of the pulmonary capillaries may be destroyed resulting in alveolar dilation and impaired blood flow in

IPF - lung tissue damaged; chronic inflammation occurs

the lungs. This reduced blood flow increases ventilation–perfusion mismatching and results in a reduction in the amount of oxygen that diffuses into the pulmonary capillaries. A common symptom of IPF is shortness of breath and a more rapid, shallow pattern of breathing. There are many different types of IPF. In many cases, the cause of the disease is idiopathic and remains unknown. In some cases, a severe viral infection can lead to IPF. Some known causes are occupational and environmental inhaled irritants such as: dust, gases, fumes or aerosols; radiation; poisons; and drugs. IPF cannot be reversed and is progressive with an expected life span of 5 to 10 years after diagnosis (http://www.nmihi.com/f/ipf.htm). In extrapulmonary RPD individuals typically have healthy, normally functioning lung tissue, but experience a reduction in the ability to expand the lungs resulting in shallow or rapid breathing. As the ability to expand the lungs is reduced, the overall capacity of air inhaled is reduced, making it difficult to meet the metabolic demands of the tissues. Extrapulmonary RPDs generally fall into three categories: neurological or neuromuscular diseases, abnormalities in the thoracic cage, and pleural disorders. Neurological or neuromuscular diseases include polio, Myasthenia Gravis, and Guillain-Barre syndrome. Nonmuscular diseases including chest wall deformities and reduced lower thoracic/abdominal volume associated with obesity, hernias of the diaphragm, and ascites (accumulation of fluid in the abdominal cavity). Pleural disorders include pleurisy and pleural thickening. All of these can lead to difficulty in expanding the thoracic cavity during inspiration and thus difficulty in generating a negative respiratory system pressure gradient.

RPD = can't expand lungs = shallow/rapid breathing

Obstructive lung diseases, on the other hand, are primarily comprised of three related conditions: chronic bronchitis, chronic asthma, and emphysema although several other diseases also fall under the heading of COPD. In each condition there is a chronic obstruction of the flow of air through the airways and out of the lungs, and the obstruction is usually permanent and may be progressive over time. These chronic lung conditions are generally classified as COPD—chronic obstructive pulmonary disease. Chronic bronchitis is characterized by a chronic, productive cough to remove mucus produced in the airways. Damage to the endothelium of the airway disrupts the respiratory escalator and the ability of the ciliated epithelium to sweep mucus and trapped bacteria and particulates out of the lungs and airway. Smoking and inhalation of toxic gases and particulates are common causes of chronic bronchitis. Inflammation of the airway and mucus secretion provides the obstructive component of chronic bronchitis. The alveoli and pulmonary capillary beds are relatively undamaged in chronic bronchitis unless infection occurs and progresses to the lung tissue. Asthma features obstruction to the flow of air out of the lungs. Usually, the obstruction is reversible and between asthma attacks or with treatment, the flow of air through the airways is normal or improves. If asthma is left untreated, the chronic inflammation associated with this disease can cause the airway obstruction to become fixed. When the obstruction is fixed the damage is permanent, and the person will have abnormal air flow between as well as during asthma attacks. Asthma patients with a fixed component of airway obstruction are considered to have COPD. Emphysema is defined by destruction of respiratory bronchioles, alveolar ducts, and alveoli. As emphysema progresses alveolar septa and the associated pulmonary capillary bed degrade. This leads to the decreased ability to oxygenate blood due to a loss of alveolar and capillary surface area. In all three cases of COPD, air flow out of the lungs is restricted and $P_{CO2}$ tends to increase, resulting in respiratory acidosis and acidemia.

COPD = chronic obstruction of air flow

For either restrictive or obstructive pulmonary diseases, chronic low oxygen levels can lead to narrowing of the arteries in the lungs. This can lead to pulmonary hypertension, increased pressure in the pulmonary arteries, and cor pulmonale or pulmonary heart disease. Because the right ventricle must work harder against higher than normal pulmonary pressure, the right ventricle may hypertrophy and the person may progress into failure on the right side of the heart.

A pulmonary function test known as Spirometry is used to determine the ability of a person to exhale air normally. There are several components to a complete pulmonary function test. These include measuring vital capacity (VC), the total volume of air that can be expelled from the lungs, FVC, the vital capacity expelled as quickly as possible and with maximal effort, forced expiratory volume FEV measured over time ($FEV_1$—vol. expired after 1 second; $FEV_2$—vol. expired after 2 seconds; $FEV_2$—vol. expired after 3 seconds), forced expiratory flow 25% to 75% (FEF at 25%–75% of air expired), forced inspiratory flow (FIF at 25%–75% of inspired air), and maximum voluntary ventilation (MVV) or maximum breathing capacity. A Flow–Volume loop illustrates most of these measurements (see Figure 1). The Flow–Volume loop includes a complete breathing cycle and measures volumes and rate for both expiration and inspiration. $FEV_1$ is the volume expired at 1 second, $FEV_2$ would be the volume expired at 2 seconds, $FEV_3$ is the volume expired at 3 seconds, and FVC is the total volume exhaled.

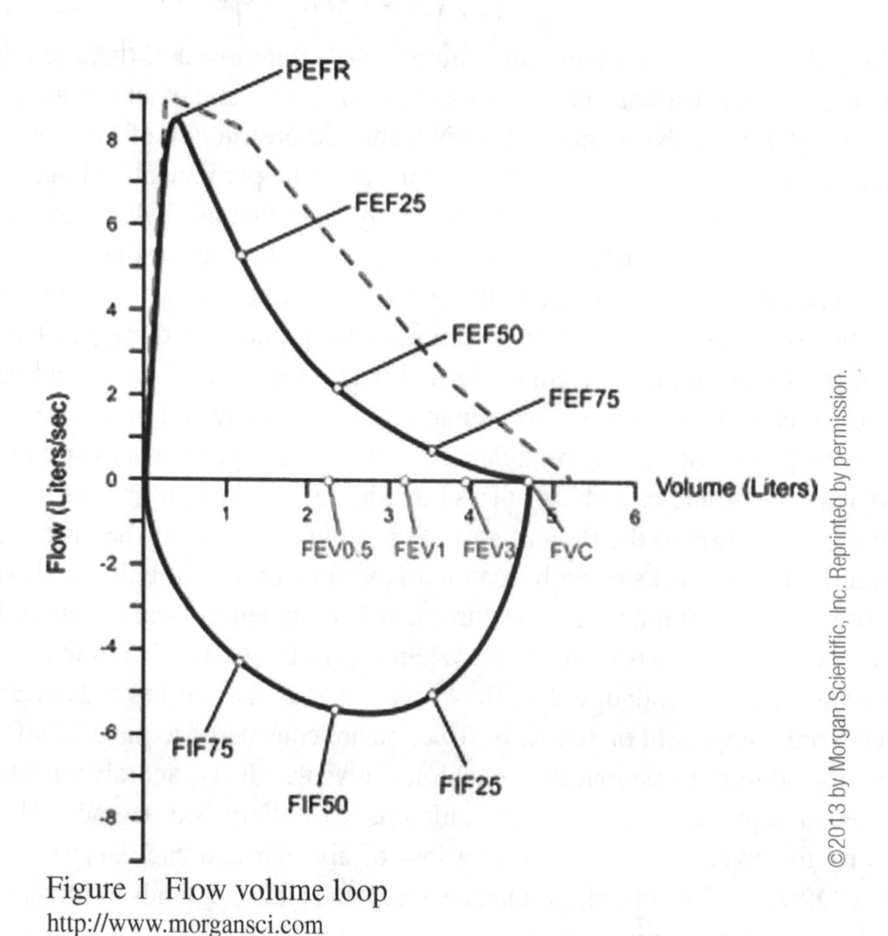


Figure 1 Flow volume loop
http://www.morgansci.com

The results of a flow–volume loop will vary depending upon the conditions of a person's lungs and can be used for diagnostic assessment of lung function (see Figure 2).

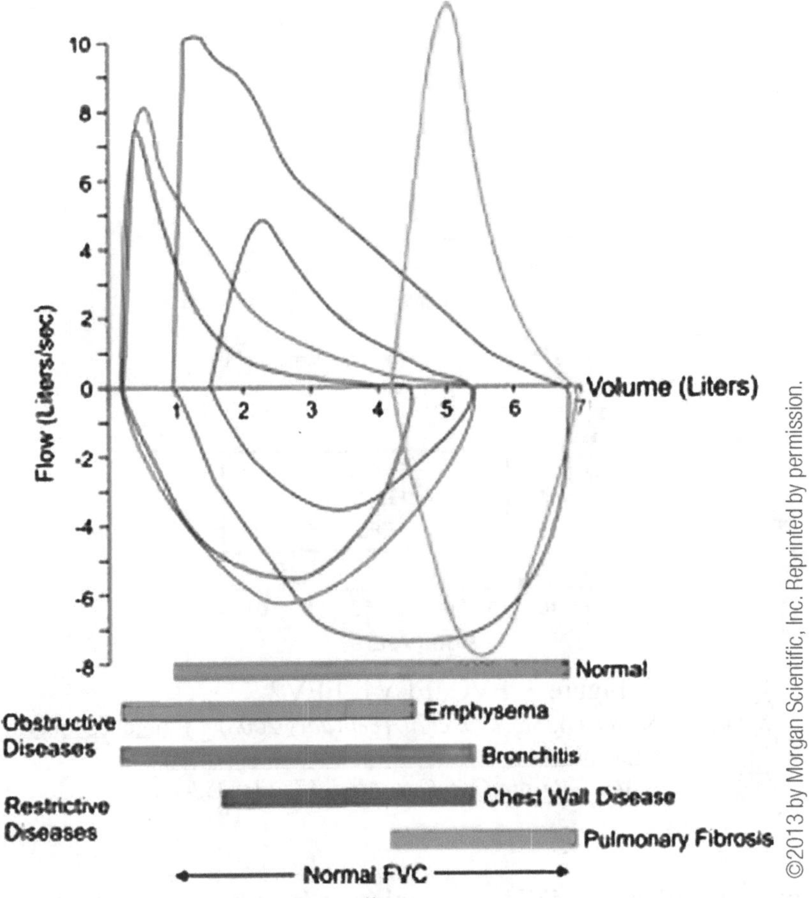




Figure 2 Flow volume loop diseases
http://www.morgansci.com

It is the ratio of $FEV_n/FVC$ that we will use to determine how well an individual is able to move the air out of the lungs.

Normal values of the FEV/FVC ratios are as follows:

$$FEV_1/FVC \bullet 100 = 66\%–83\%$$
$$FEV_2/FVC \bullet 100 = 75\%–94\%$$
$$FEV_3/FVC \bullet 100 = 78\%–97\%$$

The values of FVC, $FEV_n$, and FEF (forced expiratory flow) vary with age and gender. The values decrease with age and at the same age, are lower in females than in males (see Figure 3).

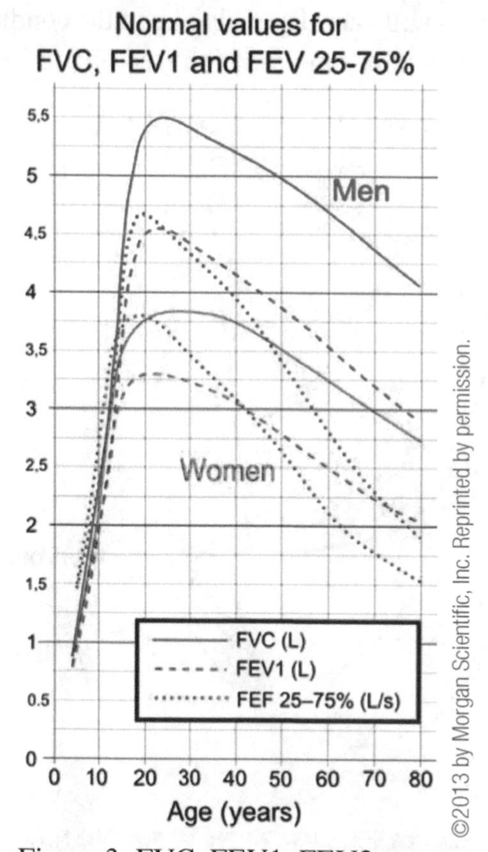


Figure 3 FVC, FEV1, FEV2
*Stanojevic S, Wade A, Stocks J, et al. (February 2008).* "Reference ranges for spirometry across all ages: a new approach". *Am. J. Respir. Crit. Care Med. 177 (3): 253–60.*

Maximum voluntary ventilation (MVV) or maximum breathing capacity may also be assessed. This is the volume of air that can be breathed in 15 seconds when a person breathes as deeply and as quickly as possible. The volume exchanged is extrapolated to one minute and is expected to be approximately equal to $FEV_1 \cdot 35$.

In obstructive airway diseases, like chronic bronchitis, asthma, or emphysema, flow rate and $FEV_1$ are reduced with $FEV_1$ less than 80% of the volume predicted for your age (see the following figure). In addition, $FEV_n/FVC$ ratios are reduced, and the $FEV_1/FVC$ ratio is usually less than 0.70. Testing individuals after the use of an inhaler or nebulizer (as for asthma) or other treatment can be used to assess the effectiveness of the treatment. The improvement as a result of treatment is referred to as reversibility and is generally better in individuals with asthma than in individuals with other forms of COPDs. COPD is often classified based on severity. The following classifications and values are based on measurements taken after a person has received a treatment and serve as a guide to the severity of COPD:

Mild—$FEV_1$ is 80% of the predicted value for the sex and age of the individual
Moderate—$FEV_1$ is 50% to 79% of the predicted value for the sex and age of the individual

Severe—$FEV_1$ is 30% to 49% of the predicted value for the sex and age of the individual
Very severe—$FEV_1$ is less than 30% of the predicted value for the sex and age of the individual

In RPDs, like fibrosis, FVC is reduced, and $FEV_1$ is often also reduced. But, because of the low compliance and high recoil of the lungs, the $FEV_1$/FVC ratio may be normal (≈0.80) or greater than normal (>0.85). Figure 4 illustrates some of the relative differences in volumes, capacities, and flow rates that are expected between a person with normal lungs and a person with either COPD or PRD.

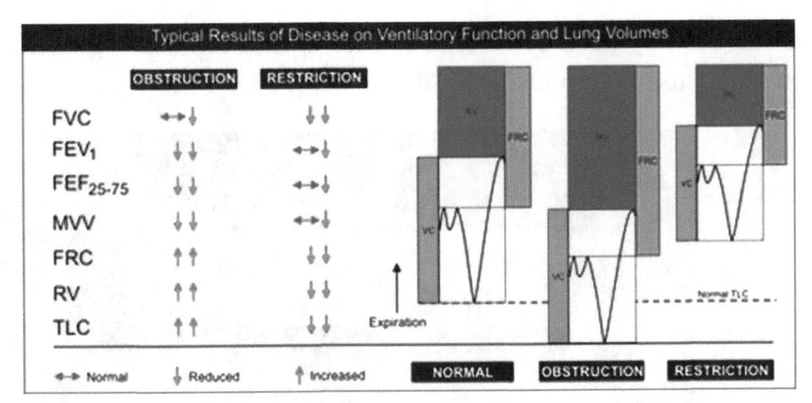


Figure 4 Disease FVC, $FEV_1$, $FEV_2$
http://www.morgansci.com

## Starting the iWorx Software

1. Be sure that the iWorx box is turned on. A small green light will be lit when the power is on. If you need to turn the power on, the on/off switch is on the back of the iWorx box.
2. Click on the **LabScribe3 shortcut** on the computer's desktop to open the program.
   You should see an information box that says "Hardware Found". Click **OK**; this will put you on the Main Window page. If the hardware is not found, check to be sure the iWorx box is turned on.
3. On the Main Window, pull down the **Settings** menu. Scroll to **Human Spirometry**, then select the **Breathing-Rest-Exercise** settings file from the "Human Spirometry" list.
   Note: If "Human Spirometry" is empty, then complete the following instructions. On the Main Window, pull down the "Settings" menu and select "Load Group." Scroll down to locate the folder that contains the settings group, "IPLMv6Complete.iwxgrp." Select this group and click "Open." At this point, you should be able to then complete step 3.
4. Instructions for the Breathing-Rest-Exercise Setup will automatically open. These instructions are not necessary. Close this page by clicking on the close button in the upper right corner of this document.
5. LabScribe will appear on the computer screen as configured by the Breathing-Rest-Exercise settings file.
6. For this lab, you do not need any information from the following files. But as a reminder, once the settings file has been loaded, clicking the **Experiment** button on the toolbar will allow you to access any of the following documents:
   - Introduction—introductory material for iWorx Human Spirometry lab exercises
   - Appendix—may have additional information and resources

- Background—may have some background information regarding the lab topic
- Labs—the lab exercise as written by iWorx
- Setup (opens automatically)—instructions for setting up this particular iWorx lab. You already closed this out.

7. If you choose to save your data, it should be saved in the **iWorx Saved Files** folder on the **Desktop**. You may create your own folder within the iWorx Saved Files folder.

## Spirometer Setup

The spirometer has already been connected to the iWorx box (see Figure 5). Do not unplug the cable. Doing so will cause you to lose your data.

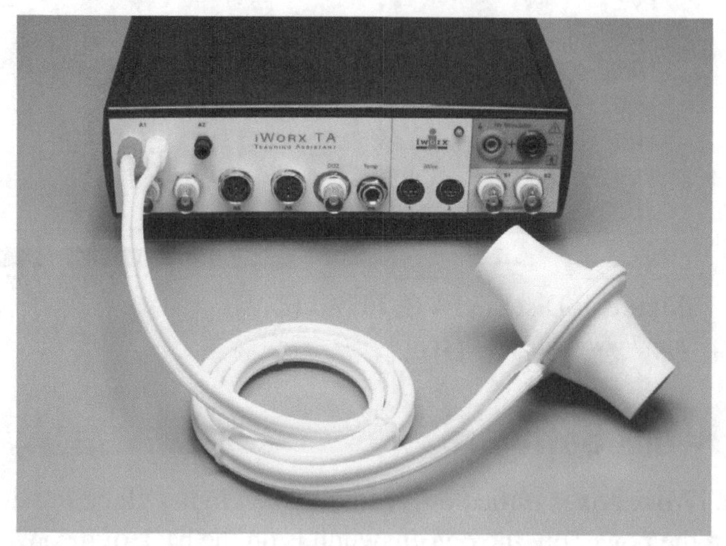


Figure 5 Equipment spirometer flowhead

You will be using a biological filter to breathe through. Simply attach the biological filter to the inlet of the spirometer (see Figure 6). The inlet side has been marked with red on the spirometer flowhead as well as on the tubing. The subject is to breathe through the biological filter NOT the open end of the spirometer flowhead. You will be using your same biological filter for all iWorx respiratory labs conducted, so you will need to label your filter and turn it in for storage at the end of each lab.

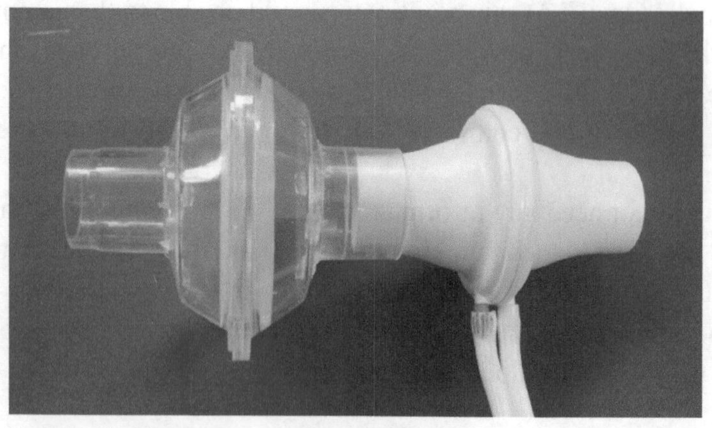

Figure 6 Equipment spirometer flowhead with biological filter

## Spirometer Calibration

You will need to check the calibration of the spirometer.

- In the Main Window, locate the "Lung Volumes" channel.
- Click on the words **Vol.Human(AirFlow)** in the heading line of the "Lung Volumes" channel. This will open a drop down menu.
- Click **Setup Function** to open the "Spirometer Calibration Dialogue."
- Check to be sure the internal spirometer being used is the **IX-TA**. If it is not, open the drop down menu associated with the spirometer and select **IX-TA**.
- Check to be sure the type of flowhead being used is the **300L**. If it is not, open the drop down menu associated with the flowhead and select **300L**.
- Check to be sure the Reset option being used is the **No Reset**. If it is not, open the drop down menu associated with the reset and select **No Reset**.
- Enter the room air temperature in degrees Celsius for the "Temperature of Inhaled Air." The "Temperature of Exhaled Air" should be **29**°C. The "Atmospheric Relative Humidity" should be **60**%.
- Click the **OK** button at the bottom of the "Spirometer Calibration Dialogue" box.
- **Allow the IX-TA to warm up for 10 minutes before recording the first data.**

## Lab Exercises

- **Exercise 1: FVC & FEV$_n$, No Restrictions**
  - Read through all instructions before beginning the testing.
  - To label the actual data for exercise 1, move the cursor to the "Mark" text box to the right of the "Mark" button and click to activate the text box. Type the subject's name followed by "resting" in the "Mark" text box. Insert the mark when you begin the recording period.

  - **Subject Instructions**
    - The subject should sit quietly and maintain good posture through the testing period.
    - It is very important for the subject to become comfortable breathing through the apparatus and that the subject breathes normally. Initially the subject will have the tendency to exaggerate tidal breathing. The subject may also fail to exhale the normal tidal volume. To become comfortable, the subject should take some practice inspirations and expirations through the apparatus. Practice ventilations should include both tidal volume ventilations and maximal volume ventilations.
    - It is also important to check that the data is being recorded correctly. Inspiration data should be displayed as an upward deflection on the "Air Flow" channel (see Figure 7; note the arrows at a tidal volume inspiration).
      - To allow the subject to practice and to check the data display, click the **Save to Disk** icon in the lower left corner of the Main Window screen. A red "X" will appear across the "Save to Disk" icon. The LabScribe software is put into preview mode and the recording system works **without saving any data**.
      - Click the **Preview** button in the upper right corner of the Main Window screen. This will start the recording.

- Have the subject begin by inhaling. Click the **Autoscale** button in both the "Air Flow" channel and the "Lung Volume" channel.
- If the flowhead is oriented correctly, the inspiration will be recorded as an upward deflection in both the "Air Flow" and the "Lung Volume" channels. If the inspiration is recorded as a downward deflection on these channels, click on the arrow to the left of the "Air Flow" channel. In the drop down menu, click "Invert."
- If it was necessary to invert the data, test the inspiration again.
- Have the subject complete as many respiratory cycles as necessary to become comfortable breathing through the apparatus.
- Click the **Stop** button.
- Click the **Save to Disk** icon in the lower left corner of the Main Window screen. The red "X" across the "Save to Disk" icon will be replaced by a green arrow. The LabScribe software is now in recording mode and the system will save your data.

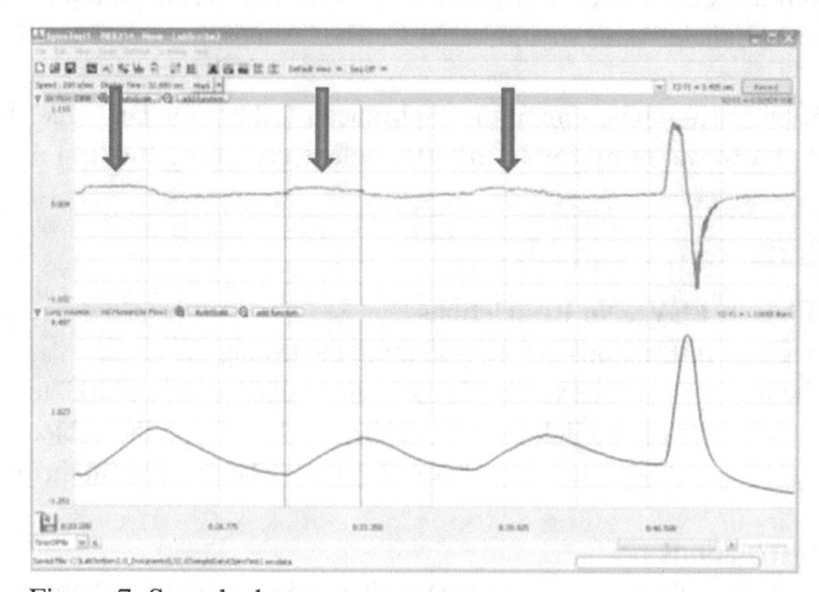

Figure 7 Sample data

- **Tester Instructions**
  - Instruct the subject to sit quietly and maintain good posture while conducting the test. Explain the testing procedure to the subject.
  - Begin the test with the biological filter and spirometer flowhead out of the subject's mouth. The apparatus should be held at mouth level in a position that does not allow air flow into the apparatus. **The flowhead outlets and tubing should be pointing down.** Click the "Record" button and allow the testing to run for 10 seconds. These 10 seconds of recording are necessary to zero the "Lung Volumes" channel.
  - After the 10 seconds required for zeroing, instruct the subject to place the open end of the biological filter into the mouth and begin breathing.
  - Click the "Autoscale" buttons in both the "Air Flow" and "Lung Volumes" channels as necessary to be able to see the data as it is being collected.
  - Record ten resting tidal volume breathing cycles (inspiration and expiration). This should take approximately one minute. More breaths may be recorded if the breaths do not look consistent.

- Immediately following the last respiratory cycle (do not interrupt the recording process), instruct the subject to **inhale as DEEPLY as possible,** then **IMMEDIATELY exhale as QUICKLY and as DEEPLY as possible. It is very important that the subject does not hesitate when beginning the maximal expiration. Maximal effort should be used at the beginning of and throughout the expiration. The subject should bend over near the end of the maximal expiration to help force air out of the lungs**. Even though the subject thinks he or she has exhaled all the air possible, ask him or her to continue exhaling until no more air can be moved (see Figure 8, note the arrow at the maximal inspiration and expiration). **The subject should exhale for at least 3 seconds even if he or she feels as though he or she has no more air to exhale.**
- Repeat this cycle of ten normal tidal ventilation cycles followed by the DEEP inspiration and expiration twice for a total of three series of ten tidal ventilation followed by maximal inspiration and expiration.
- Click "Stop."

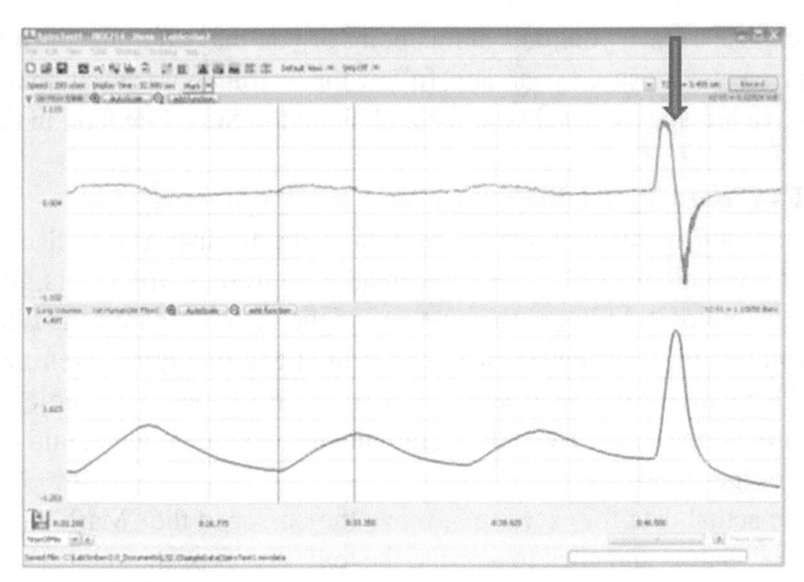

Figure 8 Sample data

- You may choose to save your data at this point or you may continue on with the experiment or data analysis. To save your data, on the Main Window, either
  1. click on the **File** menu. Scroll to **Save As,** then save your data; it should be saved in the **iWorx Saved Files** folder on the **Desktop**. You may create your own folder within the iWorx Saved Files folder. Give your file a name and designate the file type as .iwxdata, then click **Save**.
  
     Or
  2. click on the **Save File** icon in the iWorx toolbar, then save your data; it should be saved in the **iWorx Saved Files** folder on the **Desktop**. You may create your own folder within the iWorx Saved Files folder. Give your file a name and designate the file type as .iwxdata, then click **Save**.

## Exercise 2: FVC & $FEV_n$, RPDs

- Read through all instructions before beginning the testing.
- To label the actual data for exercise 2, move the cursor to the "Mark" text box to the right of the "Mark" button and click to activate the text box. Type the subject's name followed by "RPD" in the "Mark" text box. Insert the mark when you begin the recording period.
- **Subject Instructions**
  - The subject's ability to inspire air will be restricted by wrapping an ace bandage tightly around the rib cage. It is essential that the subject feels as if he or she has difficulty inhaling as compared to normal breathing.
  - It is very important for the subject to become comfortable breathing through the apparatus and that the subject breathes normally. Use the same precautions used in exercise 1.
- **Tester Instructions**
  - Repeat the measurements made in exercise 1.
- You may choose to save your data at this point or you may continue on with the data analysis. If you have not yet saved your data, follow the instructions for saving data under exercise 1. If you have previously saved your data, click on the **Save File** icon in the iWorx toolbar.

## Exercise 3: FVC & $FEV_n$, COPD

- Read through all instructions before beginning the testing. You will need to modify the aperture of the biological filter you are breathing through. To do this, place a length of a straw into the biological filter so that one end of the straw is against the filtration membrane. **Be sure you are putting the straw in the side of the biological membrane you breathe through**. Using one or two sections of tape, secure the straw to the biological filter and seal the space around the straw. The only air moving through the filter should be passing through the orifice of the straw (see Figure 9).
- To label the actual data for exercise 3, move the cursor to the "Mark" text box to the right of the "Mark" button and click to activate the text box. Type the subject's name followed by "COPD" in the "Mark" text box. Insert the mark when you begin the recording period.

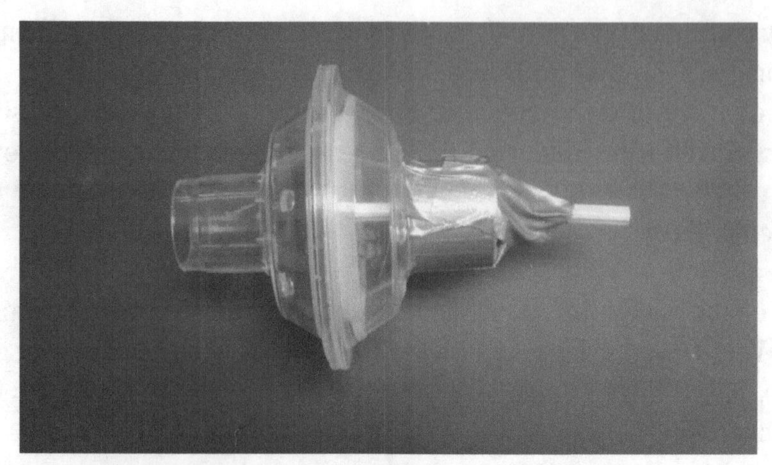

Figure 9 Equipment biological filter with airflow restriction

- **Subject Instructions**
  - The subject's ability to inspire air will be restricted by reducing the diameter of the breathing tube and breathing through a straw. This will have an effect similar to obstruction of airways with mucous or excessive bronchoconstriction (asthma) or due to the collapse of small airways (emphysema).
  - It is very important for the subject to become comfortable breathing through the apparatus and that the subject breathes normally. Use the same precautions used in exercise 1.

- **Tester Instructions**
  - Repeat the measurements made in exercise 1.

- You may choose to save your data at this point or you may continue on with the data analysis. If you have not yet saved your data, follow the instructions for saving data under exercise 1. If you have previously saved your data, click on the **Save File** icon in the iWorx toolbar.

- **Data Analysis Instructions**
  Figure 10, a diagram of icons in the iWorx toolbar may be helpful.

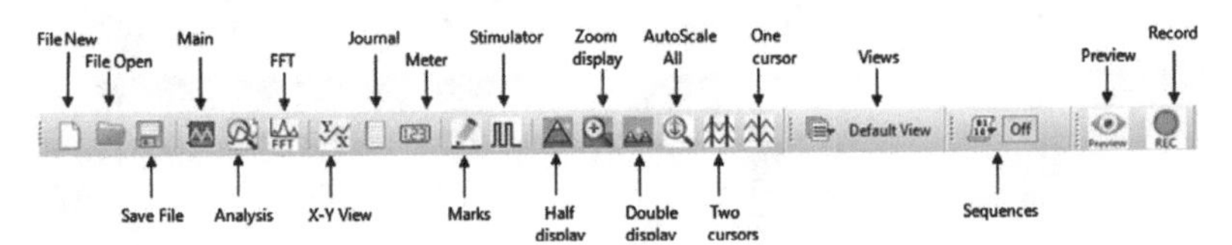


Figure 10 iWorx toolbar icons

- Click on the **Analysis** icon in the iWorx toolbar.
- Go to the beginning of your recordings. Click on the **Marks** icon to bring up the Marks Dialogue box with a table of the marks created in your lab. Once you have the table of marks, click on the number at the beginning of the row for the mark that identifies the beginning of the experimental data segment you are going to analyze. This will highlight that mark. Next, click on the **Go To Mark** button near the bottom of the Marks Dialogue box. Data at that mark will be displayed on your screen.
- Be sure that two cursors are present on your screen. If necessary, click the **Two Cursors** icon on the iWorx toolbar.
- Place the cursors on either side of four consistent breathing cycles, one cursor on the baseline just before inspiration and the second cursor just after the final expiration, then click the **Zoom Display** icon in the iWorx toolbar. This will expand the data section to the width of the Main Window. Adjust your data section as necessary with the display time icons or the zoom icon.
- Click on the **Analysis** icon in the iWorx toolbar.
- In the Analysis window, a "Function Table" will be displayed above the uppermost channel (Air Flow channel). The mathematical functions "V2 – V1," Max_dv/dt, Min_dv/dt,

and "T2 – T1" should be present. The values for each of these functions are displayed in the top margin of each channel. The "V2 – V1" function measures the difference in amplitude between cursor 2 and cursor1. The "Max_dv/dt" function is the maximum rate of air movement during inspiration in the time period between cursors. The "Min_dv/dt" function is the maximum rate of air movement during expiration in the time period between cursors. The "T2 – T1" function measures the difference in time between cursor 2 and cursor 1.

- Click on the arrow to the left of the "Air Flow" channel. In the drop down menu, click "Minimize" to reduce the "Air Flow" channel to a bar across the top of the data screen.
- Click on the arrow to the left of the "Lung Volume" channel. In the drop down menu, move the cursor to "Scale" and click "Autoscale" from the submenu. This will maximize the data displayed in the "Lung Volume" channel.
- **All measurements will be made in the "Lung Volume" channel. All data for the measurements of air flow and volume should be taken from the "Lung Volume" channel data.**
- To measure tidal volume (TV) and maximum inspiratory flow of a breathing cycle, place the first cursor at the lowest point of the breathing cycle, the beginning of the breathing cycle just prior to an inspiration. Place the second cursor at the highest point of the same breathing cycle, the end of the inspiration (see Figure 11).

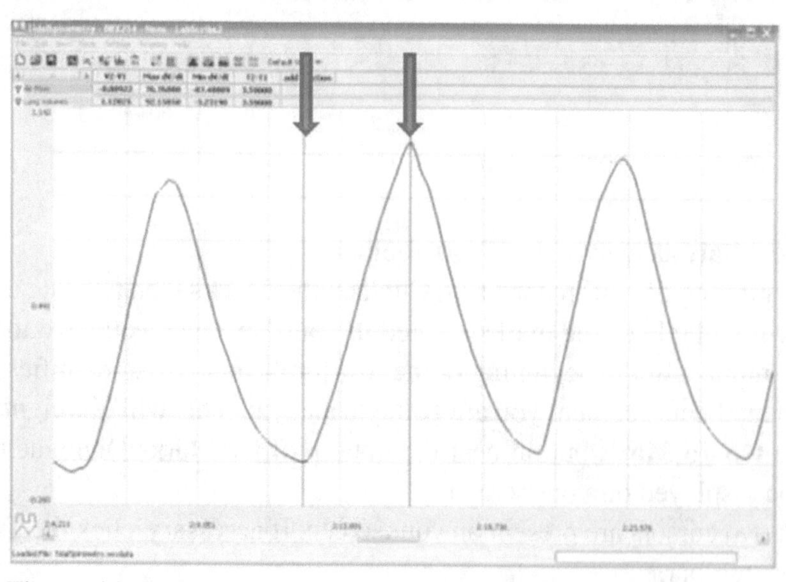

Figure 11 Data analysis TV and max inspiratory flow

- Record the "V2 – V1" data in Table 1 for the tidal volume.
- Record the "Max_dv/dt" data in Table 1 for the maximum inspiratory flow rate.
- To measure the maximum expiratory flow rate, place the first cursor at the highest point of the same breathing cycle, at the end of the inspiration. Place the second cursor at the lowest point of the same breathing cycle; the end of the expiration and the end of the breathing cycle (see Figure 12).

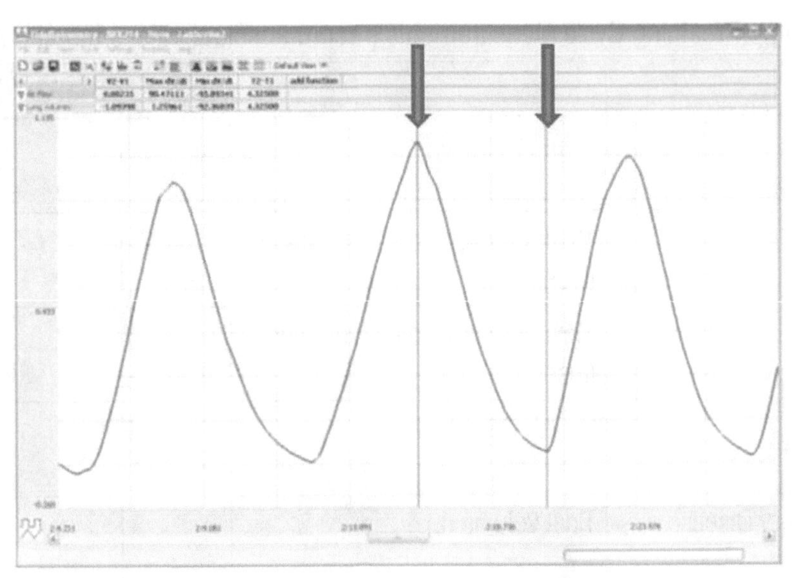

Figure 12 Data analysis Max expiratory flow

- Record the "Min_dv/dt" data in Table 1 for the maximum expiratory flow rate.
- To measure the breath period of a breathing cycle, place the first cursor at the lowest point of the breathing cycle, the beginning of the breathing cycle just prior to inspiration (as in measuring TV), and place the second cursor at the lowest point of the breathing cycle, at the end of expiration (as in measuring minimum flow rate) (see Figure 13). If the troughs are not distinct, you may measure from peak to peak (at the end of inspiration of two sequential breathing cycles).

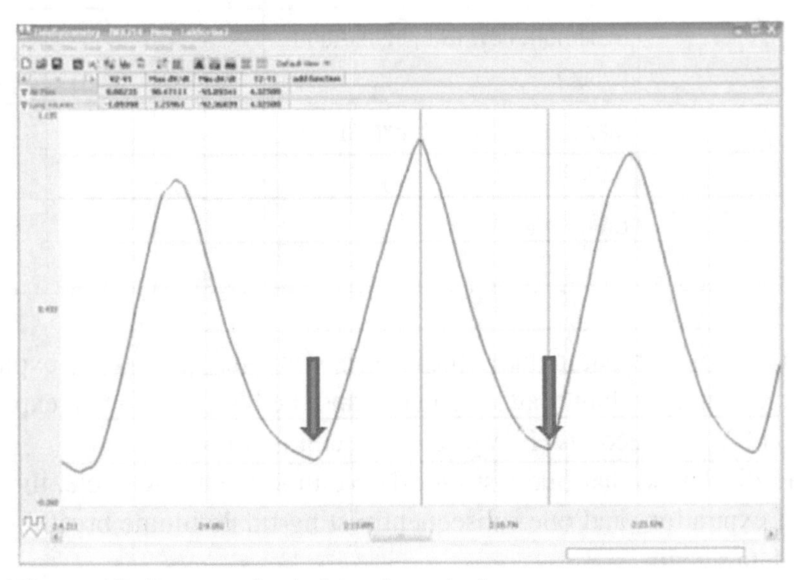

Figure 13 Data analysis breath period

- Record the "T2 – T1" data in Table 1 for the breath period.
- Repeat these four measurements for the other two cycles of tidal ventilation. Calculate the mean for each and record the value in Table 1.

**TABLE 1:** Tidal Volume Breathing Cycle Measurements (record absolute values for all numbers)

| | | Trial 1 | Trial 2 | Trial 3 | Mean |
|---|---|---|---|---|---|
| No Restrictions | Tidal volume (L) | | | | |
| | Maximum flow rate, Inspiration (L/sec) | | | | |
| | Maximum flow rate, Expiration (L/sec) | | | | |
| | Breath Period (sec) | | | | |
| Restrictive Pulmonary Disease Simulation | Tidal volume (L) | | | | |
| | Maximum flow rate, Inspiration (L/sec) | | | | |
| | Maximum flow rate, Expiration (L/sec) | | | | |
| | Breath Period (sec) | | | | |
| Chronic Obstructive Pulmonary Disease Simulation | Tidal volume (L) | | | | |
| | Maximum flow rate, Inspiration (L/sec) | | | | |
| | Maximum flow rate, Expiration (L/sec) | | | | |
| | Breath Period (sec) | | | | |

- Return to the data segment that includes the first maximal inspiration and expiration to measure the other respiratory volumes and capacities and the FVC and forced expiratory volume after 1 second ($FEV_1$), 2 seconds ($FEV_2$), and 3 seconds ($FEV_3$).
- Adjust the data display so that one resting tidal volume breathing cycle, the first maximal inspiration and expiration, and one subsequent resting tidal volume breathing cycle are all visible on the data screen.
- To measure inspiratory reserve volume (IRV) and forced inspiratory flow (FIF), place the first cursor at the peak of the resting tidal breathing cycle, the end of the tidal inspiration, just prior to the maximal inspiration. Place the second cursor at the highest point of the maximal inspiration (see Figure 14).
- Record the "V2 – V1" data for the IRV and "Max_dv/dt" data for the FIF in Maximum flow rate, Inspiration (L/sec) in Table 2.

**TABLE 2:** Respiratory Volumes, Capacities, and FVC and $FEV_x$ (record absolute values for all numbers)

| | | Trial 1 | Trial 2 | Trial 3 | Mean | $FEV_n$/FVC |
|---|---|---|---|---|---|---|
| No Restrictions | Inspiratory Reserve Volume (L) | | | | | |
| | Maximum flow rate, Inspiration (L/sec) | | | | | |
| | Forced Vital Capacity (L) | | | | | |
| | Maximum flow rate, Expiration (L/sec) | | | | | |
| | Expiratory Reserve Volume (L) | | | | | |
| | $FEV_1$ (L) | | | | | |
| | $FEV_2$ (L) | | | | | |
| | $FEV_3$ (L) | | | | | |
| Restrictive Pulmonary Disease Simulation | Inspiratory Reserve Volume (L) | | | | | |
| | Maximum flow rate, Inspiration (L/sec) | | | | | |
| | Forced Vital Capacity (L) | | | | | |
| | Maximum flow rate, Expiration (L/sec) | | | | | |
| | Expiratory Reserve Volume (L) | | | | | |
| | $FEV_1$ (L) | | | | | |
| | $FEV_2$ (L) | | | | | |
| | $FEV_3$ (L) | | | | | |

*(Continued)*

**TABLE 2:** Respiratory Volumes, Capacities, and FVC and $FEV_x$ (record absolute values for all numbers) *(Continued)*

| | | Trial 1 | Trial 2 | Trial 3 | Mean | $FEV_n$/FVC |
|---|---|---|---|---|---|---|
| Chronic Obstructive Pulmonary Disease Simulation | Inspiratory Reserve Volume (L) | | | | | |
| | Maximum flow rate, Inspiration (L/sec) | | | | | |
| | Forced Vital Capacity (L) | | | | | |
| | Maximum flow rate, Expiration (L/sec) | | | | | |
| | Expiratory Reserve Volume (L) | | | | | |
| | $FEV_1$ (L) | | | | | |
| | $FEV_2$ (L) | | | | | |
| | $FEV_3$ (L) | | | | | |

- To measure FVC and forced expiratory flow (FEF), place the first cursor at the highest point of the forced inspiration. Place the second cursor at the lowest point of the maximal expiration (see Figure 14).
- Record the "V2 – V1" data for the FVC and "Min_dv/dt" data for the FEF in Maximum flow rate, Expiration (L/sec) in Table 2.
- To measure expiratory reserve volume (ERV), place the first cursor at the lowest point of the resting tidal breathing cycle, the end of the tidal expiration. Place the second cursor at the lowest point of the maximal expiration (see Figure 14).
- Record the "V2 – V1" data for the ERV in Table 2.
- To measure $FEV_1$, place the first cursor at the highest point of the forced inspiration. Place the second cursor at the point 1 second after the forced expiration began (see Figure 14). You can use the "T2 – T1" data to determine the one second interval.
- Record the "V2 – V1" data for the $FEV_1$ in Table 2.
- To measure $FEV_2$, place the first cursor at the highest point of the forced inspiration. Place the second cursor at the point 2 seconds after the forced expiration began (see Figure 14). You can use the "T2 – T1" data to determine the one second interval.
- Record the "V2 – V1" data for the $FEV_2$ in Table 2.

- To measure $FEV_3$, place the first cursor at the highest point of the forced inspiration. Place the second cursor at the point 3 seconds after the forced expiration began (see Figure 14). You can use the "T2 – T1" data to determine the one second interval.
- Record the "V2 – V1" data for the $FEV_3$ in Table 2.
- Repeat these measurements for the second and third cycles of maximal inspiration and expiration. Record the data in Tables 1 and 2.
- Calculate the $FEV_1/FVC$ ratio, $FEV_2/FVC$ ratio, and $FEV_3/FVC$ ratio based upon the means of the three trials. Record these values in Table 2.
- Repeat the steps for data analysis for the respiratory cycle measurements in exercises 2 and 3.

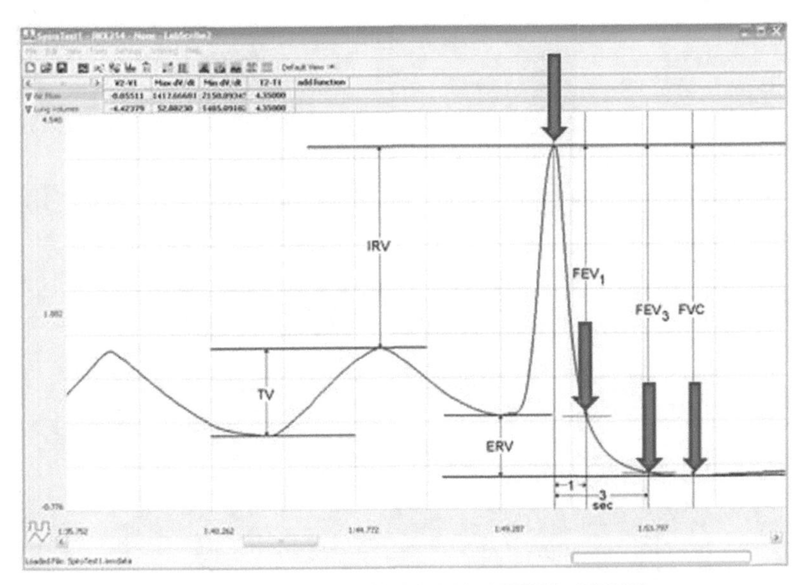


Figure 14 Data analysis FVC, FEV1, FEV2, FEV3

[illegible] the highest point of the [illegible] aspiration. He [illegible]

[illegible]

[illegible]

[illegible]

[illegible]

[illegible]

[illegible] and [illegible]

[illegible]

[illegible]

[illegible]

# Exercise 24

# Urinalysis and Nephron Function

## Objectives

- Be able to explain the function of the kidneys.
- Be able to define the term *hypoosmotic* and explain the conditions in which the kidneys would be expected to produce hypoosmotic urine.
- Be able to define the term *hyperosmotic* and explain the conditions in which the kidneys would be expected to produce hyperosmotic urine.
- Be able to explain the role of *antidiuretic hormone (ADH)*, also vasopressin, in the production of and osmolarity of the urine.
- Be able to define the term *diabetes insipidus* and explain the difference between the central and nephrogenic forms.
- Be able to explain the role of the *renin–angiotensin system* in the production of and osmolarity of urine.
- Be able to explain the role of aldosterone in the production of and osmolarity of the urine.
- Explain the difference between *aldosterone-independent* and *aldosterone-dependent* $K^+$ secretion.
- Be able to define the term *urinalysis*.
- Be able to identify the normal characteristics of urine with regard to volume produced, color, and transparency.
- Be able to identify the normal and abnormal results of a commercial test strip urinalysis.
- Be able to define the terms associated with the abnormal results of a commercial test strip urinalysis: *glucosuria, bilirubinuria, ketonuria, hemoglobinuria, acidemia, alkalemia,* and *proteinuria*.
- Be able to calculate the milliequivalents of an electrolyte.
- Be able to calculate *renal clearance* of a substance and compare it to *glomerular filtration rate (GFR)*.

## Lab Safety Reminders

- Dispose of hazardous materials properly.
- Disinfect your work area.

# Introduction

The kidneys have a number of functions in the body. The most familiar function is the production of urine. However, the kidney also acts as an endocrine gland. The kidney produces and secretes erythropoietin, a hormone used to stimulate the production of erythrocytes. The regulation of kidney erythropoietin secretion is via a negative feedback system that monitors blood oxygen content. Under normal conditions, the kidney secretes a small amount of erythropoietin that maintains normal production of erythrocytes. However, when oxygen delivery to the kidneys is reduced below normal, the kidneys increase the secretion of erythropoietin. As a result, the red bone marrow is stimulated to produce more erythrocytes. Once the oxygen delivery to the kidneys returns to normal, the secretion of erythropoietin returns to the normal level.

The kidneys also function in the maintenance of blood volume and blood pressure. When blood volume and pressure are high, the kidneys produce **hypoosmotic** or dilute urine (osmolarity of urine $<300$ mOsmoles) by allowing water to be filtered out of the plasma and not fully reabsorbed. This will result in the reduction of blood volume and pressure and an increase in the volume of urine produced. When blood volume and pressure are low, the kidneys produce **hyperosmotic** or concentrated urine (osmolarity of urine $>300$ mOsmoles) by reabsorbing some to most of the water filtered out of the plasma. This results in a decreased volume of urine due to the preservation of fluids and can help to maintain blood volume and pressure. **Antidiuretic hormone** (ADH), or **vasopressin** is produced by the hypothalamus and released via the posterior pituitary gland. Osmoreceptors in the hypothalamus monitor the osmolarity of the blood. If blood osmolarity increases above the normal 300 mOsmoles, the osmoreceptors increase firing, and ADH is released. The presence of ADH in the kidney causes more aquaporins to be inserted in the basolateral membrane of the epithelial cells of the cortical and medullary collecting ducts of the nephrons. As a result, more water is reabsorbed from the filtrate, and hyperosmotic urine is produced.

The juxtaglomerular cells produce and release **renin**, an enzyme necessary in the **renin–angiotensin system**. Renin converts the plasma zymogen, angiotensinogen, to angiotensin I. Angiotensin-converting enzyme (ACE) can then convert angiotensin I to angiotensin II. The renin–angiotensin system plays a number of roles in homeostasis. **Angiotensin II** causes vasoconstriction of blood vessels and a resultant increase in peripheral resistance. As a result, arterial blood pressure is increased. Angiotensin II also stimulates the release of **aldosterone**, a mineralocorticoid, from the adrenal cortex. Aldosterone stimulates the insertion of additional channels, transporters, and $Na^+$/$K^+$ ATPase pumps in the membrane of the epithelial cells of the cortical collecting duct. As the basolateral membrane $Na^+$/$K^+$ ATPase pumps work to remove $Na^+$ from the cell, the concentration gradient produced allows $Na^+$ to diffuse into the cell from the filtrate through $Na^+$ leak channels in the luminal membrane. As a result of reabsorbing $Na^+$, the reabsorption of water as it follows the osmotic gradient can also increase.

The kidney also regulates the secretion of potassium ($K^+$). Although $K^+$ is important for normal cell function, blood concentrations higher than normal (4 mEq/L) can be dangerous and fatal at very high levels (over 10 mEq/L). Under normal circumstances, the cortical collecting ducts will excrete the same amount of $K^+$ that was absorbed by the GI tract (minus the amount lost in sweat). If a person has a diet high in $K^+$, the plasma levels of $K^+$ may exceed the homeostatic range, and the person will experience **hyperkalemia**. If this occurs, the higher plasma concentration of $K^+$ will directly stimulate the insertion of more $K^+$ channels in the luminal membrane of the epithelial cells lining the cortical collecting ducts. These channels allow more $K^+$ to move into the filtrate and be excreted. This is called **aldosterone-independent $K^+$ secretion**. The extra $K^+$ channels will be removed when the blood concentration of $K^+$ returns to normal. The kidney can also exhibit **aldosterone-dependent $K^+$ secretion**. In this case, the high blood concentration of $K^+$ stimulates the adrenal cortex to release aldosterone. As aldosterone increases the reabsorption of $Na^+$, it simultaneously increases the secretion of $K^+$.

Aldosterone secretion can be stimulated by either low plasma concentration of $Na^+$ or high plasma concentrations of $K^+$ and is used in the regulation of both ions, and as a result of $Na^+$ reabsorption, it also plays a role in the reabsorption of water.

The kidneys regulate the plasma concentrations of many other electrolytes and molecules. The first step in urine formation is the production of the glomerular filtrate by filtration of the plasma in the renal corpuscle. The nephron tubules are responsible for the modification of the glomerular filtrate through the processes of reabsorption and secretion.

## Routine Urinalysis Tests

A routine urinalysis test checks several characteristics and components of a person's urine. These tests may be done individually or together, depending on the method and the level of accuracy desired. Simple observations of the urine can provide some basic information regarding the urine. The quantity of urine varies with diet, but the normal range in volume is 800 to 2300 mL. The color, transparency or clarity, and odor of the urine can identify the presence of substances not normally found in urine. The use of test strips is common in a routine analysis and gives good basic assessment of some of the chemical components of urine. The following are items commonly assessed in a urinalysis.

The **specific gravity** of urine is determined by the following ratio, the weight of a known volume of urine to the weight of the same volume of pure water. The specific gravity of pure water is 1.000. Because the addition of any solute to water will add weight to the solution, the specific gravity of any solution will have a specific gravity greater than 1.000. The normal range for the specific gravity of urine is 1.010 (dilute) to 1.030 (concentrated). When the body is fully or overly hydrated, the kidneys will produce **hypoosmotic** urine (dilute), and the specific gravity will be low. When the body is dehydrated, the kidneys will produce **hyperosmotic** urine (concentrated), and the specific gravity will be higher. Some medical conditions can alter the specific gravity of urine. For example, diabetes insipidus results in the production of larger-than-normal volumes of urine (**diuresis or polyuresis**) as well as hypoosmotic urine. **Central diabetes insipidus** is due to the lack of production of ADH by the hypothalamus. **Nephrogenic diabetes insipidus** is due to the inability of the kidneys to respond to ADH. Would urine from a person with either type of diabetes insipidus have a higher- or lower-than-expected specific gravity?

Many substances are filtered by the renal corpuscle, but plasma proteins, such as albumin globulins, are not filtered out of the plasma because they are too large to fit through the filtration membrane. Presence of substances that are normally not filtered indicates a potentially pathological problem with the nephrons. Other substances are filtered by the renal corpuscle but are normally reabsorbed by the tubules. If the reabsorption mechanism is not functioning properly, the substance will be present in the urine. Detection of these substances can be used to diagnose some diseases.

### General Assessment Activity

- Obtain one container of sample A urine, one container of sample B urine, and one container of sample C urine.
- Observe the color of the samples and mark the appropriate box in the following table.
- Observe the transparency or clarity of the samples and mark the appropriate box in the table.

| Color | Pale Yellow | Dark Yellow | Orange | Other Color? |
|---|---|---|---|---|
| Sample A | | ✓ | | |
| Sample B | ✓ | | | |
| Sample C | | ✓ | | |
| **Transparency** | **Clear** | **Cloudy** | **Opaque** | **Precipitates** |
| Sample A | ✓ | | | |
| Sample B | ✓ | | | |
| Sample C | | ✓ | | |

- Expected Results
  Urine is expected to have a yellow color due to the presence of urobilin (see later discussion). Abnormal colors may be the result of blood in the urine (red or brown), bilirubin (brownish-green), melanin from melanoma (brown/black), green (some bacterial infections), and others.
  Fresh urine is normally transparent. However, as urine stands, it may become slightly cloudy as epithelial cells settle and chemicals begin to precipitate and settle out of the urine. Fresh urine that is cloudy may indicate bacteria, blood cells, or pus from urinary tract infections.

### Urinometer/Specific Gravity Activity

- Obtain one urinometer consisting of a cylinder and a float. Handle the float carefully, as it is fairly delicate. Observe the markings on the float. These are the numbers you will be reading to determine the specific gravity of the urine sample.
- Fill the urinometer cylinder 75% full with sample A.
- Place the float in the filled cylinder. If the float does not float, add more of the sample until it does.

- Read the number on the stem that is at the surface of the sample.
- Record your number in the following table.
- Return the sample to the sample container.
- Carefully wash and dry the urinometer.
- Repeat the process for samples B and C.
- Keep the samples for use in the following activities.

| Sample | Specific Gravity |
|---|---|
| A | 1.015 |
| B | 1.005 |
| C | 1.025 |

- **Commercial Test Strip Urinalysis Activity**
  - Obtain three commercial reagent test strips.
  - Obtain one commercial color (results) chart.
  - Familiarize yourself with the order of test squares on your reagent strip and the order of test results on the color chart.
  - Dip one reagent strip in sample A. Be sure the test portion of the strip is completely submersed. Tap the test strip on the edge of the sample container to remove any excess sample.
  - Hold the reagent strip horizontally so the chemicals from each test do not mix and give erroneous results.
  - Following the time guideline given on the results chart, read the test results and place a checkmark in the appropriate box in the following table. Results read after 2 minutes are not valid.
  - Repeat the test for samples B and C.
  - **Note for ketones**: The commercial test strip analysis will not detect ketones in the prepared samples. A separate test for ketone bodies will be performed.
- **Ketone Test Activity**
  - Obtain three ketone test tablets and a paper towel.
  - Place one ketone test tablet on a paper towel. Put one drop of sample A on the tablet.
  - After exactly 30 seconds, compare the color change with the chart provided in the lab. No color change or a slight change to a cream color indicates a negative test and no ketones are present. A positive test and the presence of ketones is indicated by a color change that can range from light lavender to purple with an increase in the intensity of the color, indicating an increase in the concentration of ketones present in the sample.
  - Place a checkmark in the appropriate box for ketone concentration present in the sample in the following table.
  - Repeat the test for samples B and C.

| Sample A Tests | Results | | | | | | |
|---|---|---|---|---|---|---|---|
| Glucose mg/dL | Negative | 100 | 250 | 500 | 1000 | 2000+ | |
| Bilirubin | Negative | Small | Moderate | Large | | | |
| Ketones mg/dL | Negative | Trace 5 | Small 40 | Moderate 40 | Large 80 | Large 160 | |
| Blood | Negative | Nonhemolyzed trace | Moderate | Hemolyzed trace | Small | Moderate | Large |
| pH | 5.0 | 6.0 | 6.5 | 7.0 | 7.5 | 8.0 | 8.5 |
| Protein mg/dL | Negative | Trace | 30 | 100 | 300 | 2000+ | |
| Urobilinogen mg/dL | 0.2 | 1 | 2 | 4 | 8 | | |
| Nitrite | Negative | Positive | | | | | |
| Leukocytes | Negative | Trace | Small | Moderate | Large | | |

| Sample B Tests | Results | | | | | | |
|---|---|---|---|---|---|---|---|
| Glucose mg/dL | Negative | 100 | 250 | 500 | 1000 | 2000+ | |
| Bilirubin | Negative | Small | Moderate | Large | | | |
| Ketones mg/dL | Negative | Trace 5 | Small 40 | Moderate 40 | Large 80 | Large 160 | |
| Blood | Negative | Nonhemolyzed trace | Moderate | Hemolyzed trace | Small | Moderate | Large |
| pH | 5.0 | 6.0 | 6.5 | 7.0 | 7.5 | 8.0 | 8.5 |
| Protein mg/dL | Negative | Trace | 30 | 100 | 300 | 2000+ | |
| Urobilinogen mg/dL | 0.2 | 1 | 2 | 4 | 8 | | |
| Nitrite | Negative | Positive | | | | | |
| Leukocytes | Negative | Trace | Small | Moderate | Large | | |

| Sample C Tests | Results | | | | | | |
|---|---|---|---|---|---|---|---|
| Glucose mg/dL | Negative | 100 | 250 | 500 | 1000 | 2000+ | |
| Bilirubin | Negative | Small | Moderate | Large | | | |
| Ketones mg/dL | Negative | Trace 5 | Small 40 | Moderate 40 | Large 80 | Large 160 | |
| Blood | Negative | Nonhemolyzed trace | Moderate | Hemolyzed trace | Small | Moderate | Large |
| pH | 5.0 | 6.0 | 6.5 | 7.0 | 7.5 | 8.0 | 8.5 |
| Protein mg/dL | Negative | Trace | 30 | 100 | 300 | 2000+ | |
| Urobilinogen mg/dL | 0.2 | 1 | 2 | 4 | 8 | | |
| Nitrite | Negative | Positive | | | | | |
| Leukocytes | Negative | Trace | Small | Moderate | Large | | |

In addition to the commercial test strip analysis, a separate test for ketone bodies will need to be performed. Follow the instructions given in the lab for the Acetest. Note that you will record your results for the Acetest in the tables provided for the commercial test strip results.

- Expected Results

  Glucose—The normal result is negative for glucose. All glucose is filtered, but then should be reabsorbed in the proximal tubule. The normal plasma concentration of glucose is 60 to120 mg/dL. If the concentration of glucose in the plasma (and thus the filtrate) exceeds the transport maximum of glucose transporters in the proximal tubules, glucose will remain in the filtrate and will be excreted in the urine. **Glucosuria** or **glycosuria** occurs when the plasma concentration of glucose exceeds 180 mg of glucose per dL. This can occur when a person has a higher-than-normal plasma glucose concentration (**hyperglycemia**), as may be the case in unregulated diabetes mellitus or when a person has an excessive consumption of sugar. In some individuals, glucosuria occurs when the blood sugar levels are within the normal range. This is due to a genetic condition that reduces the transport maximum of the proximal tubule. Diuresis (or polyuresis) is often associated with glucosuria.

  Bilirubin—Only a trace or small amount of bilirubin should be detected. The spleen, liver, and bone marrow destroy old erythrocytes daily. Although the heme (iron) portion of hemoglobin is recycled, the globin (protein) portion is metabolized. Bilirubin is the by-product and is bound to albumin and transported to the liver. In the liver, bilirubin (as bilirubin glucuronide) is secreted in the bile. In the intestines, it is converted to stercobilin (eliminated with the feces) and urobilin (reabsorbed and excreted with the urine). A small amount of the albumin-bound bilirubin is excreted by the kidneys. A person with liver damage or blockage of the bile or hepatic ducts may excrete larger-than-normal amounts of bilirubin (**bilirubinuria**).

Ketones—Ketones are normally present in trace or small amounts in the urine. **Ketonuria** occurs when higher than trace amounts are present. Diets high in fats or low in glucose can result in ketonuria because the person is metabolizing a higher-than-normal amount of fatty acids. The urine will also be more acidic.

Specific gravity—Normal specific gravity would be within the range of 1.010 to 1.030. See the discussion of specific gravity.

Blood (hemoglobin and myoglobin)—The value of hemoglobin in the urine should not exceed 100mg/dL. If a higher-than-normal destruction of erythrocytes occurs, it may exceed the ability of the liver to metabolize the hemoglobin, and **hemoglobinuria** can occur. Improper blood transfusions, hemolytic toxins (e.g., rattlesnake venom), severe tissue damage, and other factors can lead to rapid rates of hemolysis. Myoglobin is located in the skeletal muscle cells. It should not be present in the blood and should not be in the urine. Skeletal muscle damage and heavy exercise may result in myoglobin in the urine.

pH—The pH of urine varies, usually due to diet, but is normally within the range of 4.8 to 7.5. The metabolism of most food tends to make the urine slightly acidic. **Acidemia** (blood pH $< 7.35$) will result in the production of acidic urine. The conversion of urea to ammonia can occur when urine is stored in the urinary bladder. This would result in more basic urine. **Alkalemia** (blood pH $> 7.45$) will result in the production of basic urine, as will digestion of some fruits and ingestion of sodium bicarbonate or calcium carbonate (common in antacids).

Protein—Only a trace or small amount of protein (plasma albumin and plasma globulin) should be present in the urine. **Proteinuria** (**albuminuria** for specifically albumin) occurs when large amounts are present in the urine.

Urobilinogen or urobilin—The normal range for urobilin is 0.2 to 1.0 mg/dL. Urobilin gives urine its characteristic yellow color. Other compounds are excreted that also contribute to the color of urine.

Nitrite—Normally nitrite is not present in the urine. Bacteria can convert nitrates to nitrite, so the presence of nitrites in the urine indicates the presence of some types of bacteria.

Leukocytes—Leukocytes should not be present in the urine. Bacterial infections of the urinary tract may result in the presence of leukocytes in the urine.

- Check for abnormalities in the three samples of urine.

## Milliequivalents

Ion or electrolyte concentrations in the blood are usually given as milliequivalents per liter of fluid (mEq/L). By using milliequivalents per liter, it is easy to directly compare the concentrations of all electrolytes. In systemic venous plasma, the normal concentration of anions (156 mEq/L) is equal to the normal concentration of cations (156 mEq/L). The systemic venous plasma concentration of chloride is 103 mEq/L. From this, you can see that $Cl^-$ is the major anion in the plasma. The kidneys help to maintain the normal plasma concentrations of electrolytes by regulating their reabsorption and secretion. Thus, the electrolyte concentration in urine can be extremely variable ($Cl^-$ = 61–310 mEq/L).

To calculate mEq/L, you must know the concentration, the atomic or molecular weight (MW), and the valence of the electrolyte. The MW is the sum of the atomic mass of all the atoms in the molecule (see Exercise 2: Metrics and Common Computations). The valence of an electrolyte refers to the charge. For example, $Cl^-$ has a charge or valence of 1. However, $Mg^{++}$ has a charge or valence of 2. To negate or neutralize the charge of magnesium, you need two chloride ions for each magnesium ion.

## $Cl^-$ Concentration and Milliequivalent Activity

- Obtain three tests tubes.
- Obtain four droppers.
- Measure 10 drops of urine A into one test tube. One drop is approximately equal to 0.05 mL.
- Using a separate dropper, add 1 drop of 20% potassium dichromate to the same test tube.
- Using a separate dropper, add 2.9% silver nitrate solution to the same test tube. The silver nitrate must be added one drop at a time while constantly shaking the tube to mix the solution. Count the number of drops required to permanently change the color of the solution from yellow to brown.
- Record the number of drops in the following table.
- Repeat for samples B and C.
- Determine the chloride concentration of each urine sample and record it in the table. Each drop of 2.9% silver nitrate has the ability to neutralize 61 mg of $Cl^-$ per 100 mL of sample. To calculate the concentration of chloride ions (mg/100 mL) in the sample, multiply the number of drops of 2.9% silver nitrate added to the sample by 61 mg/100 mL. (# silver nitrate drops × 61 mg/100 mL = # mg $Cl^-$/100 mL).
- Convert mg $Cl^-$ per 100 mL to milliequivalents per liter.
  - First step: Convert mg/100 mL to mM/L

$$\frac{\#\text{mg Cl}^-}{100\text{ mL}} \cdot \frac{1\text{ g}}{1000\text{ mg}} \cdot \frac{1000\text{ mL}}{1\text{ L}} \cdot \frac{1\text{ mole}}{35.5\text{ g}} \cdot \frac{1000\text{ mM}}{1\text{M}} = \text{mM Cl}^-/\text{L}$$

(MW of $Cl^-$)

  - Second step: Convert mM/L to mEq/L

mM/L • valence of ion = mEq/L

($Cl^-$ valence is 1)

12 • 61/100

| Sample | Number of Drops of 2.9% Silver Nitrate | $Cl^-$ Concentration (mg $Cl^-$/100 mL) | $Cl^-$ Milliequivalents ($Cl^-$ mEq/L) |
|---|---|---|---|
| A | 12 | 7.32 | 206 |
| B | 5 | 3.05 | 86 |
| C | 14 | 8.54 | 241 |

# Renal Clearance

**Glomerular filtration rate** (GFR) is the volume of fluid filtered per time. The normal rate is 180 L/day, 7.5 L/hr, 125 mL/min (sum of both kidneys). This is the equivalent of filtering out the volume of your plasma (approximately 3 L) sixty times per day. The **filtered load** is the rate of filtration of a particular substance from the plasma and is calculated as mass filtered per time. Filtered load is calculated as:

GFR • plasma concentration of the filterable substance (g of substance/L plasma)

**Renal clearance** is the rate of clearance (excretion) of a substance from the plasma and is measured as volume per time (L/day, L/hr, mL/min, etc.). The amount of substance cleared is the amount that is excreted by the kidneys. The renal clearance of a substance can be compared to GFR and filtered load of that same substance. If the renal clearance of a substance is equal to the GFR, the substance is neither reabsorbed nor secreted (e.g., inulin). If the renal clearance of a substance is less than the GFR, some of the substance has been reabsorbed (e.g., urea). If the renal clearance of a substance is greater than the GFR, some of the substance has been secreted (e.g., penicillin). If the substance is filtered but has a renal clearance of zero, the substance has been fully reabsorbed (e.g., glucose). Renal clearance is calculated as:

$$U_S \bullet V / P_S = C_S$$

$U_S$ is the urine concentration of substance S (mg/L or mg/100mL)
V is the volume of urine produced per unit of time (L/hr or mL/min)
($U_S \bullet V$ = the mass of substance S in the urine produced)
$P_S$ is the plasma concentration of substance S (mg/L or mg/100 mL)
$C_S$ is the clearance

## Renal Clearance Activity

- Calculate the renal clearance of substance Z given the following values.
  V = 4 mL/min
  $P_Z$ = 10 mg/100 mL
  $U_Z$ = 500 mg/100 mL
  $C_Z$ = ________
- Is the clearance of substance Z less than, greater than, or equal to the standard GFR?
- Was substance Z reabsorbed, secreted, or neither?

# Exercise 25

# Digestion

## Objectives

- Be able to define the term *mechanical digestion.*
- Be able to define the term *enzymatic hydrolysis.*
- Be able to define the term *denatured* as it applies to enzymes.
- Be able to identify the type of organic molecules or compounds acted on by *amylase.*
- Be able to identify the type of organic molecules or compounds acted on by *pepsin.*
- Be able to identify the type of organic molecules or compounds acted on by *lipase.*
- Be able to explain the specific conditions necessary for the proper function of each of these enzymes and why these conditions are necessary.
- Be able to define the term *emulsification* and compare it to the process of enzymatic hydrolysis.
- Be able to identify positive and negative results for each of the tests performed in the lab.

## Lab Safety Reminders

- Use caution when working with the concentrated acids and bases in this lab.
- The concentrated acids and bases must remain in the hoods. Do not take the dropping bottles to your lab bench.
- Dispose of hazardous materials properly.

## Introduction

The digestion of foods occurs primarily in the alimentary canal. It is here that large food particles are broken into smaller pieces with **mechanical digestion**. Mechanical digestion does not chemically change the food stuff. **Enzymatic hydrolysis** is the process that converts large macromolecules into the constituent monomers. Mechanical digestion occurs in the oral cavity with the action of the teeth and the tongue. It continues in the stomach with the mixing of the stomach. Enzymatic hydrolysis also occurs in the oral cavity with the addition of enzymes in the saliva. It also continues in the stomach, but the primary location of enzymatic hydrolysis is the small intestine. Some enzymes are produced by the stomach and the small intestine, but many enzymes are produced by the pancreas and secreted into the duodenum.

Enzymes are specific in the reactions they catalyze. Enzymes can also be specific with regard to the environmental conditions in which they function. One of the conditions is pH, something that varies dramatically through the alimentary canal. For example, the oral cavity is usually slightly basic, whereas the stomach is very acidic (near pH = 2), and the small intestines are basic (near pH = 8 or 9).

Other chemicals serve as digestive aids. For example, besides containing hydrolytic enzymes, saliva also adds liquid and helps lubricate the food. This aids in the movement of the foodstuff in the alimentary canal. The liver, via the gallbladder, adds bile to the duodenum. Most of the components of bile are waste products, but bile salts act as an emulsifying agent and aid in the digestion of triglycerides. The pancreas and liver also add bicarbonate ($HCO_3^-$) to the duodenum to buffer the acidity of the chyme that enters the duodenum.

## Carbohydrate Digestion

The digestion of starch, a polymer of glucose, begins in the oral cavity with the addition of **salivary amylase** (ptyalin), a component of saliva. Amylase is also produced by the pancreas and added to the chyme in the duodenum. Although it can begin in the oral cavity, most digestion of complex carbohydrates occurs in the duodenum. **Starch** is a polymer of glucose. The enzymatic hydrolysis of starch with amylase breaks the starch into smaller polysaccharide chains and then into maltose, a disaccharide.

A **Benedict's test** will be used to detect the presence of reducing sugars (maltose, glucose, and other monosaccharides). Benedict's solution is a basic solution that contains copper ($Cu^{2+}$, cupric ions). In the presence of reducing sugars, the copper is reduced to $Cu^+$ (cuprous ions) and forms $Cu_2O$ (cuprous oxide). Cuprous oxide is yellow-red and precipitates out of solution.

### Amylase Activity

- Obtain a test tube rack and six clean test tubes. **Note:** It is essential that the test tubes are clean when you begin this experiment.
- Obtain a 10-mL graduated cylinder.
- Obtain a glass rod.
- Label three of the test tubes 1, 2, and 3.
- Test tube 1: Add 3 mL of water and 5 mL of starch solution; mix.
- Test tube 2: Add 3 mL of α-amylase (300 units/mL) and 5 mL of starch solution; mix.
- Test tube 3: Add 3 mL of α-amylase (300 units/mL), 5 mL of starch solution, and 3 drops of HCl; mix.
- Measure the pH of the ingredients of each tube. Record the data in the following table. To measure the pH, dip the glass rod into the test tube. Transfer one drop of the solution from the test tube to the pH paper. Be sure to clean the glass rod between samples.
- Place all three tubes in the 37°C water bath and incubate for 1 hour.
- Remove the test tubes from the water bath.
- Pour half the contents of test tube 1 into another tube and label it 1A.
- Pour half the contents of test tube 2 into another tube and label it 2A.
- Pour half the contents of test tube 3 into another tube and label it 3A.
- Test tubes 1, 2, and 3: Add 5 mL of Benedict's solution to the solution remaining in the tube.
- Place tubes 1, 2, and 3 in a boiling water bath for 2 to 5 minutes. A positive test for reducing sugars is the formation of a yellow-orange precipitate. If the color remains blue, the test result is negative for reducing sugars. Rank the results according to the following scale: blue (−), green (+), yellow (++), orange (+++), and red (++++).

- Record the results in the following table.
- Test tubes 1A, 2A, and 3A: Add two or three drops of Lugol's iodine (or IKI) to each of the tubes. A positive test for starch is the development of a blue-black color. If the color is pale tan or yellow, the test result is negative for starch. Rank the results according to the following scale:
  yellow (–), blue-black (+)
- Record the results in the following table.

| Test tube # | Initial contents and treatment | pH | Reducing sugars | Starch |
|---|---|---|---|---|
| 1 | 3 mL water + 5 mL starch<br>37° incubation<br>Benedict's solution + boiling | | | NA |
| 2 | 3 mL α-amylase + 5 mL starch<br>37° incubation<br>Benedict's solution + boiling | | | NA |
| 3 | 3 mL α-amylase + 5 mL starch + 3 drops HCl<br>37° incubation<br>Benedict's solution + boiling | | | NA |
| 1A | 3 mL water + 5 mL starch<br>37° incubation<br>Lugol's iodine | | NA | |
| 2A | 3 mL α-amylase + 5 mL starch<br>37° incubation<br>Lugol's iodine | | NA | |
| 3A | 3 mL α-amylase + 5 mL starch + 3 drops HCl<br>37° incubation<br>Lugol's iodine | | NA | |

- Is amylase necessary for the hydrolysis of starch to reducing sugars?

- Explain the difference in results for the solutions from test tubes 2 and 3?

# Protein Digestion

**Proteins** are large polymers of **amino acids** held together by peptide bonds. Nonfunctional proteins are usually referred to as **polypeptide chains**. Several levels of structure are associated with a protein: Primary structure refers to the type and order of the amino acids used in the formation of the polypeptide chain; secondary structure refers to the sheet or coiling sections of the polypeptide chain; tertiary refers to the overall three-dimensional shape; and sometimes quaternary structure refers to when two or more polypeptide chains bond together to form a functional protein. The three-dimensional structure of the protein (including protein enzymes) is necessary for the proper functioning of the protein. Several factors can alter the shape of a protein; pH, temperature, and modulating chemicals. When the shape of the protein has changed and the protein is no longer functional, the protein has been **denatured**. In some cases, but not all, when a denatured protein is returned to its normal conditions, it can once again be functional. Keep in mind that proteins acting as enzymes can be denatured and as a result become nonfunctional. Do not confuse denaturing of a protein with enzymatic hydrolysis of a protein. Denaturing changes the three-dimensional shape but does not digest the protein. Enzymatic hydrolysis of a protein results in hydrolysis reactions that break the peptide bonds.

Several enzymes are used in the enzymatic hydrolysis of proteins in the digestive system. **Pepsinogen** is produced by the principal or chief cells of the gastric glands of the stomach. Pepsinogen is a **zymogen**, a nonfunctional enzyme. To function, pepsinogen must be converted to **pepsin.** This conversion occurs in acidic environments such as the stomach (pH as low as 2). Once some pepsin has been produced, the pepsin then facilitates the conversion of more pepsinogen to pepsin, an example of positive feedback. The enzymatic function of pepsin also occurs in acidic environments. Although pepsin functions well in the acidic environment of the stomach, it does not function in the basic environment of the small intestines. The pH of the chyme entering the duodenum is very acidic. However, the pH is quickly neutralized and even made basic by the addition of **bicarbonate** ($HCO_3^-$), produced in the pancreas and the liver, to the chime in the lumen of the duodenum. As a result of the dramatic shift in pH from dramatically acidic to basic, enzymes other than pepsin are necessary for the further enzymatic hydrolysis of proteins or polypeptide chains that occurs in the small intestines.

- **Protein Activity**

  **Caution:** Wearing gloves is recommended. You are using a concentrated acid (HCl) and a concentrated base (NaOH) in this exercise. Be careful to not get either on your skin. If you do, rinse the area under flowing water immediately. **Do not mix strong acids and bases**. Clean all test tubes carefully and thoroughly.

  - Obtain a test tube rack and seven clean test tubes. **Note:** it is essential that the test tubes are clean when you begin this experiment.
  - Obtain a 10 mL graduated cylinder.
  - Obtain a 250 mL beaker. Fill the beaker three fourths full with ice and add water until the solution fills the beaker one half to three fourths full.
  - Label the test tubes 1, 2, 3, 4A, 4B, 5, and 6.
  - Test tube 1: Add 10 mL of distilled water and 2 mL of biuret reagent; swirl to mix. Observe the final color of the solution and record your observations.

- Test tube 2: Add 5 mL of distilled water, 5 mL of albumin solution, and 2 mL of biuret reagent; swirl to mix. Observe the final color of the solution and record your observations.
- Test tube 3: Add 1 drop of water, 5 mL of albumin solution, and 5 mL of pepsin; swirl to mix. Measure the pH of the solution. To measure the pH, dip the glass rod into the test tube. Transfer one drop of the solution from the test tube to the pH paper. Be sure to clean the glass rod between samples. Place the tube in the 37°C water bath. Incubate for 1 hour.
- Test tube 4A: Measure 5 mL of albumin solution into a test tube, add 1 drop of concentrated HCl, and chill in an ice bath.
- Test tube 4B: Measure 5 mL of pepsin into a test tube and chill in an ice bath. Once test tubes 4A and 4B have been chilled, mix the albumin and pepsin together in one test tube; swirl to mix. Return the test tube to the ice bath for the duration of the incubation period. **Note:** It is very important that the solutions be chilled before mixing them together and throughout the incubation period. To measure the pH of the solution, dip the glass rod into the test tube. Transfer one drop of the solution from the test tube to the pH paper. Be sure to clean the glass rod between samples. Incubate in the ice bath for 1 hour.
- Test tube 5: Add 1 drop of concentrated HCl, 5 mL of albumin solution, and 5 mL of pepsin; swirl to mix. Measure the pH of the solution. To measure the pH, dip the glass rod into the test tube. Transfer one drop of the solution from the test tube to the pH paper. Be sure to clean the glass rod between samples. Place the tube in the 37°C water bath. Incubate for 1 hour.
- Test tube 6: Add 1 drop of concentrated NaOH (sodium hydroxide), 5 mL of albumin solution, and 5 mL of pepsin; swirl to mix. Measure the pH of the solution. To measure the pH, dip the glass rod into the test tube. Transfer one drop of the solution from the test tube to the pH paper. Be sure to clean the glass rod between samples. Place the tube in the 37°C water bath. Incubate for 1 hour.
- Test tubes 3 through 6: At the end of the incubation, add 2 mL of biuret reagent. Record the final color in each of the test tubes in the following table. The biuret reagent contains NaOH or KOH and copper sulfate ($CuSO_4$) and is blue in the absence of proteins. When proteins react with the biuret reagent, it turns a purple color. As proteins are hydrolyzed and peptides are produced, the color becomes pinkish-purple. For each test tube, indicate the presence of proteins, peptides, or neither in the following table.

| Test tube # | Contents | Incubation temp. °C | Initial pH | Final color | Presence of protein, peptides, or neither |
|---|---|---|---|---|---|
| 1 | 10 mL Water<br>2 mL Biuret | None | NA | | |
| 2 | 5 mL Water<br>5 mL Albumin<br>2 mL Biuret | None | NA | | |
| 3 | Drop Water<br>5 mL Albumin<br>5 mL Pepsin<br>(2 mL Biuret added after incubation) | 37 | | | |

| Test tube # | Contents | Incubation temp. °C | Initial pH | Final color | Presence of protein, peptides, or neither |
|---|---|---|---|---|---|
| 4A | Drop HCl<br>5 mL Albumin | Ice water bath | | | |
| 4B | 5 mL Pepsin<br>(2 mL Biuret added after incubation) | Ice water bath | | | |
| 5 | Drop HCl<br>5 mL Albumin<br>5 mL Pepsin<br>(2 mL Biuret added after incubation) | 37 | | | |
| 6 | Drop NaOH<br>5 mL Albumin<br>5 mL Pepsin<br>(2 mL Biuret added after incubation) | 37 | | | |

- Which test tube (3 through 6) had the highest rate of enzymatic hydrolysis? Explain your results; were your results what you expected?

- Which test tube (3 through 6) had the lowest rate of enzymatic hydrolysis? Explain your results; were your results what you expected?

- How did the pH of the solution affect enzymatic hydrolysis? Explain your results; were your results what you expected?

- How did the temperature of incubation affect enzymatic hydrolysis? Explain your results; were your results what you expected?

## Lipid Digestion

Triglycerides are the common form of lipids in the diet. Triglycerides are composed of one glycerol molecule with three fatty acids bonded to it. The enzyme **lipase** is responsible for the enzymatic hydrolysis of **triglycerides**. Although small amounts of lipase are produced by the tongue and the stomach, the majority of lipase is produced by the pancreas. As a result, most digestion of triglycerides occurs in the duodenum. Because the lumen of the alimentary canal is an aqueous environment and because triglycerides are hydrophobic, not soluble in water, the triglycerides and other lipids tend to clump together in large drops. Because enzymes must come in contact with their substrates to function, and these large droplets have less surface area than if the triglycerides were water soluble, the enzymatic hydrolysis of triglycerides is more difficult than for the other nutrients. **Bile** is a digestive aid that is produced in the liver by hepatocytes. The bile is stored in and released from the gallbladder. One of the components of bile is bile salt. The **bile salts** are amphipathic molecules, molecules that are partly hydrophilic and partly hydrophobic. The action of bile salts is the **emulsification** of triglyceride droplets. As mechanical digestion breaks the large triglyceride droplets into smaller droplets (still triglycerides), bile salts (and phospholipids) surround the small droplets with the hydrophobic edges of the bile salts and phospholipids facing inward and in contact with the triglycerides and the hydrophilic edges of the bile salts and phospholipids facing outward and in contact with the aqueous environment. This prevents the small droplets of triglycerides from coalescing back into large droplets. Breaking the larger droplets into small droplets provides a much larger surface area for contact between the lipase and the triglycerides and enzymatic hydrolysis to occur.

Lipase acts by hydrolyzing triglycerides into two free fatty acids and a monoglyceride (glycerol molecule with one fatty acid bound). As the enzymatic hydrolysis of more triglyceride molecules occurs, more free fatty acids are produced, and the pH of the solution decreases. This production of acids and the corresponding reduction of pH can be used to monitor the relative amount of triglyceride hydrolysis.

- **Bile Salt and Lipase Activities**
  - Obtain a test tube rack and six clean test tubes. **Note:** It is essential that the test tubes are clean when you begin this experiment.
  - Label the test tubes 1, 2, 3, 4, 5, and 6.
  - Test tube 1: Add 3 mL of oil and 5 mL of water and mix vigorously.
  - Test tube 2: Add 3 mL of oil, 5 mL of water, and a pinch of bile salts and mix vigorously.
  - Once again vigorously mix the contents in test tube 1 and in test tube 2. Observe the solutions in tubes 1 and 2 over the next couple of minutes and record your observations in the following space. If there were no differences, add another pinch of bile salts to tube 2; remix both test tube 1 and test tube 2 and continue your observations.
  - Add 3 mL of litmus cream to test tubes 3, 4, 5, and 6.
  - Test tube 3: Add 3 mL of litmus cream and 5 mL of water; mix.
  - Test tube 4: Add 3 mL of litmus cream, 5 mL of water, and a pinch of bile salts; mix.
  - Test tube 5: Add 3 mL of litmus cream and 5 mL of 1% pancreatin solution (enzyme compound); mix.

- Test tube 6: Add 3 mL of litmus cream, 5 mL of 1% pancreatin solution (enzyme compound), and a pinch of bile salts; mix.
- Place test tubes 3, 4, 5, and 6 in the 37°C water bath and incubate for 1 hour.
- Litmus powder has been added to the cream to act as a pH indicator. At a basic pH, the color of the litmus is blue, but at an acidic pH, the color is pink. The more acidic the solution, the more color change will be observed. Observe your samples after incubation and check the appropriate box in the following table to indicate the color of the solution in each of the test tubes.

  Note: Pancreatin solution is a combination of enzymes produced by the exocrine cells of the pancreas. It includes the enzyme lipase.

Test tubes 1 and 2:

- What happened when you mixed the solutions in test tubes 1 and 2?

1. it bubbled, but oil is slowly seperating & rising
2. bubbled, & fizzed

- Were there any differences in what happened in test tubes 1 and 2 after you allowed them to sit for a brief period of time?

2 oil is foggy. but then both seperated
2. seperated, oil is slightly bubbly

| Results table | Tube 3 litmus cream and water | Tube 4 litmus cream, water, and bile salts | Tube 5 litmus cream and pancreatin | Tube 6 litmus cream, pancreatin, and bile salts |
|---|---|---|---|---|
| Blue | ✓ | | | |
| Purple | | ✓ | ✓ | |
| Pink +, ++, or +++ | | | | Some digestion ✓ |

Test tubes 3, 4, 5, and 6:

- Why is pH used as a measure of enzymatic hydrolysis of triglycerides? Explain your results; were your results what you expected?

To test the level of acidity.

- Which test tube had the highest rate of enzymatic hydrolysis? Explain your results; were your results what you expected?

#6; purple

- Which test tube had the lowest rate of enzymatic hydrolysis? Explain your results; were your results what you expected?

#3; blue

# Exercise 26

# Metabolic Rates

## Objectives

- Be able to identify the different forms of metabolism, catabolism, and anabolism.
- Be able to explain when energy would either be stored or used in the body based on energy intake and energy use.
- Be able to define the terms *calorie* (small calorie) and *kilocalorie* (large Calorie).
- Be able to distinguish between energy produced per liter of oxygen consumed and the energy per gram of a type of organic molecule.
- Be able to define the terms *direct calorimetry* and *indirect calorimetry*.
- Be able to define the term *basal metabolic rate (BMR)* and identify the conditions that must be met to measure BMR.
- Be able to define the term *metabolic rate*.
- Be able to determine when BMR and/or MR would change.
- Be able to calculate BMR and MR and percent deviations.
- Be able to identify hormones that affect BMR and their effects on BMR.
- Be able to explain the use of a respirometer.
- Be able to explain the purpose of soda lime in the respirometer.

## Introduction

Many chemical reactions take place within the body. As a whole, these reactions are referred to as **metabolism**. Although we often assume metabolism means the breaking down of molecules, it also includes the building of molecules. A reaction that begins with large molecules or macromolecules and breaks them down to smaller subunits is a **catabolic reaction**. Catabolic reactions often release energy that may then be used to synthesize ATP (energy). The hydrolysis of a protein to amino acids would be an example of catabolism. In this case, the amino acids may be further broken down by cellular respiration to produce ATP, or they may be used as building blocks to produce other types of molecules. An **anabolic reaction** uses small subunits or monomers to produce large molecules often with the addition of ATP (energy). The process of translation in protein synthesis is an example of anabolism.

In the body, there is always a combination of catabolic and anabolic reactions. When you eat food, many of the organic macromolecules in your food are hydrolyzed to their monomers or subunits within the digestive tract. These are then absorbed into the bloodstream and carried to the cells of the body, where they may be further broken down in cellular respiration or where they may be stored for future use. Some of them may be used by the cell to build a different type of organic molecule. The new molecules that are synthesized may be stored (lipids in adipose tissue, glycogen in liver and skeletal muscle), or they may be used to build new tissue or perform some function. In general, when you take in more energy (food) than your body uses, you will have an excess of anabolic reactions, and you will gain energy (usually stored as lipids). When you take in less energy (food) than your body uses, you will have an excess of catabolic reactions as you break down tissue (initially glycogen and lipid) to produce energy for body functions, and you will lose energy that was stored. Anabolism and catabolism are usually balanced when there is no net energy gain or loss. The energy balance of the body can be illustrated mathematically: energy stored = energy taken in – (energy used in work + energy used in heat production). More intuitively, we know that when we have a balanced energy equation, we neither gain nor lose weight. However, when we take in excess energy, we tend to gain weight, and when we take in too little energy, we tend to lose weight.

Food energy units are referred to as calories. The definition of a **calorie**, or the **small calorie** (c), is the amount of energy required to raise the temperature of 1 g of pure water 1°C, from 14.5° to 15.5°C as standard temperature and pressure (STP). The unit we are more familiar with and that is found on food labels is actually the **kilocalorie** (kcal), or the **large calorie** (C). The definition of a kilocalorie is the amount of energy required to raise the temperature of 1000 g of pure water 1°C, from 14.5° to 15.5°C at STP. So, 1 kilocalorie is equivalent to 1000 calories.

The process of calorimetry can be used to determine the exact amount of energy present in a particular type of molecule and how much energy is released for every liter of oxygen consumed (see the following table). For example, 1 mole of glucose contains 673 kcal of energy. It takes 6 moles or 134.4 L of oxygen (6 moles $O_2 \times 22.4$ L/mole = 134.4 L oxygen) to completely break down 1 mole of glucose. So the catabolism of glucose produces 5.01 kcal per liter of oxygen consumed (673 kcal/134.4 L $O_2$). A more familiar way of comparing metabolic values of different organic molecules is to determine the amount of energy produced per gram of material.

| Foodstuff | Energy (kcal) produced per L of oxygen consumed | Energy (kcal) produced per g of food consumed |
|---|---|---|
| Glucose | 5.01 kcal/L $_{o}2$ | 4.0 kcal/g |
| Lipids | 4.70 kcal/L $_{o}2$ | 9.0 kcal/g |
| Proteins | 4.60 kcal/L $_{o}2$ | 4.0 kcal/g |

Which foodstuff is most energy dense?

Which foodstuff has the lowest energy yield per volume of oxygen?

## Metabolic Rates and Respirometry

Humans measure their metabolic rates in two ways, basal metabolic rate (BMR) and metabolic rate (MR). The distinction between the two rates is very important with regard to a person's energy budget. **BMR** is the minimal amount of energy used by the body to maintain itself. It is usually measured under the following conditions: upon waking after sleeping 8 hours, no food ingestion for 12 hours prior to measurement (postabsorptive state), mentally and physically relaxed, no fever, testing must be in a room of comfortable air temperature, generally within the thermoneutral zone (18°C to 22°C or 64.4°F to 71.6°F, depending upon a variety of factors), subject rests for 30 minutes prior to testing, and subject should be tested in a reclining position in a semi-dark room. These conditions minimize factors that may artificially raise the oxygen consumption of the subject. Multiple factors can result in a change in BMR. One common factor is altering the amount of lean muscle mass a person has. Different types of tissues have different energy demands. At rest, skeletal muscle has an energy demand of 13.0 kcal/kg (54.4 kJ/kg), and adipose tissue has an energy demand of 4.5 kcal/kg (18.8 kJ/kg). This is why one of the recommendations for weight loss is to increase muscle mass through strength training. Although the density (mass/unit volume) of skeletal muscle tissue (≈1.06 kg/L) is approximately 15% greater than the density of adipose tissue (≈0.92 kg/L), skeletal muscle tissue has a higher per mass energy demand even at rest, so replacing fat with muscle increases the BMR and can help increase overall weight loss. Several hormones can also increase BMR: thyroid hormone, growth hormone, and other hormones stimulated by the sympathetic system. Any condition that increases the concentration of these hormones above the normal plasma concentration would be expected to increase BMR (e.g., hyperthyroidism). A decrease in the plasma concentration of these hormones below normal would result in a decrease in BMR (e.g., hypothyroidism). Age and gender also have an influence on BMR. In general, males have a higher BMR than females due to the lower percent body fat and higher percent lean muscle mass in males relative to females. BMR decreases with age once an individual reaches adolescence. For every decade after age 20, BMR is expected to drop by 2% to 3%. These are generalizations, and there are many other factors that influence a person's BMR including genetics. The following table indicates the overall effect on BMR of the physiologic conditions listed.

| Physiologic Condition | Effect on BMR |
|---|---|
| Hyperthyroidism | + |
| Acromegaly | + |
| Gigantism | + |
| Polycythemia vera | + |
| Pregnancy | + |
| Hypothyroidism | – |
| Anorexia nervosa | – |

BMR can be predicted or estimated for an individual based upon the person's height, weight, age, and gender.

For females:

655 + (4.35 • weight in lb) + (4.7 • height in inches) – (4.7 • age in yrs) = kcal/24 hrs

For males:

66 + (6.23 • weight in lb) + (12.7 • height in inches) – (6.8 • age in yrs) = kcal/24 hrs

**MR** is the total energy expenditure of the body regardless of the conditions, so it is at least the BMR and usually the BMR plus any additional expenditure. How would MR be affected by the conditions listed in the previous table?

**Respirometry** is used to determine BMR and MR. It measures the amount of oxygen consumed by an individual in a given time. With a standard respirometer, the subject inhales from a sealed chamber containing pure oxygen. The subject also exhales into the same chamber. As the subject inhales, the volume of the chamber decreases, and the volume change is recorded on a chart. As the subject exhales, the volume of the chamber increases, and the volume change is recorded on a chart. Under normal circumstances, the volume of air inhaled and exhaled is the same, but the chemical composition of the air is different. Although the inhaled air is 100% oxygen, the exhaled air is combination of oxygen and carbon dioxide (and a very minor amount of other gasses). A chemical called soda lime is also present in the chamber. The soda lime rapidly absorbs the carbon dioxide exhaled into the chamber. As a result of absorbing the carbon dioxide, the volume of air remaining in the chamber after exhalation is less than before the subject inhaled (volume of $O_2$ before inhalation – volume of $CO_2$ absorbed by the soda lime). The difference in volumes before inhalation and after exhalation is the volume of oxygen consumed by the subject. The change in volume is measured over a period of minutes and is then used to calculate MRs. The rate is measured as kcal/m$^2$/hr.

## Respirometry Activity 1

- Obtain a correction chart for STPD (standard temperature, pressure, and dryness).
- Based on the information given in the testing information that follows, find the correction factor for STPD. Record the value in the following table.
- Based on the information given in the testing information, determine the body surface area for the subject. Using the nomogram provided, draw a line from the subject's height to the subject's weight. Read the square meters at the point the line crosses the column of body surface numbers. Record the value for body surface area (m$^2$) in the following table.

  Testing information:
  - Spirometer temperature: 25°C
  - Barometric pressure: 750 mm Hg
  - Female subject
  - Height: 5′8″
  - Weight: 113.5 lb
  - Age: 20

655 + (4.35 • 113.5) + (4.7 • 68) – (4.7 • 20) =
655 + 493.725 + 319.6 – 94 =
1374.325 kcals/24 hrs
BMR

- Using the following spirometry tracing, Figure 1, determine the amount of oxygen consumed (L) in 3 minutes for the BMR recording. Convert this value to the amount of oxygen consumed in 1 hr (L $O_2$/hr). Record the value in the following table.

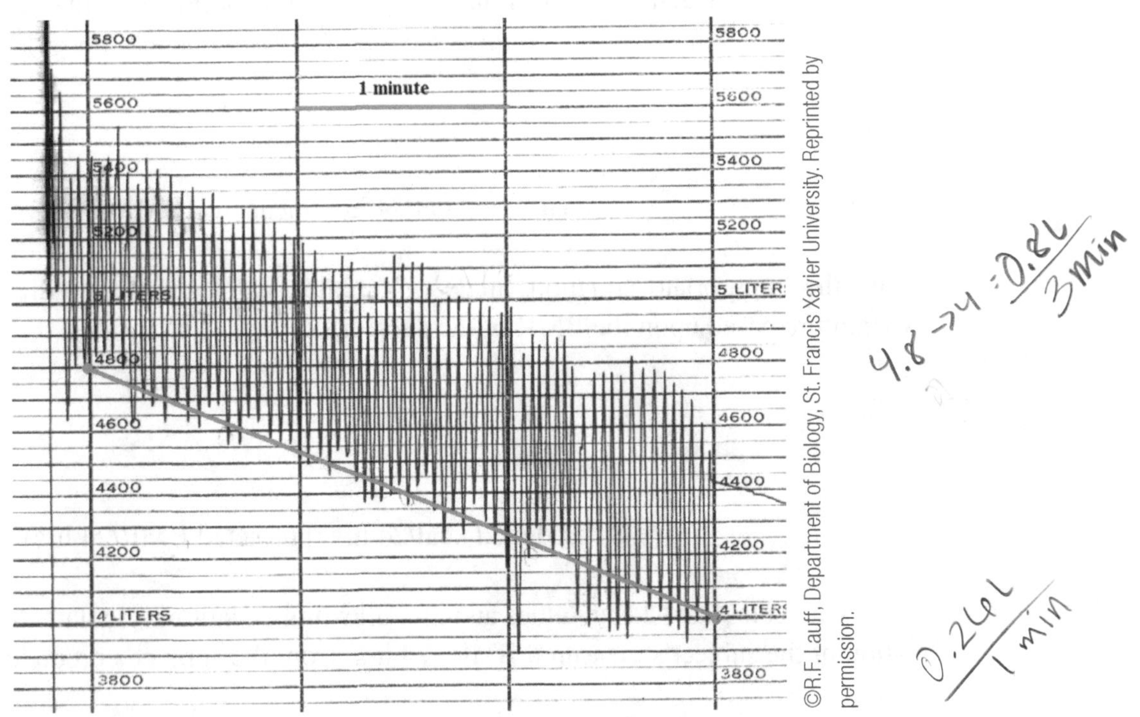




Figure 1 Spirometery tracing

| | $O_2$ consumed (L/hr) | STPD correction factor | Avg. kcal/L $O_2$ | Body surface area ($m^2$) |
|---|---|---|---|---|
| BMR | 16L/hr | | 4.825 | |

- Using the data, testing information, and the equation given, calculate the BMR for the subject. Note: This calculation is for kcal/$m^2$/hr.

$$\text{Calculation: } \frac{(\text{L } O_2 \text{ consumed/hr}) \bullet (\text{STPD correction factor}) \bullet (\text{Avg. kcal/L } O_2)}{\text{body surface area } (m^2)}$$

__________ = kcal/$m^2$/hr

(16 L/hr) ·

- Calculate the number of kcals utilized by this person in 24 hrs if she maintained the same constant level of activity as during her testing.

Calculation: kcal/m$^2$/hr • body surface area (m$^2$) • 24 hrs

__________ = measured kcal/24 hrs (measured BMR/24 hrs)

- Using the appropriate equation (male or female), calculate the predicted 24-hr caloric expenditure based upon the BMR of the subject.

__________ = predicted kcal/24 hrs (predicted BMR/24 hrs)

- Using your calculated 24-hr values and the following equation, calculate the percent (%) deviation of the subject's measured BMR (24 hrs) from the subject's predicted BMR (24 hrs).

$$\frac{\text{measured BMR} - \text{predicted BMR}}{\text{predicted BMR}} \cdot 100 = \%\ \text{deviation}$$

__________ = % deviation

## Respirometry Activity 2

- Obtain a correction chart for STPD.
- Based on the information given in the testing information that follows, find the correction factor for STPD. Record the value in the following table.
- Based on the information given in the testing information, determine the body surface area for the subject. Using the nomogram, draw a line from the subject's height to the subject's weight. Read the square meters at the point the line crosses the column of body surface numbers. Record the value for body surface area (m$^2$) in the following table.

  Testing information:
  - Spirometer temperature: 24°C
  - Barometric pressure: 743 mm Hg
  - Female subject
  - Height: 5′3″
  - Weight: 124 lb

- Age: 28
- BMR test: oxygen consumed in 6 minutes = 0.8 L
- Postexercise MR test: oxygen consumed in 6 minutes = 3.6 L

- Determine the amount of oxygen consumed in 1 hr (L $O_2$/hr) for the BMR recording. Record the value in the table.

- Determine the amount of oxygen consumed (L) in 1 hr (L $O_2$/hr) for the postexercise MR recording. Record the value in the table.

| | **$O_2$ consumed (L/hr)** | **STPD correction factor** | **Avg. kcal/L $O_2$** | **Body surface area ($m^2$)** |
|---|---|---|---|---|
| BMR | | | 4.825 | |
| MR | | | 4.825 | |

- Using the data, testing information, and the equation given, calculate the BMR for the subject. Note: This calculation is for kcal/$m^2$/hr.

$$\text{Calculation: } \frac{(\text{L } O_2 \text{ consumed/hr}) \bullet (\text{STPD correction factor}) \bullet (\text{Avg. kcal/L } O_2)}{\text{body surface area } (m^2)}$$

__________ = kcal/$m^2$/hr

- Calculate the number of kcals utilized by this person in 24 hrs if she maintained the same constant level of activity as during her testing.

Calculation: kcal/$m^2$/hr • body surface area ($m^2$) • 24hrs

__________ = measured kcal/24 hrs (measured BMR/24 hrs)

- Using the appropriate equation (male or female), calculate the predicted 24-hr caloric expenditure based upon the BMR of the subject.

__________ = predicted kcal/24 hrs (predicted BMR/24 hrs)

- Using your calculated 24-hr values and the following equation, calculate the percent (%) deviation of the subject's measured BMR (24 hrs) from the subject's predicted BMR (24 hrs).

$$\frac{\text{measured BMR} - \text{predicted BMR}}{\text{predicted BMR}} \bullet 100 = \%\ \text{deviation}$$

_________ = % deviation

- Using the data, testing information, and the equation given, calculate the postexercise MR for the subject. Note: This calculation is for kcal/m²/hr.

$$\text{Calculation: } \frac{(\text{L } O_2 \text{ consumed/hr}) \bullet (\text{STPD correction factor}) \bullet (\text{Avg. kcal/L } O_2)}{\text{body surface area } (m^2)}$$

_________ = $\text{kcal/m}^2\text{/hr}$

- Calculate the number of kcals utilized by this person in 24 hrs if she maintained the same constant level of activity as during her testing.

Calculation: $\text{kcal/m}^2\text{/hr} \bullet \text{body surface area } (m^2) \bullet 24\text{hrs}$

_________ = measured kcal/24 hrs (measured MR/24 hrs)

- Using your calculated 24-hr values and the following equation, calculate the percent (%) deviation of the subject's measured MR (24 hrs) from the subject's measured BMR (24 hrs).

$$\frac{\text{measured MR} - \text{measured BMR}}{\text{predicted BMR}} \bullet 100 = \%\ \text{deviation}$$

_________ = % deviation

# Exercise 27

# Body Temperature Regulation

## Objectives

- Be able to define the term *homeotherm.*
- Be able to define the term *endotherm.*
- Be able to define the term and explain the function of *shivering thermogenesis.*
- Be able to define the term and explain the process of *radiation.*
- Be able to define the term and explain the process of *conduction.*
- Be able to define the term and explain the process of *convection.*
- Be able to define the term and explain the process of *evaporation.*
- Be able to define the term *hypothermia.*
- Be able to discuss mechanisms of detecting and correcting hypothermia.
- Be able to define the term *hyperthermia.*
- Be able to discuss mechanisms of detecting and correcting hyperthermia.
- Be able to explain the role of the hypothalamus in body temperature regulation.

## Introduction

Humans are **homeotherms**, meaning that we maintain a relatively constant internal or core body temperature (36°C–37°C or 97°F–98.6°F), regardless of the environmental temperature. This relatively constant body temperature allows chemical reactions or metabolism to occur at a fairly predictable rate. We are also **endotherms**, meaning that we internally generate the majority of our heat. This heat is generated as a by-product of metabolic reactions that occur in our bodies. For example, you know that when you exercise, your body temperature increases due to the increased activity of your skeletal muscles. Although when consciously using your muscles, the primary reason for doing so is movement, one of the by-products is heat. In some cases, that heat is considered a waste product, and the body has to work to dissipate the heat to the environment in to maintain a normal body temperature. At other times, we involuntarily cause our muscles to contract specifically for heat generation, as in **shivering thermogenesis**.

## Body Temperature

Body temperature can vary from one part of the body to another. The **body core**, tissue and organs of internal cavities, is usually 2° to 3° warmer than tissues of the extremities (body shell).

The temperature of the surrounding environment can have a large impact on this difference, and temperature may vary by several degrees. Poor peripheral blood circulation may also result in this same effect. Body temperature is regulated by the **hypothalamus**. **Peripheral thermoreceptors** detect temperature in the extremities, and **central thermoreceptors** detect temperature in the body core. Detection of temperature changes by the peripheral thermoreceptors results in activation of a feedforward system to attempt to return peripheral temperature to normal before it has a chance to affect core temperature. Detection of temperature changes by the central thermoreceptors results in activation of a negative feedback system to return body temperature to normal. The hypothalamus responds to the sensory information from the thermoreceptors by adjusting metabolic rate via hormones, peripheral vasodilation or vasoconstriction, sweating, panting, shivering, and piloerection. The cerebral cortex also responds to the sensory information from the thermoreceptors by altering voluntary responses of skeletal muscles. Body core temperature follows a **circadian rhythm** and fluctuates approximately 0.5°C through the day, with temperature being the lowest early in the morning and highest in the midafternoon.

The hypothalamus normally maintains body temperature, and specifically core temperature, within an acceptable range (36°C–37°C or 97°F–98.6°F). Body temperatures below the norm result in **hypothermia**. Hypothermia decreases metabolism and results in a decreased ability to think rationally and decreased ability to move skeletal muscle, thus decreasing the ability to generate more heat. Body temperature above the normal range can be the result of a fever or hyperthermia. There is a very important distinction between these two. In the case of a **fever**, the hypothalamus establishes a new temperature setpoint, and the hypothalamus activates heat gain mechanisms to purposefully raise body temperature. In the case of **hyperthermia**, the core body temperature increases without a change in the temperature setpoint. Although hyperthermia often occurs in more strenuous exercise, the body will ultimately match heat loss with heat gain to prevent core temperature from becoming dangerously high. A core temperature of 41°C (106°F) will usually result in some brain death, and a prolonged core temperature of 45°C (113°F) will result in death.

## Body Temperature Activity

- Obtain a mercury-free oral thermometer.
- Shake the alcohol down into the bulb, and cover the thermometer with a disposable probe cover.
- Insert the bulb under the tongue with the tip at the base of the tongue.
- The mouth should be kept closed and the tongue held down.
- Remove the thermometer after 15 seconds and record the temperature in the following chart.
- Immediately return the thermometer to the mouth and continue the procedure at 15-second intervals for each of the times and record the temperature in the following chart.
- Take a drink of cold water and immediately repeat the entire procedure beginning with shaking the alcohol down into the bulb.
- Repeat the entire procedure beginning with shaking the alcohol down into the bulb, but take axillary (underarm) temperature by placing the thermometer bulb in the closed axilla.

| Time (sec) | Oral temp (°C) | Oral temp (°C) after drinking cold water | Axillary temp (°C) |
|---|---|---|---|
| 15 | | | |
| 30 | | | |
| 45 | | | |
| 60 | | | |
| 75 | | | |
| 90 | | | |
| 105 | | | |
| 120 | | | |
| 135 | | | |
| 150 | | | |
| 165 | | | |
| 180 | | | |

- How long did it take to observe duplicate or stable temperature readings for the first readings?

- How long did it take to observe duplicate or stable temperature readings after drinking cold water?

- How long did it take to observe duplicate or stable temperature readings for the axillary readings?

- How did the axillary temperature data compare to the first oral temperature data?

- Expected Results
  If positioned correctly, the oral thermometer is measuring the temperature of the lingual arteries. This gives a good indication of the core body temperature. However, oral temperature can be influenced by the drinking of fluids and by the temperature of the air in mouth breathing.

  If positioned correctly, the axillary temperature is measuring the temperature of the axillary artery. Although this should give a good indication of core body temperature, it is difficult to get correct placement of the thermometer and is more likely to be influenced by environmental conditions.

  Between oral and axillary thermometry, measuring oral temperature is generally the preferred method of thermometry when using traditional thermometers (mercury, alcohol, and digital). Axillary temperature is generally 0.3°C to 0.6°C lower than oral temperature. Infrared thermometers are becoming more precise and more prevalent as well. These are used to measure thermal radiation (tympanic membrane or ear canal thermometer and forehead temporal artery thermometer). Tympanic membrane temperature is generally 0.3°C to 0.6°C higher than oral temperature. Rectal temperature may be taken in young children. It is generally 0.3°C to 0.6°C higher than oral temperature. Plastic strip thermometers measure only skin temperature not true body temperature, and they are not recommended for home use.

## Mechanisms of Heat Loss and Gain with the Environment

Although metabolism is the means by which our body produces its own heat, we also exchange heat with our surrounding environment. This is especially important in involuntary heat loss and situations in which we purposefully dissipate heat to stay within our homeostatic temperature range. **Radiation** is the radiant (through air) gain or loss of heat following a thermal gradient. If the air temperature surrounding the body is cooler than body temperature, the thermal gradient will lead to heat loss from the body. However, if the temperature surrounding the body is warmer than body temperature, the thermal gradient will lead to heat gain by the body. Think about standing close to a campfire to warm up.

- **Thermal Radiation Activity**
  - Obtain a 500-mL beaker and fill it with hot tap water.
  - Ask the subject to place his or her hands around the beaker but he or she should not touch the outer surface of the beaker.
  - Could the subject feel heat radiating from the beaker?

**Conduction** is the gain or loss of heat following a thermal gradient due to physical contact. Whereas radiation transfers heat through air, conduction transfers heat via direct contact between objects. Think about warming your hands by holding a cup of coffee. An object that serves as a good conductor of heat will easily pass heat on to another area. An object that does not conduct heat well will not pass the heat on but rather localize it and warm itself. We generally think of the materials that are poor conductors of heat as insulators. Insulating materials can be used not only to hold heat in but to keep heat out as well. For insulation against heat loss think of a coat on a cold day, and for insulation against heat gain think of the effect of coolers and can wraps. In the human body, adipose tissue serves as an insulator to help keep our core body temperature within the homeostatic range.

## Conduction Activity

- Obtain a glass rod and a brass rod of equal sizes.
- Obtain a 500-mL beaker and fill it with hot tap water.
- Ask the subject to place his or her hands around the beaker filled with hot water, touching the outer surface of the beaker.
- Could the subject feel the hands warming as a result of making contact with the beaker?
- Have the subject place his or her arms on the lab bench with the inner aspects up, place the glass rod at a 90° angle (perpendicular) across one forearm and the brass rod at a 90° angle (perpendicular) across the other forearm.
- What was the immediate sensation in each of the arms?

  Arm with glass rod:

  Arm with brass rod:
- Over a little time, did the subject notice any difference between the two arms? If so, what was the difference?
- Expected Results

  Initially both rods should feel cool. The brass rod is a better conductor of heat and should have continued to feel cool for a longer time than the glass rod.

**Evaporation** is the loss of heat as water is converted from a liquid to a vapor. It takes 600 kcals to convert 1 L of liquid water to water vapor. That energy is taken from the body. As the water gains energy and evaporates, the body loses energy (heat) and cools. **Active water loss** (and thus heat loss) occurs with sweating, but **passive water loss** (and thus heat loss) occurs with respiration and through our water-permeable skin. One factor that affects the rate of evaporation of water is the concentration of water vapor in the air. Absolute humidity is a measure of water vapor in the air without regard for air temperature. A much more useable number with regard to evaporation is

**relative humidity**. It tells you how much water vapor is in the air compared to how much the air could hold at that temperature. Relative humidity is given as a percent and is calculated as:

$$RH = (\text{actual vapor density}) / (\text{saturation vapor density}) \cdot 100$$

Evaporation can only cool; it is not a mechanism for heat gain.

### Evaporation Activity

- Have the subject wet a hand in water that feels lukewarm or roughly body temperature.
- Expose the wet hand to the air.
- Could the subject feel the wet hand cooling as the water evaporated?
- What does evaporation have to do with dehydration on a hot summer day?
- How would relative humidity affect evaporative cooling?

**Convection** is the heat loss or gain as a result of moving air or water. A convection oven works on this principle. As warm air passes over the food, some of the heat follows the temperature gradient and is absorbed by the food. For the body, we often use moving air to cool ourselves either by fanning ourselves with a piece of paper or by sitting in front of an electric fan. For the body, convection is usually a means of heat loss, but it can also result in heat gain. If you have ever been in a hot dessert breeze with temperatures over 100°F, you may have noticed that the breeze did not cool you but made you feel hot. Piloerection is used to reduce heat loss as a result of convection.

### Convection (and Evaporation) Activity

- Have the subject wet both hands in water that feels lukewarm or roughly body temperature.
- Wave one of the wet hands in the air, but hold the other hand still.
- Did the subject notice a temperature difference between the hand the subject waved and the hand the subject held still? What was the difference?

### Combination Activity

- Obtain two chemical thermometers, each with a detection range of 0 to 100°C.
- Wrap the bulb of one thermometer in a thin layer of cotton and secure the cotton with a rubber band.

- Place the bulbs of both thermometers in 40°C water. Allow the thermometers to equilibrate, then remove them from the water.
- Secure both thermometers to a clamp and suspend them from a stand.
- Record the temperature of both thermometers in the following chart.
- Continue to record the temperatures of each thermometer at 1-minute intervals.

| Time (min) | Temp. unwrapped (°C) | Temp. wrapped (°C) |
|---|---|---|
| 0 | | |
| 1 | | |
| 2 | | |
| 3 | | |
| 4 | | |
| 5 | | |
| 6 | | |
| 7 | | |
| 8 | | |
| 9 | | |
| 10 | | |
| 11 | | |
| 12 | | |
| 13 | | |
| 14 | | |
| 15 | | |

- Did the temperature of the two thermometers fall at the same rate?

- Did the thermometers stabilize at the same temperature?

# Exercise 28

# Glucose Tolerance Test

## Objectives

- Be able to identify the normal range of blood glucose.
- Be able to define the term *hypoglycemia.*
- Be able to define the term *hyperglycemia.*
- Be able to define the term *glucosuria* and explain its cause.
- Be able to define the term *polydipsia.*
- Be able to explain the functions of *insulin* and *glucagon* in the regulation of blood glucose concentrations.
- Be able to explain what an *A1c test* measures and how *glycosylated hemoglobin* can indicate uncontrolled diabetes mellitus.
- Be able to explain the expected changes in blood glucose in a *glucose tolerance test.*
- Given values for a glucose tolerance test, be able to identify a person as normal or diabetic.
- Be able to explain the difference between *glucose-dependent diabetes mellitus* and *glucose-independent diabetes mellitus.*
- Be able to interpret a glucose tolerance curve.
- Be able to explain why the glucose tolerance curve is elevated in a person with diabetes mellitus.

## Introduction

The normal blood glucose concentration ranges from 60 mg/dL to 120 mg/dL. When the value falls below this range, the person is suffering from **hypoglycemia**. Symptoms associated with hypoglycemia may include hunger, trembling, feelings of nervousness, sweating, weakness, and some individuals may have trouble speaking. When the value exceeds this range, the person is suffering from **hyperglycemia**. Symptoms associated with hyperglycemia may include diuresis, dry mouth and excessive thirst, headache, and vomiting. The ability to regulate blood glucose level involves many tissues and systems within the body. The primary hormones responsible for the regulation of blood glucose are **insulin**, produced by the beta (β) islet cells of the pancreas, and **glucagon**, produced by the alpha (α) islet cells of the pancreas. In general, insulin causes cells to take up glucose from the blood and thus lower the blood glucose concentration. Glucagon causes the liver cells to release glucose to the blood and thus raise the blood glucose concentration. The inability of the pancreas to produce insulin or the failure of cells to respond to insulin results in **diabetes mellitus**. Because a person with uncontrolled diabetes mellitus can no longer regulate their blood glucose levels, they will be characterized by hyperglycemia (blood glucose > 120 mg/dL) and **glucosuria** (blood glucose > 180 mg/dL exceeds glucose transport maximum in the kidneys and results in glucose in the urine). The person will also have a higher-than-normal metabolism of lipids and proteins and may experience ketoacidosis. The person may also be characterized by **polyuria**, production of large quantities of dilute urine. To compensate for the excessive loss of water, the person may experience **polydipsia** or excessive thirst.

There are two types of diabetes mellitus, **insulin-dependent diabetes** (type I) and **insulin-independent diabetes** (type II). Both result in the same problem, inability of the individual to properly regulate blood glucose concentration. Insulin-dependent diabetics cannot produce adequate amounts of insulin. As a result, their cells are not adequately directed to take up glucose from the plasma, and blood glucose concentration often remains higher-than-the-normal homeostatic range. Insulin-independent diabetics can initially produce adequate amounts of insulin, but their cells will not respond to it appropriately. As a result, the cells do not take up glucose from the plasma at a normal rate and blood glucose concentration often remains higher–than-the-normal homeostatic range. As insulin-independent diabetes progresses, the individual may lose the ability to produce adequate amounts of insulin.

Other medical conditions interfere with the body's ability to regulate blood glucose levels. These conditions may be related to the functioning of the thyroid, anterior pituitary, digestive system, and other homeostatic imbalances.

## A1C Test

Some hemoglobin molecules may have a glucose molecule bound to an amino acid in the protein portion of the hemoglobin. This form of hemoglobin is called **glycosylated hemoglobin** (Hb A1C). In most individuals, the percent of glycosylated hemoglobin is low, usually 3% to 5%. In individuals who tend to routinely experience hyperglycemia, as in uncontrolled diabetes mellitus, the percent of glycosylated hemoglobin may be 2 to 3 times higher. The percent of glycosylated hemoglobin is used as an indicator of how well blood glucose levels have been controlled over the past 2 to 3 months. A person with a hemoglobin A1C test result of 6.5% or above is considered diabetic.

Not all people have the same type of hemoglobin. Hemoglobin A is the most common form of hemoglobin (2 alpha globin chains and 2 beta globin chains), but some people have alternate forms of hemoglobin: S, C, or E. Hemoglobin S is the form of hemoglobin associated with sickle cell disease. In these individuals a single amino acid substitution in the beta globin chains results in the abnormal hemoglobin. Hemoglobin C is an alternate form of hemoglobin resulting from a mutation in the gene that codes for the beta globin chains. Individuals with hemoglobin C disease tend to have smaller and fewer erythrocytes and may be slightly anemic. Hemoglobin E is an alternate form of hemoglobin resulting from a mutation in the gene that codes for the beta globin chains. In these individuals a single amino acid substitution in the beta globin chains results in the abnormal hemoglobin. Individuals with hemoglobin E disease may suffer from slight hemolytic anemia.

Having any of these hemoglobin variants can affect the results of an A1C test. In some cases the A1C may be artificially low, and in other cases the A1C may be artificially high. As a result, the A1C test should not be used in isolation to diagnose a blood glucose abnormality in an individual with alternative forms of hemoglobin.

## Glucose Tolerance Test

A **glucose tolerance test** measures the ability of an individual to deal with glucose. A person is required to fast overnight. A blood and urine sample is taken from the person to obtain a baseline value of glucose in the blood and urine. The person then drinks a solution containing a specific amount of glucose (the solutions can range from 50 to 100 g of glucose, depending upon the test). Blood and urine samples are taken at 30-minute or 1-hour intervals for 1-3 hours. For a normal individual, the blood glucose level is expected to rise above the 120-mg/dL level and may even exceed the transport maximum of 180 mg/dL. However, the blood glucose level is expected to begin decreasing 60 minutes after glucose consumption. Blood glucose levels are expected to stabilize 2 to 3 hours after glucose consumption. For a person with diabetes mellitus, the baseline blood glucose may already be above 120 mg/dL and will rise as a result of glucose consumption. As the blood glucose rises above 180 mg/dL, the person will also exhibit glucosuria. The blood glucose may remain higher than normal for several hours. A person with diabetes mellitus will exhibit a reduced tolerance to glucose. As a result, the glucose tolerance curve will be elevated as compared to a normal glucose tolerance curve (see Figure 1). Some women experience **gestational diabetes**, a higher than normal blood glucose concentration that starts or occurs during pregnancy.

The following blood glucose concentrations would be considered impaired or abnormal results for a 3-hour, 100-g oral glucose test:

Fasting—greater than 95 mg/dL
1-hour—greater than 180 mg/dL
2-hour—greater than 155 mg/dL
3-hour—greater than 140 mg/dL

The following blood glucose concentrations would be considered impaired or abnormal results for a 2-hour, 75-g oral glucose test:

Fasting—greater than [illegible] mg/dL
1 hour—greater than [illegible]00 mg/dL
2 hour—greater than 140 mg/dL (impaired), greater than 200 mg/dL (diabetes)

The following blood glucose concentrations would be considered impaired or abnormal results for a 1-hour, 50-g oral glucose test:

Fasting—greater than 110 mg/dL
1-hour—greater than 140 mg/dL

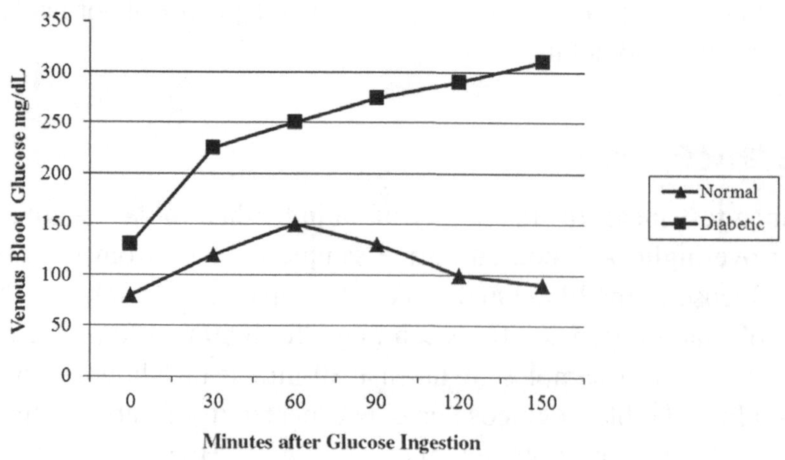


Figure 1 Glucose tolerance curve

## Blood Glucose Testing

A person who needs to monitor or track her/his blood glucose will use a meter such as a TRUEresult® meter system to measure blood glucose (see Figure 2). Routine blood glucose testing does not require fasting and drinking the glucose solution. An auto-lance is used to prick the skin to obtain a small drop of blood, usually 0.3–0.6 μL. Blood is usually taken from a finger, because this gives the best results. Blood can also be taken from alternate sites such as the palm, forearm, upper arm, thigh, or calf, but the particular glucometer being used must be able to support blood samples drawn from these areas. Capillary action draws the blood into the test strip in the glucometer. The glucose in the blood reacts with the chemicals in the test strip. An electrical current is passed through the sample and from this, the concentration of glucose in the sample can be calculated. There are other types of blood glucose monitors. A continuous glucose monitor uses a sensor implanted under the skin. This allows readings to be taken routinely 24 hours a day and at more regular intervals. A continuous glucometer must be prescribed by and the sensor implanted by a medical doctor. A non-invasive glucometer uses a laser light and Raman spectroscopy to determine the blood glucose concentration in blood-rich areas of the skin. As a result, the non-invasive glucometer doesn't require a blood draw.

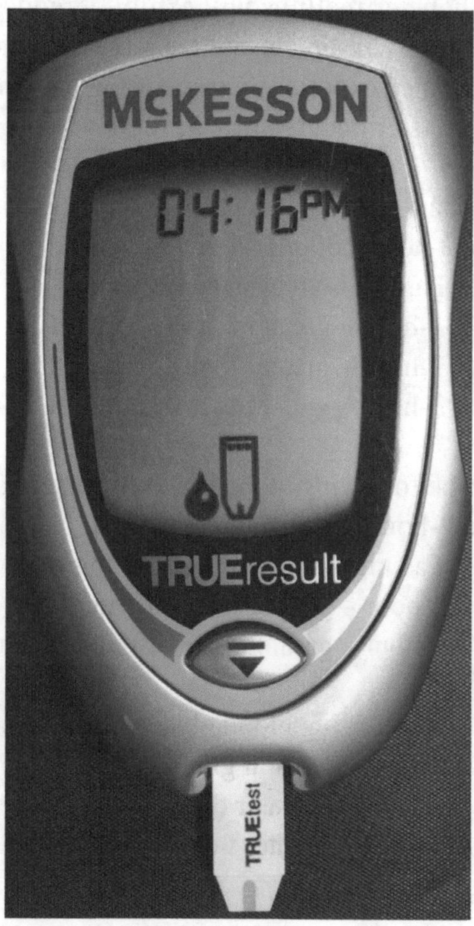


Figure 2 Glucometer with test strip

Multiple factors can influence the results of a blood glucose test. Some of these factors are mechanical or environmental: improper storage of meters and test strips, expiration of test strips, poor calibration of meters, humidity, and high altitude. Other factors are behavioral: improper cleaning of the test site, small blood sample volume. Other problems are physiological: dehydration, anemia, and interference by vitamin C and Tylenol.

# REFERENCES

Crowley, L. *An Introduction to Human Disease Pathology and Pathophysiology Correlations*. 6th ed. Sadbury, MA: Jones and Bartlett, 2004.

iWorx Systems, Inc – iWorx. Advancing Physiology. https://www.iworx.com/teaching-landing/human-physiology/

*New Medical Information and Health Information*. April 2015. http://www.nmihi.com/f/ipf.htm

Pflanzer, R. G. *Experimental and Applied Physiology*. 8th ed. New York: McGraw-Hill, 2007.

Silverthorn,D.U. *Human Physiology An Integrated Approach.* 7th ed. USA: Pearson, 2016.

*U.S. National Library of Medicine & National Institutes of Health* https://www.nlm.nih.gov/

Widmaier, E. P., H. Raff, and K. T. Strang. *Vander's Human Physiology:The Mechanisms of Body Function*. 12th ed. New York: McGraw-Hill, 2011.